D1524502

No. 1211
$14.95

The Complete Handbook of Videocassette Recorders

Second Edition

by Harry Kybett

TAB BOOKS Inc.
BLUE RIDGE SUMMIT, PA. 17214

SECOND EDITION

FIRST PRINTING

FEBRUARY 1981

Printed in the United States of America

Library of Congress Cataloging in Publication Data

Kybett, Harry.
The complete handbook of videocassette recorders.

"Tab book #1211."
Includes index.
1. Video tape recorders and recording—Handbooks, manuals, etc. I. Title.
TK6655.V5K87 1981 621.388'33 80-28237
ISBN 0-8306-9658-X
ISBN 0-8306-1211-4 (pbk.)

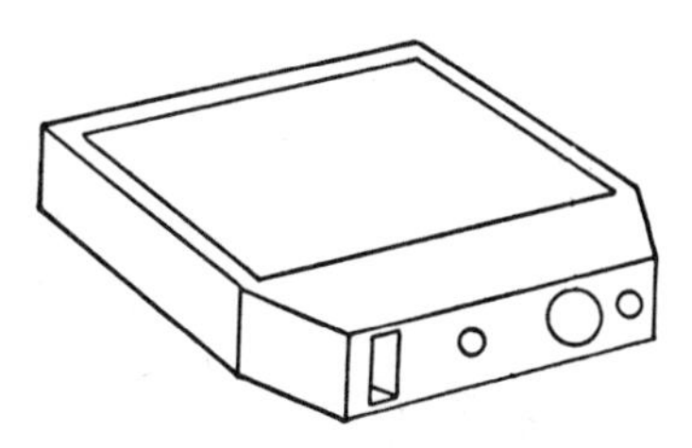

Contents

Preface

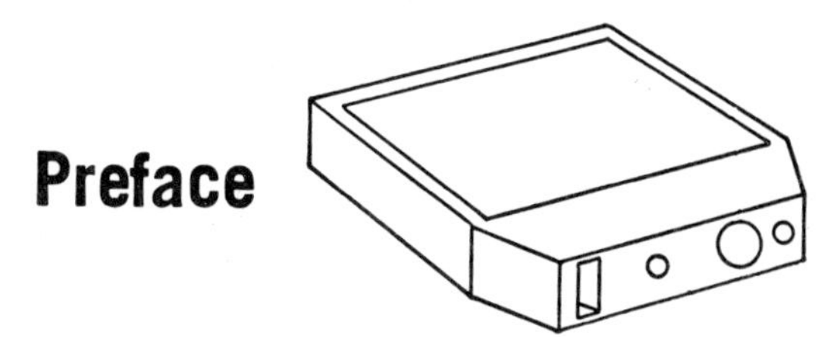

Since this book was first written the videocassette scene has undergone two major changes. The first change has been the upgrading of the original U-matics to the quality of a broadcast machine, where they are now used extensively for news inserts, commercials, promos, and other short pieces. This was brought about by CBS News, who took the Sony V0-2850 machines to Moscow to cover President Nixon's visit. Their performance proved beyond doubt that they could produce pictures good enough for broadcasting. Shortly after this the first broadcast U-matics appeared.

The second change has been in the other direction. Starting with the Betamax, several excellent small machines have appeared which are inexpensive enough and reliable enough to have become a viable consumer item.

Consequently, this second edition is slightly different from the first. It covers briefly the broadcast U-matics, but does not go into detail about these, as these tend to be used by professionals and all the information necessary on their use and servicing is available from the manufacturers.

However, there is a real need for information about the consumer level machines. These will be serviced in local centers and TV service stations where the TV service man will have had little or no opportunity to work with VTR's. Hence, much of this second edition is concerned with the newer consumer models and the accessories made available for them.

Harry Kybett

1 Introduction to Videocassettes

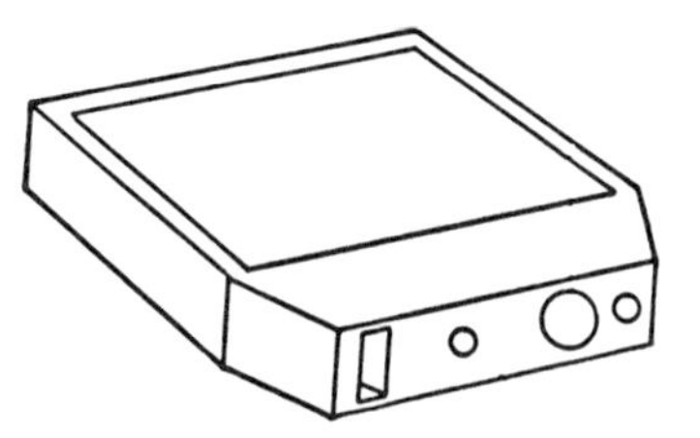

The videocassette machine has been designed to be used by nontechnical persons in everyday situations. It requires no special training or knowledge of video or television on the part of the user, and is ideal for use in a variety of educational, entertainment and information exhange situations.

It was approached by the designers primarily as a playback device for the nontechnical user; and this simple fact makes it different from every other videotape machine, all of which were designed and built as recording machines to be used by an engineer in a TV studio.

The simplest use of the videocassette machine is playing back a prerecorded tape. It is easier to do than drive a car, learn to type, or use a small calculator; it has even been compared to placing a coin in a slot or mailing a letter. This ease of operation has been responsible for removing a psychological barrier to using high-level technology, and it has caused the wide and ever expanding proliferation of these machines.

They are ideal for the nontechnical person and are now found in places such as homes, schools, advertising agencies, banks, sales offices, and hospitals. These are all spots where information exchange, entertainment, or education are vital, but where other video techniques have not been fully adopted due to the technical difficulties of using television. One demonstration of the videocassette machine in these areas has usually been enough to overcome or remove the objections of any skeptic.

A videocassette machine is basically a helical video tape recorder (VTR).

A helical VTR is a tape machine in which the tape is wrapped around a head drum of large diameter, and in travelling around the drum the tape path is in the form of a helix, or a single turn of a spring. This is shown in Fig. 1-1. The tape leaves the drum at a level different from that at which it joined the drum, and it can be wrapped halfway or completely around the drum.

It differs from other helicals in its specialized mechanical construction, its appearance, and its mode of operation. The major differences are that the tape is completely enclosed in a plastic container which is simply inserted into the machine, and that the tape threading is automatically performed by the machine instead of by the operator.

The general appearance is a neat box with a few indicator lights and controls on the front panel which are kept to a minimum, and are as unobstrusive and pleasing to the eye as possible. There are no unsightly plugs or cables in view (these are hidden at the rear), and the tape reels, head drum, and the other familiar mechanical items of the open reel machines are inside and are not exposed to view during normal operation. They all look like a piece of attractive furniture in the same way a phonograph or TV set does, and none looks like an item of professional TV equipment or military surplus gear. These machines can be placed side by side with anything found in a living room or expensive office, and not look at all out of place.

THE BASIC OPERATIONAL ASPECTS

The most important requirement the manufacturers and designers set out to meet was that of extreme operational simplicity. The actual operational controls and facilities vary slightly from model to model, but all have been made so simple to operate that a young child can easily insert a tape and see the picture.

The simplest mode of operation is that of viewing a prerecorded tape on a domestic color receiver. Only two connections are required: the power connection to the wall and the cable from the machine to the antenna terminals of the TV set. The set is then turned to the correct channel and possibly fine tuned. The cassette is pushed into an opening in the front of the machine, and the **play** button is pressed. The machine then automatically threads the tape, and no further operations are required from the user.

A large number of tapes are now available in cassettes and cartridges in the form of prerecorded lessons, programs, and other material supplied from parent organizations and commercial sources. Thus this simplist of playback modes is one of the most important uses of these machines and in many places it covers the full extent of use. Other playback modes are possible, as are several recording modes, and these are covered in greater detail in later chapters, but in all these cases this extreme operational simplicity has been maintained.

At this point it should be understood that operational simplicity and neat appearance completely mask the internal design and construction, which must be placed among the most complicated mechanical devices and sophisticated pieces of electronics ever presented to the public. They have a mechanical precision and construction far above anything else widely available—even a Swiss watch—and they have a complexity of moving parts which is quite astounding. Although the operation of these machines has been reduced to the lowest level, the care and maintenance has not undergone such a radical transformation. In spite of this, a reliability of over 99% has been reported by the U.S. Army, which is one of the largest users.

BASIC CONSTRUCTION AND TAPE PATH

At this time it is useful to include a short description of the tape path and the transport system in the PLAYBACK and RECORD modes. A full description of the individual systems is given in later chapters.

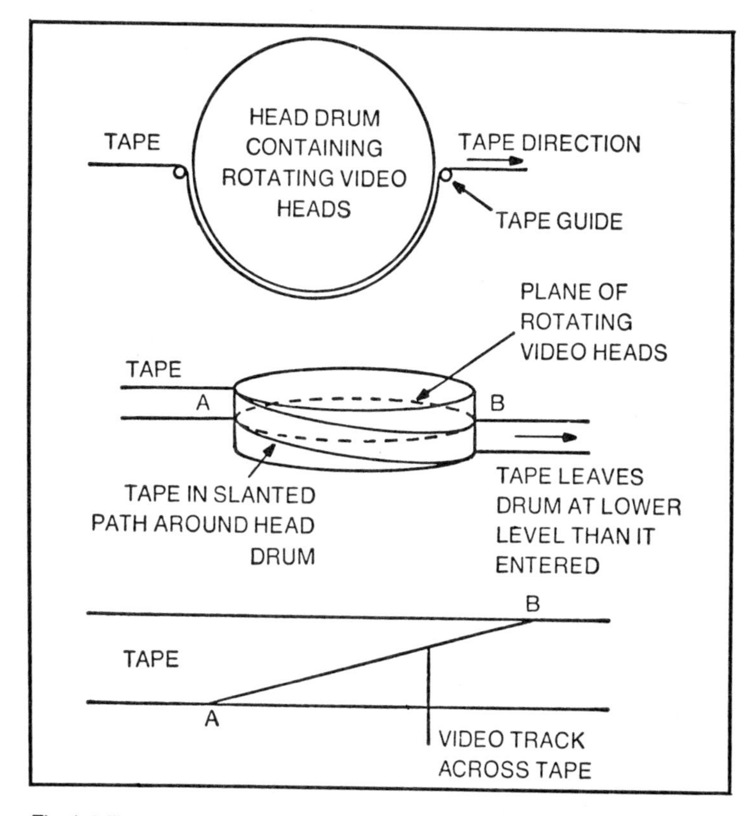

Fig. 1-1. Tape path on a helical VTR.

When the PLAY button is pressed the tape is pulled out of the cassette and automatically threaded around its path. This is shown in simplified form for the U-matic in Fig. 1-2. The other machines are similar in principle. When the tape is fully run out the pressure roller closes against the capstan and the tape is played. The take-up spool is puck and belt driven by the same main motor, and back tension is provided by a friction clutch in the feed spool. The tape guides keep the tape correctly aligned at all points.

When the **stop** button is used, the pressure roller releases and the tape is unthreaded by a process which is the reverse of the threading. The feed spool is braked, and the tape is pulled into the cassette by the rotation of the take-up spool. Once it is fully retracted, an automatic mechanical action causes it to be rewound for a few seconds back onto the feed spool. It then stops, the main motor is depowered, and the heads stop turning.

If the end of the tape is reached in the playback mode, a plastic leader on the end of the tape allows light from a small lamp to fall onto a phototransistor which automatically stops the machine, retracts the tape, and initiates a complete rewind.

Fast forward and **rewind** of the tape are conducted with the tape inside the cassette, and it is never run out for these purposes.

This whole process of tape threading and transport can be seen by removing the top cover of the machine and observing the internal mechanism. Doing this will make the operation of these machines clear to the newcomer. However, nothing inside the machine should be touched or interfered with in any way.

TYPES OF MACHINES

In the past about 20 manufacturers introduced several different videocassette machines, but gradually the 3 following types began to dominate, while the others were phased out.

Umatics

This is the most widely used format in the United States for professional uses; such as broadcasting, industry and education. The main manufacturers are Sony, Panasonic, JVC and Shibaden. Other companies have marketed these manufacturers products with their own name tag.

Betamax

One of the two most widely used consumer cassette systems, it is made in 1 hour and 2 hour versions. Some machines are single speed only while others are dual speed. Basically it is a Sony format but it is also available from Sanyo, NEC and Zenith.

VHS

This is the other widely used consumer format, which is made in 2 hour and 4 hour versions. The two main manufacturers are JVC and Panasonic, but it is also available from RCA and Magnavox.

Other systems have been introduced but have not become widely used in the United States, and are not covered in this book. These are:

Philips

This is the most widely used European system, especially in the consumer field, where it is manufacturered by several companies in several countries (including the USSR). It has found only limited use in the US.

EIAJ

This uses a ½ inch tape in a one reel cassette or cartridge, and conforms to the EIAJ open reel standards. It is manufacturered by Panasonic.

Sanyo V-Cord

This is an early system discarded in favor of the Betamax.

Quasar

This is an early system marketed by Magnavox until discarded in favor of the VHS.

Several other systems appeared but never became widely used. Examples are Cartrivision, IVC, Ampex Instavision, various TCS Selectavisions, and the 8 mm film EVR system.

The Kodak Super 8 flying spot scanner is an 8mm film system which permits home movies to be played over an ordinary TV set.

Four basic models are offered by the manufacturers at this time: the playback only deck, the video record deck, the TV record deck, and the portable deck.

The Playback Only Deck

This is the simplest and most basic of the machines and is capable of playback only. It is ideal for use in schools, sales offices, and other places where the prime requirement is the playback of tapes supplied from some parent organization or central source and recording is not likely to be undertaken. These machines are smaller, lighter, and less expensive than the recording machines and afford protection to valuable tapes by their inability to erase or record over the programs (Fig. 1-3).

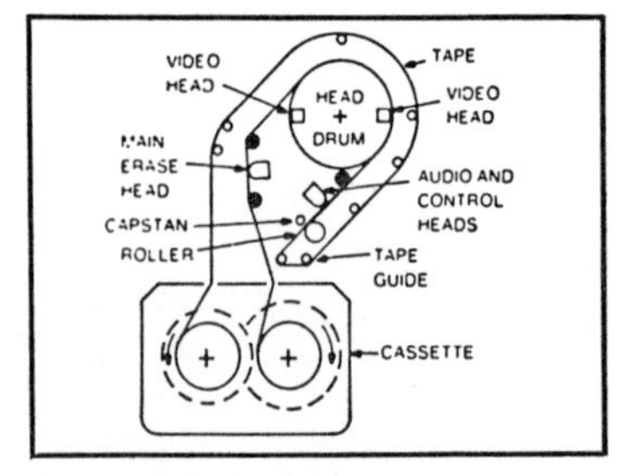

Fig. 1-2. Tape threading on a Umatic.

The Video Record Deck

This is similar in many respects to the normal open reel machines. It can record from the usual video sources and can play back in the same manner as the above decks. These machines are ideal for small industrial or educational uses where a fairly simple video system is in use and **off air** programs do not form a major part of the required material (Fig. 1-4).

The TV Record Deck

The main characteristic of this machine is the TV tuner. This is a normal TV tuner as found in a domestic TV set, and it is usually mounted on the right-hand side of the machine. It was included so the machine would be accepted in the home (Fig. 1-5). An antenna is connected to the rear terminals of the machine, and TV programs can be received just as with a normal TV set. These programs can now be recorded directly onto the tape and viewed by either a monitor or a TV set which is plugged into the machine. These machines also will record from standard video sources and play back in the same manner as the machines listed above.

The record machines are easily distinguished from the playback decks because they are longer, due to the inclusion of the record circuits and the tuner at the right-hand side.

Both the record machines have a continual electronics-to-electronics signal (E-E) and radio frequency (RF) output at all times, so the program can be viewed both prior to and during recording. Several audio modes of recording are possible: two audio tracks are provided which can be recorded upon, either separately or simultaneously, and postproduction dubbing is possible onto one of the tracks. Separate playback of the tracks is possible, giving the capability of stereo music or dual language commentary.

The Portable Deck

Several types of portables have been introduced with a cassette-type format. One of these is a rather large mechanism

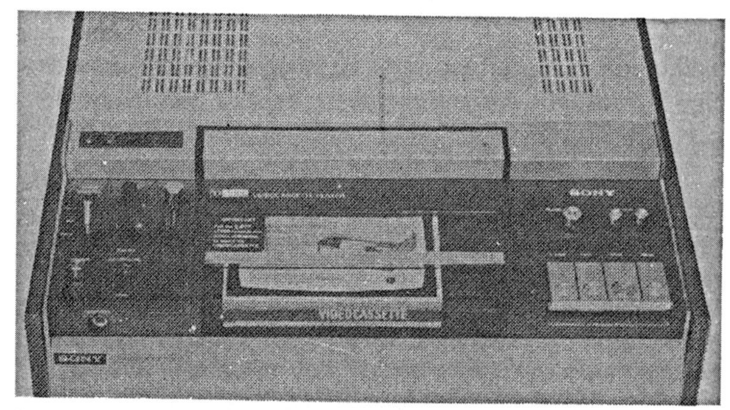

Fig. 1-3. Playback deck (courtesy of Sony).

designed to use the U-matic type cassette, and another is a smaller Betamax unit.

MASS DUPLICATION

The videocassettes were designed as a total system rather than as another isolated helical VTR, and a major part of the overall thinking was the provision of mass duplication facilities. This is both crucial and necessary if the videocassette is to gain acceptance as a major information transfer and distribution medium. This intention was recognized in the early phases of design and has heavily influenced some of the decisions in the development.

The photograph in Fig. 1-6 shows a typical duplicating center. The main playback VTR is a large helical with near broadcast specifications, and the record machines are modified versions of the standard decks. The whole system can be remotely controlled

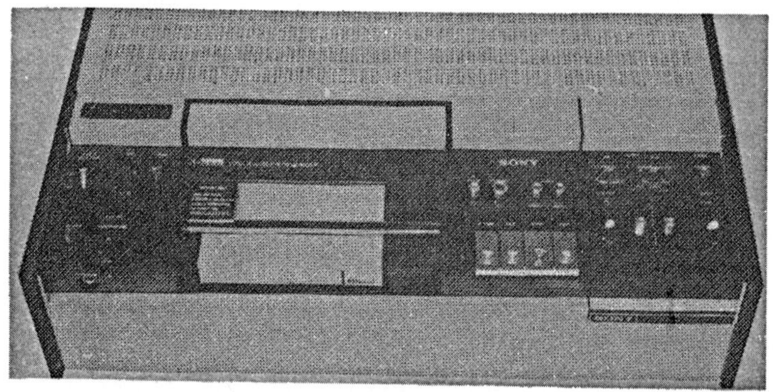

Fig. 1-4. Record deck (courtesy of Sony).

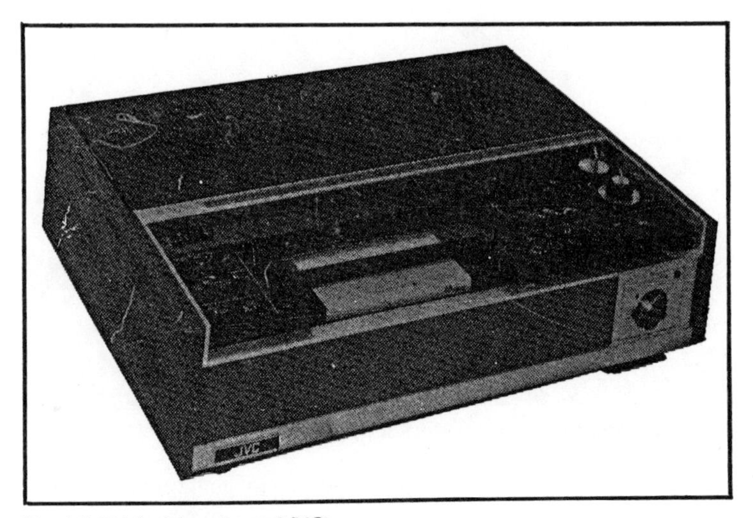

Fig. 1-5. TV deck (courtesy of JVC).

to stop and start simultaneously, and, with the provision of quick load and unload, many copies can be made in a short period.

A smaller level of duplication is possible with the standard decks, as shown in Fig. 1-7.

CONCLUSION

All of the videocassette manufacturers have provided a user oriented, foolproof, automatic threading videotape machine, with most of the controls interlocked to prevent misoperation and damage to both machine and tape. All have been devoted to the principle that the operator should not handle the tape. All have about the same facilities, and the ease of operation is best seen by examining the photographs of the front sections, with the controls clearly visible and marked with their exact functions. They all exhibit a very stable mechanical system, and a high degree of interchange of tapes is possible within each of the formats.

No monitor or a TV set is provided with these; in all cases this must be obtained separately. But no other equipment is needed to hook these items together and get a picture.

Most manufacturers provide a sturdy case, often with casters, so the machine can be transported with ease and not be exposed to damage. These cases contain padded sections for all the general purpose accessories which may be needed, and they are ideal for long distance shipping or use in rugged terrain.

The manufacturers are of the opinion that the basic design has been finalized and that no major changes are expected in the future.

Minor modifications and additions of facilities will naturally be introduced, but these will be internal to the machines to effect greater reliability and ease of tape handling and transport. These changes will not alter the tracks on the tape, the type of tape or cassette needed, or any of the other basics. There will be no major format changes and, in short, they are a future-safe system open only for improvements in manufacture and overall reliability. Every manufacturer has accessories which will permit facilities like still frame, remote control, automatic rewind, backspace, and pause to be included in machines which were not originally built with these capabilities.

The cassette will probably replace the open reel machines in several areas, and will be introduced with success into many places where the open reel could never gain a foothold. However, it is doubtful that cassettes will ever be dominant in the production studio, because basically they are post-production machines.

Three things permitted the development of the videocassette from a device with no precedent to an almost overnight success. First, the audiocassette had transformed the public attitude toward tape recording: it no longer required preparation and technical knowledge, and it was now something which was easy. This is a profound psychological change which is not well understood, but it did mean that the marketing of the videocassette was comparatively easy—user education to sell the idea was not necessary.

Fig. 1-6. Mass duplication center (courtesy of Phillips).

Fig. 1-7. Small multiple copy center (courtesy of University of Michigan).

Second, the technology of the open reel the VTR had progressed to the point where all the major development work has been done, and thus all the major problems had been resolved one way or another. Thus the cassette development work involved more applications engineering and not the research of unknown quantities and new principles. This shortened the work considerably: the developers knew where they were going.

Third, the advances in electronics in the mid-1960s, especially in miniaturization and integrated circuits, had matured to the level of producing reliable general purpose and special devices on a commercial basis. Without these modern semiconductors, modern small video technology would not exist.

The appearance and acceptance of the videocassette ranks alongside the photocopying machine, the microcomputer, and the data terminal as a new social communication tool and information exchange medium, and like these other items its effects are likely to be widespread and not fully understood for a long time. Videocassettes can now be found everywhere. There is hardly a major establishment with a use for video information which does not have cassette as well as or instead of open reel VTRs.

In conclusion, perhaps the reason for their astonishing success is that they represent an idea whose time has come—and for once the technology and the financing of the research and development were in step with public feeling at all levels.

2
Simple Cassette Playbacks

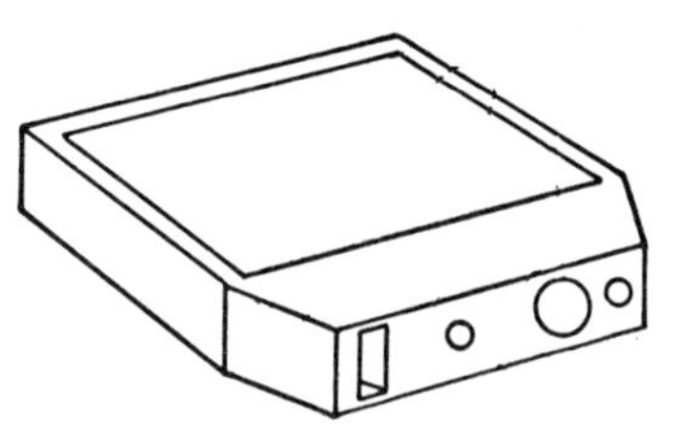

At this time it is expedient and useful to introduce the simplest playback modes. This will help to gain familiarity with the machines and to prepare the new user for the more complex modes needed for maximum use.

There are two modes of playback with which all videocassette machines are provided. The first is to play back a tape into an ordinary TV set, and the second is to use a TV monitor and loudspeaker amplifier. All the more complex playback modes are the same as these two basic situations, only the equipment connected to the videocassette machine changes. These two basics are covered below, and the difference between them is explained. Videocassette insertion and removal also are briefly covered.

PLAYBACK INTO A TV SET

All videocassette machines will play back into a domestic TV set and produce a color picture, just as if it was a normal TV channel in the viewer's area.

The first step is to ensure that the machine is plugged into the wall power outlet. The power switch can now be depressed, and the green indicator light should come on. If this does light up, then it is safe to proceed. The TV set can be either plugged into the wall outlet or into the auxiliary power outlet on the rear of the machine.

The only other connection needed is between the TV antenna terminals and the radio frequency (RF), or very high frequency (VHF), output at the rear of the machine (Fig. 2-1).

This output uses an F fitting, which is quite different from all the other sockets on the machine, so confusion is impossible. A special RF cable should be used, type RG 59/U or a similar one, and an F fitting should be attached to both ends. One end of the cable attaches to the machine, and the other end should be connected to a small antenna transformer similar to that in Fig. 2-2.

Both the cable and transformer are supplied with the videocassette machine. They can also be purchased at any electronics or TV store. Connection to the TV antenna can be made without this transformer but is not advised, for the reasons explained later.

The complete correct connections are shown in Fig. 2-3. When these connections have been made the TV set can be tuned to either Channel 3 or 4—whichever is not used in the area—and the channel select switch on the rear of the videocassette machine is switched to the same channel.

The cassette can now be inserted and the PLAY button pressed. The user need do nothing else to get a picture on the screen. The tape can be viewed and listened to just like a normal TV program, and, if privacy is required, headphones can be plugged into the machine and the TV sound turned down.

PLAYBACK INTO A TV MONITOR

To play back into a monitor the power connections must be made as before, but now two other connections are needed: video and audio. The **video out** of the machine must be connected to the **video in** of the monitor. Again, an RG 59/U-type cable can be used, but now either the PL 259 UHF plugs or the BNC-type plugs must be used. For the audio, either the **line out** or the **monitor** positions can be used, and these are connected to the input of the audio amplifier or monitor. The audio amplifier may be built into the TV monitor or may be separate, but the connections are the same in each case. Figure 2-4 explains these situations more clearly. The cassette is inserted and played back as above, and the TV monitor and audio monitor are adjusted as needed.

TV Set and TV Monitor. A TV set has a tuner which enables it to pick up programs transmitted by the local stations; usually it does not have video and audio inputs and outputs. A TV monitor

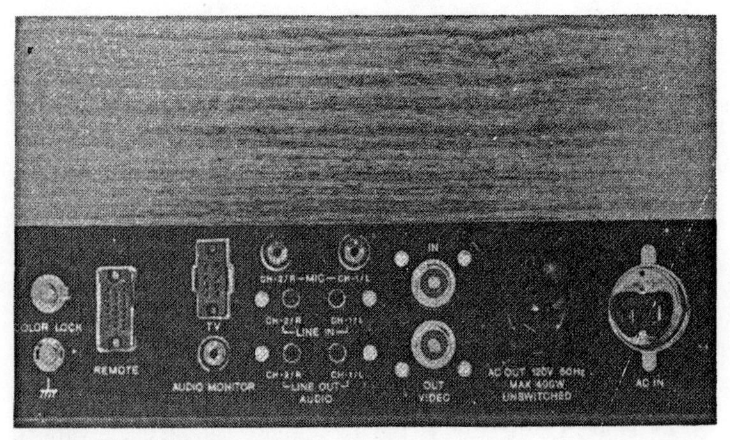

Fig. 2-1. Rear of machine (courtesy of Sony).

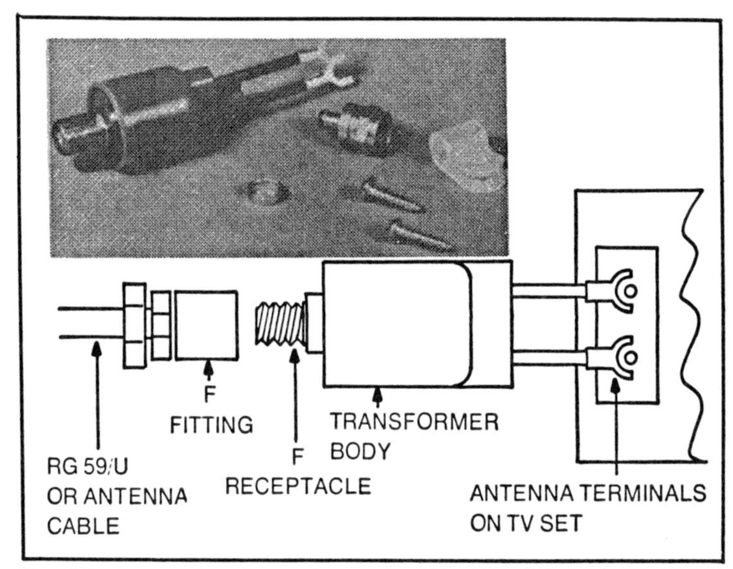

Fig. 2-2. Antenna transformer (courtesy of Sony) and drawing of antenna transformer.

does not have a tuner, so it cannot pick up broadcast stations, but it does have video inputs and sometimes video outputs. TV monitors usually do not have an audio section built in, and so audio amplifiers are in common use; videocassettes have been designed with the ability to play back into both. Several manufacturers in the industrial and educational fields have modified TV sets so that they also can be used as monitors. They are provided with a **video out** and an **audio out** so that **off air** recording is possible. These are often called "jeeped" TV sets.

CASSETTE INSERTION

The insertion of a cassette into a machine is one of the easiest things to accomplish, and all manufacturers have reduced this to a level where it can be done by a small child. It is even easier than the audiocassette.

To insert the cassette, the loading slot or carriage is raised by operating the **eject** or **insert** lever. In some machines this is not necessary: the cassette is simply pushed into an opening. The cassette must be correctly oriented, and it is shaped so that it will only enter the machine one way. When it is pushed fully in, the loading slot drops to the threading position, and then it is safe to use the function buttons. The tape will automatically thread when the **play** or **forward** button is pressed.

Prior to insertion it is first necessary to check that the tape has been fully rewound onto the feed spool. This is easy. The cassette is turned upside down, and the center section behind the protective flap is examined to see if the clear plastic leader is visible. If tape is seen, then a rewind is indicated. This is shown in Fig. 2-5. If the tape is to be used for a playback, the small red safety cap on the underside of the cassette should be removed. This disables the record functions and prevents inadvertent program destruction.

With the one-reel cartridges the tape is always fully rewound when not in the machine, and thus the program always will start at the beginning of the tape.

A typical set of instructions similar to those found in the cassette boxes is shown below. This covers the main points regarding cassette insertion.

OPERATING INSTRUCTIONS FOR CASSETTES

To get the most out of this videocassette, follow these precautions (see Fig. 2-6):

Keep the videocassette away from high temperatures, excessive dust, and moisture.

Keep the videocassette away from strong magnetic fields.

Do not leave the videocassette exposed directly to the sun.

Do not attempt to open the videocassette.

Before inserting, take up slack in the tape by turning the reels manually in the direction indicated by the arrows. (See Fig. 2-6A.)

Avoid repeated insertion and removal of the videocassette without operating the machine.

If the machine fails to function as desired when the **play, rewind** or **forward** button is pressed, check that the tape is not at either end of its travel.

You may monitor the tape on the supply reel through the translucent window. (See Fig. 2-6B.)

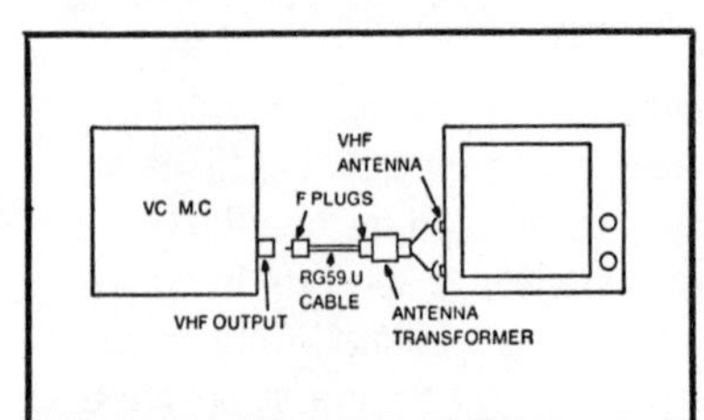

Fig. 2-3. V.C machine to TV set connection.

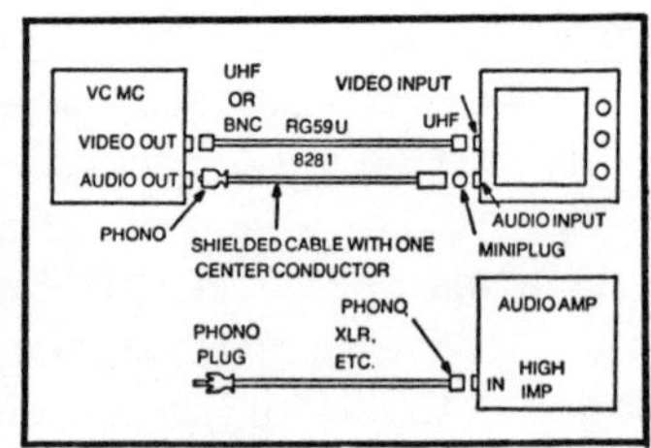

Fig. 2-4. To TV monitor.

Since the full width of the tape is used for recording pictures and sound, the tape cannot be reversed. (See Fig. 2-6C.)

If the red safety cap at the bottom has been removed, recordings cannot be made. To protect a recording from accidental erasure, remove the cap.

Rewind the tape onto the supply reel. Always store the videocassette in the cassette case when it is to be carried or mailed or when it is not to be used for a long time.

CASSETTE REMOVAL

Removal is just as easy as insertion. With the two-reel type of machines, the **stop** button is pressed, and this retracts the tape into the cassette. When the tape is fully retracted the cassette can be removed from the machine. With these it is not necessary that the tape be fully rewound to the beginning. To actually remove the cassette, the **eject** lever or button is used, and the cassette pops or when it is not to be used for a long time.

With the one-reel cartridge types, the cartridge removal is automatic. Pressing the **rewind** button will cause the tape to be fully rewound to the beginning, at which time it is fully inside the cartridge. It then automatically ejects from the machine. Unlike the previous type, cartridge removal implies full rewind.

FURTHER PLAYBACK OPERATIONS

When a tape is played back it is possible to get a perfect picture without using any controls other than the **play** button. However, during playback three main problems can occur to upset the picture. To obviate these, three controls are provided.

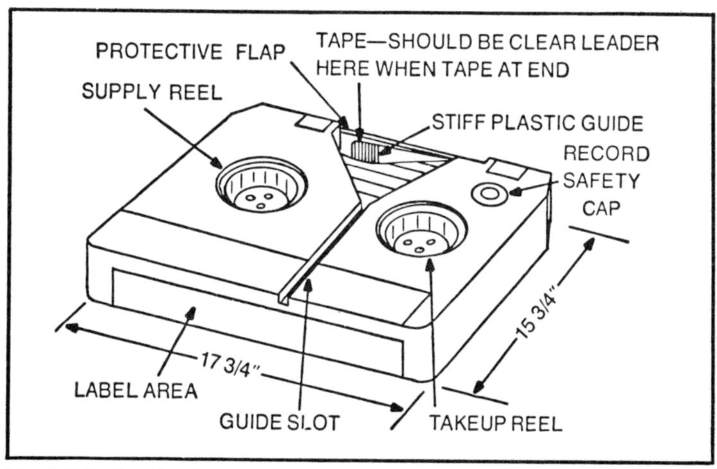

Fig. 2-5. Underside of a cassette.

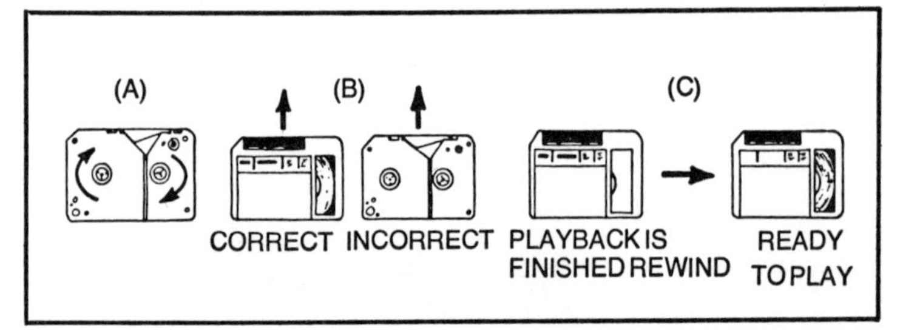

Fig. 2-6. Cassette instruction sheet diagrams.

● **Mistracking.** Mistracking a tape is caused by the rotating video head scanning two video tracks during one revolution. This effect appears as a few lines of noise across the picture, and they appear to run up or down the screen, causing a shuddering effect. To restore a perfect picture the **tracking** control should be slowly turned until this effect disappears.

● **Hooking.** A hooking or bending of the picture often appears at the top of the screen, and it usually is not stable. The effect is often called *flagging*. It is caused by the playback tension in the tape differing from that during record. Many factors can cause this, ranging from a bad tape to a maladjusted machine to atmospheric conditions. It is corrected by using the **tension** or **skew** control. Often the range of this control is inadequate to correct the fault.

● **Color problems.** In a well-adjusted machine playing back a good tape the color will be automatically controlled and will need no adjustment. However, two problems can be observed: either color bands will be seen across the screen or the actual colors will be wrong, for example, green faces. To correct these the **color lock** control is pulled and then turned. When the tape has been completed, this control should be pushed back in, because it will affect the next tape.

The three controls above work in the **playback** modes only and have no affect in **record.**

SUMMARY

Although the two playback modes described here are the simplest uses of the machines, they do serve as a good starting point for learning the correct and full use. The extensions of these two modes and the other possibilities of use are covered in the next few chapters, which explains all the essentials needed for using videocassette machines.

3
Simple Cassette Recording

This chapter offers a simple introduction to the operational procedures needed to record a tape.

In its simplest modes, recording is just as easy as playback. The cassette is inserted into the machine as previously described, but now it is essential that the protective cap is in place. Once the cassette has dropped into the threading position this cap holds back a small lever or microswitch, which allows power to be applied to the record circuits and associated relays. To accomplish recording all that is necessary is to see that the required connections are made and that the input switch is in the correct position.

The first step is to check for an electronics-to-electronics signal at the output of the machine. (This is usually referred to as an E-E or E-to-E signal.) A video monitor connected to the **video out** will provide this check. An E-E signal is an input signal which has passed through all the electronics of the machine and appears at the output: it is thus an indication that everything in the machine is working perfectly. It is not an indication that the recording has actually been put onto the tape—only a playback can give this assurance.

To get an E-E signal the cassette must be in place and the **record** button must be pressed. This button will lock down but will not run the tape out or transport it—only the **playback** button will do this. When the **record** button is pressed it allows an E-E picture to be seen at both the **video out** and the **vhf out** and means that a recording can take place. Without the cap or cassette inserted the record electronics cannot function and an E-E signal cannot be obtained. If recording is attempted, the playback mode will be initiated.

In order to view the E-E picture and to hear the incoming sound, a monitor or TV set must be hooked up to the machine just as in the playback mode, and this must be done in addition to the record connections. Any of the playback connections can be used, either separate audio and video or composite RF (audio and video) to the TV antenna. If an E-E picture is observed and the audio signal is heard on the speaker, then a short test recording should be made. To record on the tape, the **record** button is held down and the **play** button is pressed. This will lock the **record** button down and thread the tape. If the **record** button is not held down it will release and the playback mode will be entered.

When the tape is fully threaded, the record mode will be initiated and the incoming signal will be recorded onto the tape. About 30 seconds should be recorded and then the **stop** button pressed. The tape will retract back into the cassette and then-rewind for a few seconds. After the warning light goes out, the **play** button is pressed and the recorded section is played back. If a satisfactory recording has ocurred then it is safe to proceed with the main programs. The tape should now be rewound to the start.

On most machines no controls are provided in the record mode. Provided the video signal is of a correct level and the audio has been connected properly, then the signals will be automatic gain controlled (AGC) and no further operations will be necessary.

RECORDING EXAMPLES

Three simple recording situations are shown in the following examples, and these represent the most common and basic of the recording modes normally encountered. Although the operations and interconnections are somewhat inter-dependent, these examples concentrate on the operations; connections are covered in detail later.

The main difference between playback and recording is one of attitude rather than technique. Playback is a passive procedure: it is the viewing of something already created. Recording, on the other hand, is an active procedure: it is creative and requires participation. Even in its simplest modes it requires thought, preparation, and work. The meaning of this will become apparent as experience is gained.

Recording Off Air With The Internal TV Tuner

The TV antenna or the CATV cable is hooked up to the antenna terminals at the rear of the machine in the same way they are connected to the TV set. The **input select** switch is set to the TV position, and the channel desired is selected and tuned just as with a TV set. Figure 3-1 shows the simple connections.

The **record** button is pressed, and the incoming signal is viewed in the E-E mode on a monitor or a TV set. For best reception the fine tuning ring is used, and the tuning indicator pin is made to give a peak reading on the dial. The recording can be started whenever desired.

Recording From A Camera And A Microphone

Figure 3-2 shows the easiest of the recording setups. This suffices to cover many situations and is the simplest of all to accomplish. It is a setup which can work anywhere, such as in an office or a classroom. A proper studio is not needed.

The camera must produce a fully composite video signal, and it is connected to the **video in** of the cassette machine.

The microphone should be connected directly to the **mic in** of Channel 1. Channel 2 can be used if desired. The difference between these is covered in detail later.

The main operational problem with this simple setup is that there is no way of fading into the sound and picture at the start and no way of fading out at the end, because no controls are provided. This is the main drawback to the videocassettes as simple production machines.

This means that the talent must sit still and silent until the tape is threaded and ready. Once the warning light goes out, about 10 seconds should be allowed for the machine to fully stabilize. Then a hand cue must be used to start the talent. At the end of the program the talent must sit still and silent again for about 30 seconds before the stop button is pressed. This sitting still and silent at the beginning and end allows for fading in and out as the tape is played back in a studio through a mixer. It ensures that the

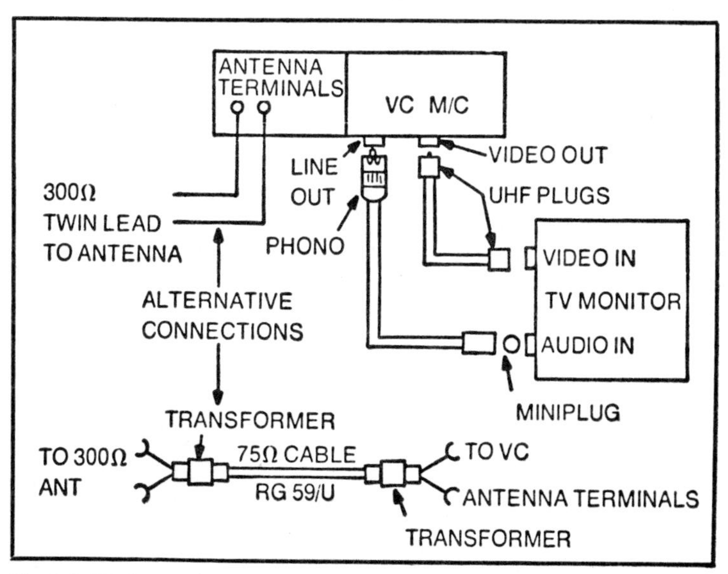

Fig. 3-1. Off air recording connections.

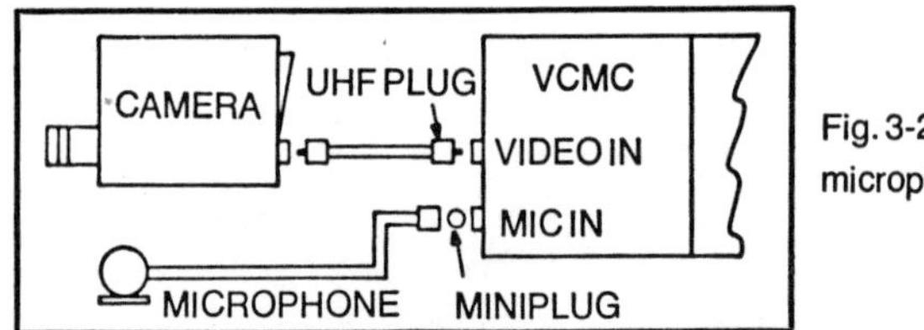

Fig. 3-2. Camera and microphone connections.

beginning of the program will not be clipped in recording later in copying sessions.

Although this simple procedure is inadequate for a professional production, it does give practice in simple recording and playback and allows the new operator to build confidence. Once this is gained, the more complicated procedures may be attempted.

The main problem with this is the impossibility of keeping the tape black and the sound down until the start, when ideally both should be faded in together. A black opening and end can easily be achieved by capping the lens, but the talent must still be instructed to remain silent.

COPYING ONTO ANOTHER VIDEOCASSETTE

One of the most common and most used recording modes with cassettes is that of copying a master tape onto a cassette format. Once the correct connections have been made it becomes only a matter of pressing the **record** and **play** buttons on the recording machine and then pressing the **play** button on the playback machine.

If the master tape has a sufficient amount of silent black, then these two buttons can be pressed simultaneously. When the machine begins to record it will start off with a section of the silent black before the program begins. If silent black is recorded at the end of the program, then it is easy to copy about 10 to 30 seconds of this before pressing the stop button.

If there is no silent black on the master tape, then difficulties can be experienced with both the opening and the closing of the program, and the starting point will have to be chosen and rehearsed carefully prior to the actual copying.

The connections for the various copying modes are covered in detail in a later chapter.

CONCLUSION

In general the recording of a cassette is as easy as playback, requiring that only one extra button is pressed and one additional precaution is observed. The simplest modes should give no difficulty to anyone, even the new user of a videocassette machine.

4 Operational Controls and Facilities

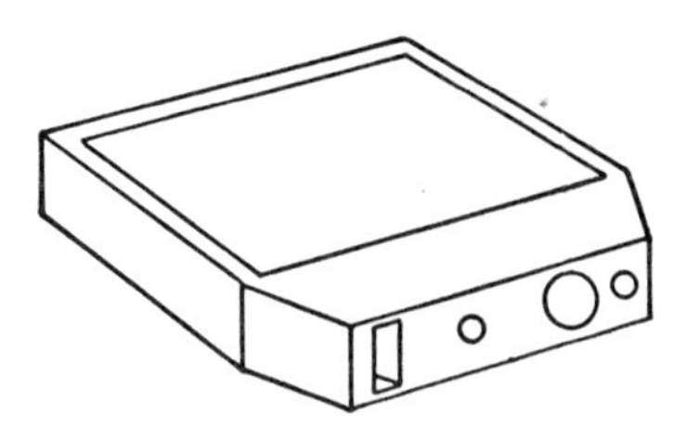

It is appropriate at this time to cover the operational controls provided on the videocassette machines and to explain the correct use of these controls. It makes sense to explain what the controls are and what they do before they are used, but since it is likely that many operators will have some experience with these machines, this chapter can be read later or used solely for reference purposes. It is put here for convenience in developing the operational procedures.

Because the cassette machines have been designed to be the easiest of all VTRs to use, and with the idea that all the important parts should never be touched by hand, or system of knobs, buttons and levers is provided which initiates the various functions by powering motors and solenoids inside the machine.

There are many ways in which these machines can be used, but the more advanced uses require some knowledge of their capabilities and how to perform a few semitechnical operations. The exact details of each model are given in the operations manual, and this should always be consulted. Everything that is possible with cassette machines is within the capabilities of the average person. This is the strength of these machines, because much can be accomplished with a minimum of technical training.

Three general warnings should be observed with all cassette machines at all times:

1. Due to their automatic nature it is imperative that all operating instructions be obeyed explicitly. Failure to do so can and will cause damage to both the machine and the tape.

2. Do not touch anything while the warning light is on.

3. Never force any of the controls. They should all operate easily and smoothly. If they do not, then maintenance is indicated.

For easy reference this chapter is divided into four main sections:

1. The controls—This covers the actual knobs, levers, and buttons. It explains what they are for and what they do, and it explains correct use and when not to use.

2. The plugs and sockets—Explains their functions and correct use.

3. The automatic functions—Several facilities are provided which are completely automatic in their operation, with no controls for the operator. These are covered very briefly.

4. Operating advice—A general list of do's and don'ts about machine and tape care is provided.

This chapter is fairly complete in its list of the controls and functions. Obviously not all are found on every machine, and certain slight variations in operation also occur from model to model. To determine which are actually found and which is the correct procedure, the individual operator's manual must be consulted.

A complete physical description of the internal mechanics and electronics of the machines and the construction of the cassettes has been left until later in the book, which covers the technical aspects more fully.

THE CONTROLS

The controls which are most used are naturally placed in the easiest and most convenient locations. This is either on the top of the machine near the front or on the front panel. The actual layout varies from machine to machine, but within a widely used format a given arrangement is closely adhered to. This is shown in Fig. 4-1. The actual number of controls also varies, but the following sections cover those found in most machines and the comments can be applied to all.

The General Controls

Power switch. This must be used by the operator before any other function can be selected, because the power is not turned on automatically by the insertion of a cassette. This switch is most often a push-on/push-off button which is mounted away from the rest of the controls, where it will not be accidentally touched. Associated with it is a light—usually green—to show when the power is on or off. After use, the power must be turned off by hand, because power is not automatically turned off by removal or ejection of the cassette.

The main function keys. The most important and most used controls are those dealing with the tape transport and the selection of the major operating functions. These are grouped together for ease of operation, and most are in the form of "piano keys" which are mechanically interlocked to enable only one

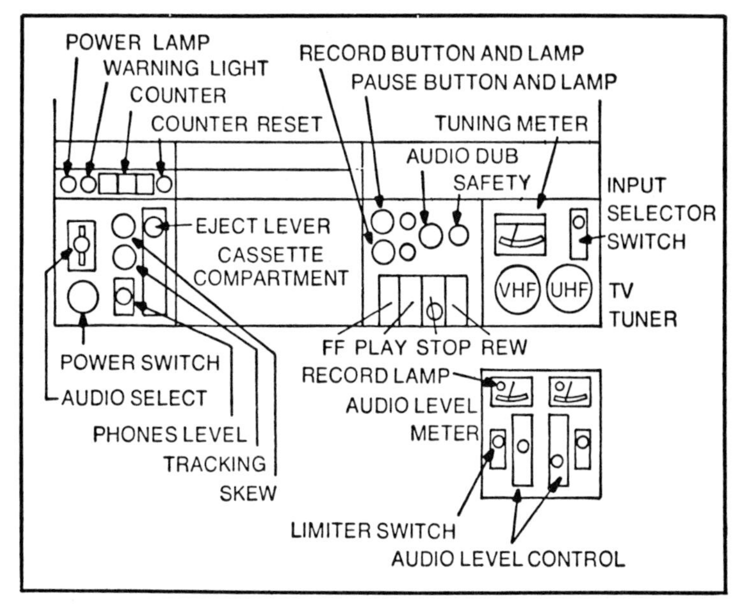

Fig. 4-1. Layout of front panel controls.

function to be selected at a time. They are built into a single unit, which also contains a small printed circuit board, switches, a release solenoid, and several levers which extend into the mechanical works of the machine. They select the mechanical functions, route power to the appropriate places, and serve to protect the tape and the machine from wrong or inadvertent operation once a function has been selected. These controls should be operated one at a time, and only at the correct time. They should never be forced, and if they give trouble, then a qualified serviceman should be consulted.

The **stop** button is the only one which is mechanically interlocked out when a function is chosen, and it must be used between the selection of all other functions. No buttons should be operated when the warning light is on.

An important feature is that if a tape reaches its end—either end in any function or mode—then the plastic leader actuates an automatic stop mechanism. This powers the solenoid on the key mounting which releases the keys, depowers the electronics, and stops the selected function, leaving the machine in the **stop** mode.

Associated with these controls is a warning light, which comes on when the tape is being threaded or retracted. Whenever this light is on, no keys should be depressed or even touched. They are mechanically locked out during this time; attempting to use them or slightly forcing them can cause damage to both the machine and the tape.

In most machines these controls are mechanical in their action and thus are not easily adapted to remote control. A few machines use a different type of function button. These are a nonmechanical action type: a gentle touch causes the key to operate a microswitch placed underneath. This is directly connected to an electronic control circuit, which consists of latches and gates, and these now control the functions required. The advantage of this is that it enables remote control with a very simple hand-held unit containing duplicate switches.

On some machines indicator lamps are associated with each of these buttons, other models have indicators only for record, and a few have no such lamps.

The individual buttons or controls are now covered. Figure 4-2 is a typical layout.

Play or forward. Both these legends are used for this most important control button. Depressing this button has the major effect of running the tape out of the cassette and around the threading path and the heads. It is the *only control* which does this and must be used in both the **play** and the **record** modes.

When pressed, the warning light comes on while the threading is in process, and the light stays on until the threading is complete. At this time power also is applied to the playback and servo electronics and the control circuits. Thus the key stays down and the playback mode remains active, and it is impossible to use any other control or select any other mode without first using the **stop** button. The exceptions to this are the **audio dub** and the **edit** controls.

If power is removed when this button is depressed, then the tape will stop in its run-out position, and it will be impossible to remove the cassette from the machine. Once power is returned the play mode will continue.

If the tape reaches the end, then the plastic leader causes an automatic stop, followed by a retraction of the tape back into the cassette, a short rewind (in some models a full rewind), and the automatic release of the **play** button. In some machines the mode is ended with an automatic cassette ejection from the machine.

Record. When this button is pressed it will stay down, the red record light will come on, and an E-E picture will appear at the video output and the VHF output of the machine. This will only occur if a cassette is in place and the safety cap is included. With no cassette inserted or with the cap omitted the record will not function. This button will not thread the tape; only the **play** button will do this. Momentarily pressing any of the other main functions keys will release the **record** button, but the correct one to use is the **stop** key.

To effect a recording the **record** button should be held firmly down and the **play** button pressed. This will now lock the **record** button in place and cause the tape to be threaded. Recording is ended by using the **stop** button; this releases the **record** and **play** buttons and retracts the tape.

The **record** button can be mechanically locked down with the **play** button while the machine is depowered. When power is applied the tape will run out, and recording will begin when the threading is complete.

With a cassette inserted and the **record** button pressed an E-E picture of the incoming video can be seen on a monitor. At this time power has not been applied to the recording amplifiers or to the erase circuits, so this is a safe viewing procedure which will not erase a small part of the tape. Power to these circuits is not applied until the tape is threaded.

As with all other tape recorders, use of the **record** function will wipe away all previous material; this includes the video, the control track, and both audio tracks. Hence recording video over an existing audio commentary is impossible.

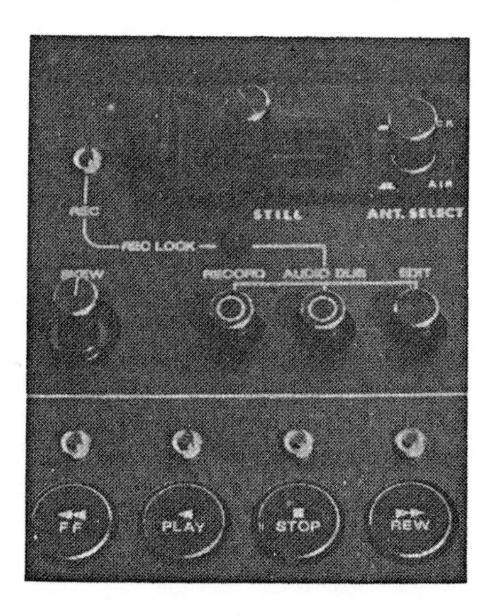

Fig. 4-2. Function buttons (courtesy of JVC).

Stop. This is a most important button or key. It must be used between all other modes of operation to stop whatever mode the machine is in. Its operation causes the warning light to come on, it releases all other buttons and functions, and no other function must be attempted until the warning light goes out.

In the playback and record modes, using the **stop** button on a cartridge machine will simply arrest the tape in its run-out mode at the time the button is pressed. It will stay in this position until another function is selected.

With cassette machines, the **stop** button causes the tape to be fully retracted into the cassette and the rewound for about 10 seconds of program time. The total time for this process is about 6 seconds. During retraction and rewind the **stop** button is locked down and automatically releases when the short rewind is over. Simultaneously the warning light goes out, and only now may the next function be selected.

Rewind and fast forward. With the cassette machines both of these functions are conducted with the tape entirely within the cassette. (Note: This is not true of the larger editing machines.) If the tape has been played up to some point and either of these functions are required, the **stop** button must be used first. This retracts the tape into the cassette, and when fully retracted either of these functions may now be selected. The tape is then transported from one reel to the other without being run out around the threading path. A belt and puck drive is used from the main motor to the appropriate spool, the other being left to rotate freely.

If the **rewind** or **fast forward** is continued to the end of the tape, then the clear plastic leader will cause an automatic stop. (At this point the same function should not be selected again, because an automatic lockout is not provided, and the mechanism will just continue to run and wear out the drive belts.) When the end of the function is reached, the keys are automatically released and will pop up to the rest position.

To go from one of these modes to the other, the **stop** button must be used, and it must be used prior to selecting any other mode. Each of these modes can be stopped anytime and anywhere in the tape.

With the single reel cartridges, the **rewind** and **fast forward** modes can be used once the tape is run out around the threading path. Neither can be initiated immediately after cartridge insertion. The **play** mode must be allowed to be automatically entered first, then **stop** is used, and the required other function is selected.

At the end of FF the tape will automatically stop and jump into **rewind.** At the end of rewind the tape is fully inside the cartridge and the cartridge will reject. For this reason the **rewind** button is sometimes labeled the **rewind eject** button.

Those machines which have timers and auto-repeat functions will also automatically **fast forward** and **rewind** over the time or section of tape selected for continuous repeats.

Controls Associated With the Main Function Keys

Several other controls are provided on some models, and these act as extensions of the main functions.

Audio dub. This facility is provided on all recording machines. In the normal recording mode, audio is recorded with the video onto either or both tracks. Further audio can be added onto Channel 1 only at a later time by using this control. To use it, the tape must be running in the **play** mode, and when the new audio is required the control is pressed. The input to Channel 1 can be from any source. This is provided as a useful feature for adding a second language, additional commentary, a verbal cue track, etc. Note that TV audio is recorded onto Channel 2 and that this cannot be overdubbed.

Audio safety. This is a mechanical interlock which protects the **audio dub** control. It must be pressed and held down, and then the **audio dub** control is pressed.

Phase. Not all machines are provided with this useful control. It momentarily stops the tape transport without changing the operating mode, and when it is released the tape runs again in the same mode. It is provided for use in the **record** and **play** modes but can be used in the **dub** and **edit** modes.

If power is applied with both the **play** and **pause** buttons pressed, then the tape will run out and thread, and the machine will remain in **pause** until the **pause** control is released. Then normal operation will begin. Because the use of this button does not cause the tape to retract, it is most useful in setting a tape to a cue point. Individual models should be checked to see if this gives a **still frame** picture, because it is not really the same function as **still frame.** If such a picture is produced it may have tracking noise lines running horizontally across it at some places. The only way to adjust the tape position to obviate these is to release the button and immediately press it again. Repeat this until a picture free of interference is obtained.

Edit. The earlier models were not provided with this function, but it has been incorporated into some of the later models. Each model should be checked to see if **assemble** or **insert edit** has been provided, because these are quite different in their operation. To use the **edit** button, the tape must be in the **play** mode, and it is simply pressed at the time the new program material is required. This switches the machine over into the record mode, and the new material is put onto the tape. The necessary connections and the operational procedures are covered in detail in chapter 6.

Eject lever. The functions of this have been covered adequately under cassette insertion and removal. Some machines have an automatic ejection, and therefore this control may not be provided.

Record lock. This is not found on all machines, and its functions are covered in the following section.

Memory counter. This is an **on-off** switch located near the tape counter and it is similar to the **normal-repeat** switch. In the on position, the tape will rewind to the point where the counter shows 000 and then stops.

This is useful for repeated playbacks of the same section of tape. When the required start point is found the counter is reset to 000, and the tape is played to the desired end point. The **rewind** is used and the tape automatically stops at 000, and is ready for a replay.

Eject Button. This is an automatic cassette removal function. The cassette holding slot is mechanically interlocked to the threading ring and cannot rise until the tape has been fully retracted. Notches in the threading ring ensure this, and mechanical levers are used to lift the tape slot at the correct time.

Timers, Counters, and Auto Repeats

All machines are provided with a timer or a counter to indicate how much tape has been used, but the type and the use for which it is intended vary considerably. It is convenient to include a brief description of the auto repeat facility because it is controlled by the more advanced timers. Figure 4-3 shows a typical dial.

Counter. The simplest of the indicators is the belt driven counter. This is three decade wheels which reset manually to zero and then count up to 999 before overspilling into 000 for a second time around. It is a very inaccurate method because the drive is taken from a spool and is only an indication of how much tape has been used.

Tape indicator. This is basically the same as the above counter in that it is driven from a spool, but it is more complex in its

Fig. 4-3. Timer dial (courtesy of JVC).

construction and functions. It uses a dial face marked in minutes from 0 to 30, and has a single pointer which indicates elapsed time. A second (usually red) pointer is used to set the point at which **playback** or **record** is required to start. Once this is set, the FF button is used, and when the timing pointer reaches the red pointer, the machine stops and enters the pbk mode. These are quite accurate in their timing of the tape.

Clock. Some machines have a battery driven clock with two sets of hands. The clock runs continuously, and the second set of hands provide for automatic turn-on at a preset time of day. This is different from the tape counter function. A 1.5 V UM-2 battery is used, which will run for about a year, and can drop to as low as 1 V before it affects the operation of the clock. Easy replacement is possible without affecting any other machine operations.

Timing Unit. Some manufacturers provide a separate timing unit. This works similarly to the common "coffee-pot" or "swimming pool" turn-on/turn-off timers, but has the turn-on function only. The desired turn-on time is set on a dial and the end of the tape provides the machine turn off.

The timer is plugged into the wall outlet and the cassette machine and TV set (if used) are plugged into the power outlets on the timer. The timer is switched to off, and this provides power to the cassette machine and TV set. The cassette is loaded, the tuner set to the desired station, and the machine is put into the record mode. The timer is then turned ON. This interrupts the power to the machine and TV set and supplies power only at the desired time.

The time set dial and the actual time dial may be conventional clock faces or digital readout.

Normal repeat playback select switch. In the **normal** position the tape will play in the normal manner. In the **repeat** position the tape will repeat indefinitely a section which has been selected by the timer or the tape indicator.

The red pointer is set to the desired time at which the repeated playback is desired to begin, and the select switch is set to **repeat.** The tape will now play from this point to its end—that is, the end of tape where the plastic leader starts, not to the end of the program section. The plastic leader initiates the **rewind,** which takes the tape back to the selected start point instead of to the beginning of the cassette. The playback of this section is now repeated. This process continues until it is stopped manually.

It is also possible to **record** from a preset point on the tape and then to have this new recorded section play back repeatedly. Instructions for accomplishing this tend to differ from model to model, and the operator's manual should be consulted before attempting this mode.

Record lock. This is for use when a timed, unattended recording is required. The clock controls can be set to start a recording at some time when the machine is unattended. If the **record lock** button has been pressed, it will lock the **record** button down when it is pressed. When the selected time on the clock arrives, the machine will then automatically depower and shut off. It is essential that the **auto repeat** switch is left in the normal position; otherwise the tape will rewind and record again and again.

The Playback Only Controls

Certain controls are active in the **playback** mode only. These have been reduced to a bare minimum and in most cases do not need to be touched at all. If excessive use of any of these is required, then either a defective tape is being played or the machine requires maintenance and realignment.

Tracking. This is a normal helical VTR control, which should be adjusted for correct tracking of the tape, because it is viewed on the monitor screen. A tracking meter or a video level meter is not usually found on these machines and thus cannot be used to aid in tracking adjustments.

Skew. This is a normal tension control. It works in playback only, being preset in the record mode. It should be used to provide minimum or no hooking of the picture at the top of a monitor screen.

Color/mono switch. This is found at the rear of the machine. All machines will automatically record a color signal in color and a monochrome signal without the color. Because most tapes are in color this switch should be left in the **color** position.

However, if a monochrome monitor is used for viewing or a monochrome tape is played, then it should be set to the **mono** position.

Color lock control. This normally should not be touched. It is preset at the time of manufacture or alignment for correct color playback and should be used only if a tape will not lock up on color when played. To be used, it must be pulled and rotated. After use it should be pushed back in; otherwise it will affect other tapes.

The Record Controls

The cassette machines are completely automatic in their record functions, and it is possible to make a perfect recording without touching any controls. However, it is necessary to select the source to be recorded, and a switch is provided for this. Also, several controls are provided with the tuner section to facilitate its use.

Input select switch. This selects either the external audio and video sources which are plugged into the back of the machine or the internal TV tuner. The video and audio are switched simultaneously. Whatever is selected by this switch also appears at the outputs in the **record** and E-E modes. Its two positions are labeled ANT or TV, and EXT.

TV tuner. The controls with this are exactly the same as those found in normal domestic TV set and should be used as such, with the picture viewed on a TV set or monitor plugged into the output of the machine. The tuner is described in more detail later. Up to four other controls may be provided with the tuner.

Fine tune meter. Sometimes called TV meter or tuning indicator, this is set to a peak reading when fine tuning the tuner.

AFT switch. This switches in an automatic fine tune circuit.

Antenna select switch. This selects either the incoming program or the output of the tape to be fed to the outputs for viewing. It is very similar to the input select switch.

Local/DX switch. This adjusts the input level from the antenna to the TV tuner. It reduces the level from a strong local signal and does not attenuate that from a distant weak station.

Note: Two antenna terminals are provided for use with the TV tuner. Both UHF and VHF antennas are connected separately and are selected by the tuner as with a normal TV set. When the machine is depowered and not in use the VHF antenna is fed to the antenna terminals of a TV set which is connected to the VHF output of the machine. The UHF is not fed through. This subject is covered more fully later in the book.

The Audio Record Controls

Some models of machines are made without a TV tuner, and in its place is provided an audio control panel. This allows control over the audio to be recorded and makes the machine suitable for music and other types of studio recording. A level control, level meter, and limiter switch are provided for each channel. To use it, the **limiter switch** should be set to its OFF position and the **level controls** adjusted so that the audio peaks do not exceed the arrowhead mark on the meter. The **limiter switch** is now put to the **on** position and will prevent the audio peaks from exceeding this preset point without introducing distortion. The meters are similar to normal volume unit (VU) meters, and they are in the output circuit of the audio amplifiers, so they will work in both the **record** and **playback** modes. A small record lamp in the corner of each indicates that the **record** mode has been selected for that channel.

Controls Common to Record and Playback

Audio select switch. This selects the audio which will appear at the audio monitor jack on the rear panel. This same audio is fed to the VHF modulator and so will be heard on a TV set used for viewing and listening. It selects either Channel 1, Channel 2, or a mixture of both. The headphones are fed audio from this switch, but the feed is different from that just described: In position 1, Channel 1 is fed to both ears; in position 2, Channel 2 is fed to both ears; and in MIX, Channel 1 is fed to one ear and Channel 2 is fed to the other.

Headphones switch. This controls the level heard in the headphones. Only 8 or 16 Ω (ohms) headphones are specified. High impedance phones can be used, but the level control will not now function.

Channel select switch. This selects the TV channel used to view the output of the machine. Some machines use Channels 3 and 4, and other prefer Channels 5 and 6. This should be checked with the operator's manual.

THE PLUGS AND SOCKETS

One of the most characteristic features of these videocassette machines is the variety of plugs and sockets used, which is quite deliberate. Audio, video, and RF require different types of sockets and cables, and the correct one has been used in each case. This also provides the operational advantage that makes it impossible to plug the wrong cable into the wrong input or output. All of these have been mounted on a rear panel, which is generally divided into input and output sections, as in Fig. 4-4.

The Outputs

Video out. Only one direct video output is provided, and this is a standard video signal suitable for use with any other piece of video equipment. It must be terminated in 75Ω at all times during use by a monitor or some other video input. In the **playback** mode the video out from this plug is the video on the tape; in the **record** and E-E modes the video input to the machine appears at this plug. A PL 259 UHF-type plug or a BNC-type plug is used.

VHF or RF out. Both terms are used to describe this output. Most often this is a standard F plug used in the MATV and CATV industry. When the machine is depowered, the output of this socket is the antenna connected to the VHF antenna terminals. When the machine is powered the output is the E-E of the input video and audio, and in the **playback** mode it is the signal from the tape. This is a 75Ω output, which must be connected through a matching transformer to the antenna terminals of a TV set.

In some machines the same UHF plug used for the video out is used for the RF.

Line out left. This is the audio output of the left channel, or Channel 1. An output is always present whenever the machine is powered, and it is either the input signal or the audio from the tape.

Line out right. This is the audio output of the right channel, or Channel 2. It is identical to the above otuput.

Audio monitor. The output here is selectable with the use of the audio select switch. It is either Channel 1, Channel 2, or a mixture of both. Each of these outputs are unbalanced, approximately 0 dB (decibels) level, and are rated as 10 kΩ output impedance.

Headphones. This is a stereo ¼ in. phone plug which provides listening facilities to both tracks. A set of 8Ω or 16Ω phones should be used. Two preset listening levels are available with the phones select switch.

The audio outputs are covered in more detail later in this book.

The Inputs

Video in. This is a standard UHF PL259 or sometimes a BNC socket. It will accept the output of any standard-type TV equipment provided it is a composite video signal.

Antenna inputs. Two sets of screw terminals are provided, one for VHF and one for UHF. These are normal 300Ω inputs such as those found on a domestic TV receiver. If a 75Ω antenna or a CATV lead-in is used, then a matching transformer must be used.

MIC in right. This is a microphone input to the right channel, or Channel 2.

MIC in left dub. This is microphone input to the left channel, or Channel 1. This channel should be used for dubbing an audio commentary onto an existing tape.

A wide range of microphones can be used with these machines; the input level is rated from −72 dB up to −30 dB. However, it is advisible to check a microphone with a test recording before committing oneself to it for a program.

Line in right. Also called **aux in r.** This accepts a relatively high level and is suitable for most output found on electronic equipment.

Line in left dub. This is the same as the above input, but it is also active during the **audio dub** mode.

Both line inputs will accept a wide range of levels and are rated to work between −20 dB and +10 dB approximately. The input impedance is around 100K and will not load down any input signal.

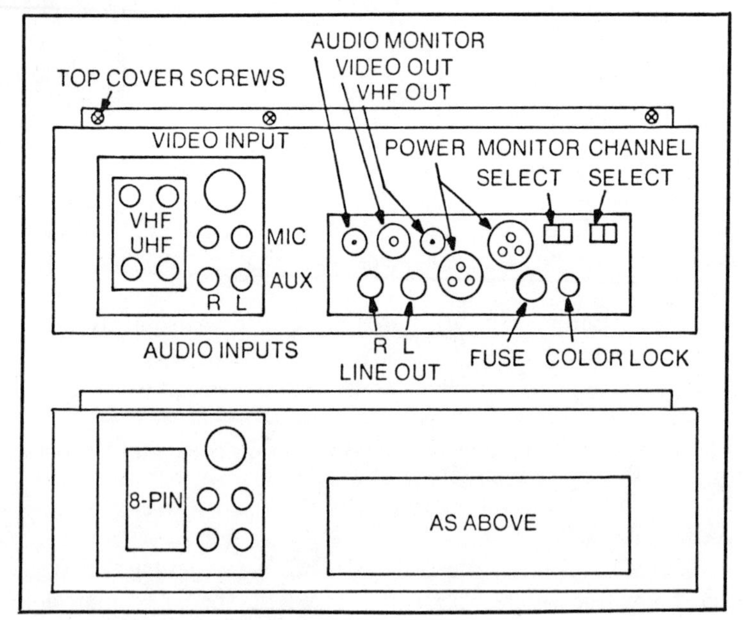

Fig. 4-4. Rear panel of machine.

General Purpose Sockets

The 8-Pin. This is a most important connector within the EIAJ-1 format of VTRs. It is provided on some cassette and cartridge machines to enable easy connection to the TV monitors made with this series of equipment. Its use is fully covered in Chapter 6.

AC power sockets. Most machines have two of these. One accepts the output from the wall power plug, and the other is a loop through so that a monitor can be plugged into the rear of the machine. This is useful when only one wall socket is available. Never plug more than one monitor into this output.

Fuse. This is in the AC line input. Its size is specified on the rear of the machine, and only this size should be used. It is dangerous to use a larger fuse. Usually 1.5 or 2 amp (amperes) is the approved size.

THE AUTOMATIC FUNCTIONS

Several automatic controls and facilities are provided in the videocassette machines, and these aid in the recording and playback of the video signal. They are a feature of most of the machines available today, but the term automatic should be understood for what it implies in practice. This tends to vary from one manufacturer to another and is often used more as a selling point than as a hard technical fact.

An automatic function is something which the machine does for the operator and which the operator does not actually have to do. Tape threading is perhaps the best example. However, the tape is not threaded just because the cassette is inserted into the machine; in most machines it requires that the operator push a **play** button. So the definition of automatic can vary.

The automatic functions can be divided into three main areas in which they work: mechanical, operational, and electronic.

Mechanical. All the major tape transport functions are conducted by the mechanics of the machines with no operator involvement. At the correct time a combination of mechanical and electrical interlocks selects the correct solenoids, circuits, etc. Although the functions are started by the operator pushing a button, many of the subsequent functions are started and stopped within the machine by the action of a light from a small lamp falling on a phototransistor. The opaque tape keeps the light from the transistor, but the clear plastic at each end allows it to fall on the transistor. In many machines it is the change of light level which is important and not the actual level, which is why a function—such as **rewind**—will not automatically stop, if used after it has reached the full rewind position.

Operational. In many respects these are not easy to distinguish from the mechanical functions. Although the auto repeat playback must be set up and selected by the operator, its implementation is fully mechanical. The same is true of the auto rewind and eject. The facility of being able to view one OFF AIR program while recording another requires little operator work, because most of the interconnections are performed within the machine.

Electronic. A range of automatic functions are provided within the electronic circuits of all machines. The most obvious is the automatic gain control—or AGC—for both the video and audio in the record mode. Provided the input signals are within the ranges specified, a sufficient signal will be recorded onto the tape to ensure a perfect playback.

Depending upon the complexity of the machine, certain other automatic controls are provided, which often are mentioned in the manufacturer's literature and have no provision for operator adjustment. These are:

1. *Automatic color control.* This controls the level and sometimes the phase of the color on playback.

2. *Automatic phase control.* This controls the phase—or hue—of the picture on playback.

3. *Automatic color killer.* This will disable the color circuits when a monochrome signal is used.

4. *Color noise canceler.* This will adjust and minimize the noise in the color signal.

5. *Drop out compensator.* This replaces the bad sections of the tape with a repeat of the previous line of video, thus avoiding annoying flashes on the screen caused by an old or damaged tape.

AGC and limiters. Because both of these are now incorporated into many models a brief explanation of the difference is in order. An AGC circuit controls the gain of the amplifier so that it presents a narrow range of output levels regardless of the input level. Thus a low level input is amplified more than a high level input. Normal dynamic ranges of input which are very close to each other in time during a program passage are preserved relative to each other, because it takes an AGC circuit a short time to react to input changes. In this way the artistic quality of music and speech is not destroyed.

The main objection to AGC circuits is that during periods of extended silence, the AGC will gradually work wide open, and this tends to introduce unnecessary background noise. Then, when the sound does begin, the opening few seconds can overload due to the circuits reaction time.

A limiter is not an AGC. A level is set above which the circuit will not permit further amplification. All dynamic ranges below this level are left alone. But as this level is approached the amplification is gradually reduced, preventing both signal overload and distortion. Several types of limiter circuits exist, and they work slightly differently, but those used in most cassette machines are quite simple.

OPERATING ADVICE

All machines have a set of operating instructions which should be understood and followed at all times. A videotape recorder is no different, and the advice given here is applicable to all VTRs and videocassette machines. These statements are not esthetic but are based on good practical reasons learned from operational experience and by paying attention to the way the machine is designed. The result of following this advice, which is given by all manufacturers, is beneficial to any videotape system in terms of ease of operation, long life, minimum downtime, and reliability.

Machine Care and Handling

● **Do** the following with the machines:

Keep the machine away from strong magnetic fields, air conditioners, heaters, dust, and moisture.

Avoid unnecessary shocks or impacts to the machine, especially while it is in operation.

Operate it in horizontal position only; this is the only position in which it is designed to operate.

When inserting a cassette, push it in as far as it will go, but never force it.

Only remove cassettes with the eject lever.

Remove the cassette when transporting the machine.

Always use the **stop** button between all operating functions.

Always wait until the warning light goes out before trying to eject the cassette or selecting another function.

When the **play** mode is discontinued before the end of the tape, fast forward the tape to the end and then rewind it fully. This prevents uneven winds which can damage the tape and affect its stability.

Beware of the automatic **repeat** function, especially when using the automatic timer for recording. It can cause continuous repeats of the recording mode, and thus valuable material can be lost.

When receiving a UHF program, view the incoming program on a TV set to check the fine tuning.

If the power is interrupted, press the **stop** button. This will cause the tape to unthread and retract when the power is reapplied. Normally the machine will unthread when the power is reapplied, and occasionally the warning lamp will stay on. If it stays on for longer than 10 seconds, press the **play** key.

● **Do not do** the following to the machine:

Do not attempt to load the cassette with the loading slot lowered.

Do not eject the cassette while it is playing.

Do not force the eject lever. If difficulty is experienced, then depower the machine.

Do not depower the machine when the warning lamp is on.

Do not touch the eject lever until the warning light goes out, even during or after rewind or fast forward.

Do not use the **rewind** key after the tape has been fully rewound and automatic shutoff has occurred. It will continue to run without a further automatic shutoff.

Do not stop playback or recording with the **power** switch.

Do not connect the VHF output to anything other than an approved RF device, such as the antenna terminals of a TV set or a proper distribution amplifier.

Do not put objects on top of the machine while it is in operation. This will block the ventilation holes and cause overheating.

Do not transport the machine with a cassette inside. Remove it and transport it in its own protective box.

Do not throw away the packing carton. This is a piece of advice given by all manufacturers in their operating manuals.

Do not attempt to service these machines unless you have some basic technical training.

Tape Care and Handling

Correct tape care and handling is most important and is necessary for the long life and reliability of the tape. Much of this means giving attention to small details and often appears to be a lengthy, fussy ritual which can impair the smoothness of playback sessions in a nontechnical atmosphere. It was consideration of this factor which made cassettes so attractive. Because of their fully enclosed and protected nature, many of the problems which plague the open reel tapes are largely offset; but this does not mean that the cassettes are immune to bad handling and bad storage. If the following advice is adhered to, then the cassettes will remain trouble-free.

● **Do** the following with cassettes:

Always rewind the tape onto the supply reel when it is finished.

Always store and ship the cassette inside its plastic protective case.

Keep all tapes away from strong magnetic fields, excessive dust, and moisture.

Always mark or write clearly on the label, not on the plastic cassette case. (Use a different color from that previously used, and erase previous titles with the same color used for the latest title.)

Always store the cassette so that it is on end, never lying down.

● **Do not do** the following with the cassette:

Do not put it on top of air conditioners or heaters.

Do not repeatedly insert and remove it without playing the tape.

Do not attempt to put the cassette in upside down. Unlike audio cassettes there is no "other side," because the full width of the tape is used for the video signal.

Do not pull the tape out of the cassette or cartridge. This invites damage and lessens its useful life.

Never touch the tape with the fingers.

The Cleaning Cassette

Due to their entirely closed construction, normal head and guide cleaning is not advised by the manufacturers. For this a cleaning cassette is available. It is played for about 30 seconds, or from 000 to 010 on the counter, and then removed *without rewinding*. Thus when it is next used a new section of the cleaning tape is run through the machine. When the end of the tape is reached it is then rewound and used again. It can be used several times before a new tape is required. Excessive use of this will shorten the life of all the heads, especially the rotating video heads, so its frequent use is not recommended. With care, the top of the machine can be removed and normal cleaning procedures can be followed. This procedure is covered in Chapter 15.

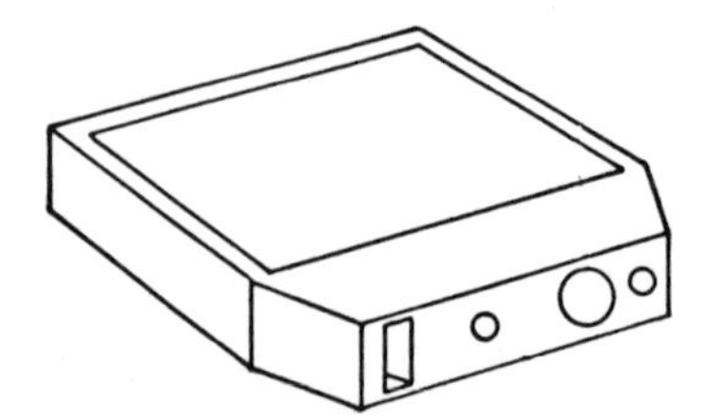

5 Common Interconnections

This chapter covers the most common of the interconnections between videocassette machines and other items of equipment. It concentrates on situations found in practice and illustrates them with examples. All equipment must be connected to some other item to produce a tangible, usable system for both playback and recording.

In general, with both playback and recording, the video connections are quite easily understood and are almost always the same. The audio is much more complicated and is treated at greater length.

The examples given do not exhaust all the possible situations, but they do provide guiding principles and form a good basis for just about every practical situation which could occur. If some other setup not shown here should be encountered, then no difficulties should arise when dealing with it.

Playback is treated first, and then recording is covered. The connections for each mode are entirely separate, and any recording situation can occur with any of the playback situations. So treating them separately allows for clarity of discussion and does not create the impression that a particular recording mode must be associated with or accompany a particular playback mode. They are entirely separate and independent.

Two particular special case connections are encountered with most VTRs. These are the use of the "8-Pin" multiplug and the RF connections. They are different in several respects from the above cases and are covered later.

To fully understand interconnections and how to make full use of the machines, some knowledge of video and audio circuitry is essential. However, much is possible without this, so the circuit descriptions have been left until later in the book. Because the different models of videocassette machines have slightly different inputs and outputs switching arrangements, plugs, and sockets, these are discussed in the corresponding chapters to avoid unnecessary confusion now.

Complete coverage of audio inputs and output levels, impedances, and plug wiring is found later in the book and can be consulted for specific details and used in conjunction with this chapter.

With regard to audio connections, three pieces of advice can always be observed.

1. When in doubt, try the connection. Equipment is seldom destroyed due to audio connections. The only caution to be observed is not to short or ground any outputs. This can destroy equipment very quickly.

2. No audio interconnection is perfect and has not been designed to be so, thus a wide latitude of levels and impedances will be found which will work when connected together.

3. Experimentation with audio connections is always in order, and often necessary. So when in doubt experiment.

INPUTS AND OUTPUTS

Before covering these as separate items, a few general operational points common to both should be observed.

The video and audio channels are active all the time power is applied to the machine. All outputs are inhibited or muted during threading, rewind and fast forward operations but are active at all other times, even when the tape is not running.

In most cases the video inputs and outputs are made with a PL 259 UHF or a BNC type connector. The RF is almost always an F socket. In both cases the proper video and RF cable must be used. This is usually the RG 59/U or the 8281 types, both of which are 75Ω cable. At all times the outputs should be terminated in a 75Ω input or load. The input to all VTRs is a terminating input.

Unlike the video, audio is not standardized with respect to levels, impedances, and connectors. These all vary from model to model, and each must be treated on its own merits. A wide range of adaptors is available to facilitate audio interconnections, but the wiring connections within these must be treated with some care. Shielded cable should always be used with the audio.

In most machines no level controls are provided, either for the inputs during record or for the outputs during playback. The input levels are specified to be within a certain range, and this is wide enough to accommodate most situations found in practice. If the signals are within this range, then the audio and video AGC circuits will record a satisfactory signal onto the tape and a satsifactory playback will be assured. The playback preamplifiers are set during alignment procedures and normally do not drift. They only need to be readjusted with a change of tape type, with age, or with a change of components due to failure.

Those machines which do not have a TV tuner have audio level controls, a meter, and a limiter switch instead. This enables some control to be exercised over the recording, which is desirable for music and some speech situations.

Outputs and Playback Connections

The video outputs. Two types of video outputs are provided on most machines. One is a standard pure video output, which is a fully composite video signal of standard amplitudes and is suitable for connecting directly to a monitor, a TV system, another cassette machine, or any other VTR.

The other output is an RF signal, with the video and audio modulated onto a carrier at the frequency of a standard TV channel. This is suitable for connecting to the antenna terminals of a domestic TV set or for plugging into an RF distribution system.

When the tape is in the playback mode a video signal is always present at both the **video out** and the **rf out.** So when a tape is being played, the picture will appear on whatever is connected to these outputs. In the **record** and **E-E** modes, the input signals to the machine are seen at both outputs.

Both outputs are affected by the tracking, color lock, color mono switch, and the skew controls in the playback mode only. These do not function in record, and they are video controls only. Both these outputs always must be terminated in 75 Ω.

The audio outputs. In most of the cassette machines two identical audio channels are provided, one for each of the tracks on the tape. Each has an independent line output socket, which always has the output of the channel present whenever power is applied to the machine. Many models have an audio monitor jack, the output of which is selected by a switch on the front panel and can be either Channel 1, Channel 2, or a mix of both. This same audio is fed to the input of the RF modulator.

To accomplish audio playback, it is mainly a matter of selecting the best or correct output from the machine and connecting it to the input of the other item of equipment. This can vary enormously, depending upon the location and use and the range of inputs likely to be found in practice is very wide.

The audio output from both channels comes from either the tape audio in the playback mode or from the inputs to the machine in the **record** and **e-e** modes.

All machines use the same heads for both record and playback, so in **record** mode the audio signals at the outputs are the input signals which have passed through the entire electronics of the machine. It is not a playback from a separate head, as in professional audio machines.

Note that **no** audio outputs will directly drive a speaker. An amplifier is required at all times.

Examples of Output Connections

These examples are intended to illustrate the range and type of audio equipment one can expect to encounter and the range and type of inputs one can expect to deal with.

Although certain types of inputs have been ascribed to definite pieces of equipment in these examples, it should be clearly understood that the given item may have a very different input when found in practice and that the described input can be found on many other items. No hard and fast rules can be laid down with audio equipment that has not been built for broadcasting or communications use.

Each of the following situations can be realized with any model or make of cassette machine, and thus the plugs, levels, and impedances may vary if another model is substituted for that in the example. The exact details of the plugs and sockets used on the machines are given in Chapter 4. The illustrations here use a simplified form of drawing which makes the connections obvious without complicating the drawing.

TV Monitor

The **video out** of the cassette machine is connected to the **video in** of the monitor; if there are two video inputs to the monitor, then either can be used. The **term off** switch should be set to the **term** or 75Ω position. The TV monitor will have an audio input, and this is connected by a separate cable to the **line out** of the machine. Use the output of the channel with the audio recorded on it. If no sound is heard, then try the other line out socket. In most cases the audio input to the TV monitor is unbalanced, so no difficulties arise with this connection.

The ANT VTR switch on the set should be put in the VTR position. Figure 5-1 shows this connection.

TV Monitor with Separate Audio Amplifier

This connection is used when the TV monitor has no internal amplifier and speaker. It is essentially the same as the previous mode, but now the audio is connected to a separate audio amplifier with its own speaker. Ideally this amplifier should have a high impedance input, and the **line out** of the cassette machine should be used. If an amplifier with a low impedance input is used, then the output level from the cassette machine will be dropped a great deal, but it still may be usable. It is necessary to have a level control or volume control on the amplifier because there is none on the machine. Depending upon the amplifier used, the audio connections will be subject to some variations and experimentation, and each situation must be treated on its own merits. Figure 5-2 shows typical situations.

Several Monitors and Separate Audio

This is an extension of the previous example and is typical of the situation found in auditoriums, lecture halls, and other public places.

The monitors are usually connected in a "loop through" mode, as in Fig. 5-3. The video is fed into the first one, and all the termination switches are set to the **off** or **high** position. Only the last one in line is set to **on** or 75Ω.

The audio can vary and must be treated on its own merits, as many situations can occur. The most common are the following:

1. Several audio amplifiers with their own speaker. In this case each must have a high input impedance, and they are parallel. Use the **line out** of the cassette machine, and connect it to one of the inputs.

2. One amplifier feeding all the speakers. This may have a high or low level input, and a range of impedances could be encountered.

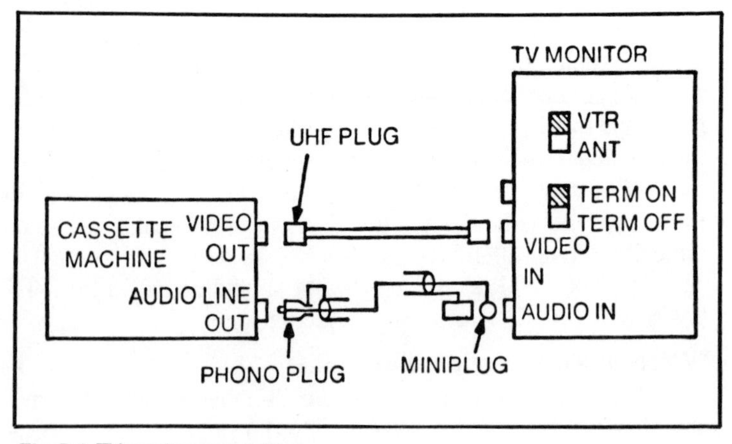

Fig. 5-1. TV monitor connections.

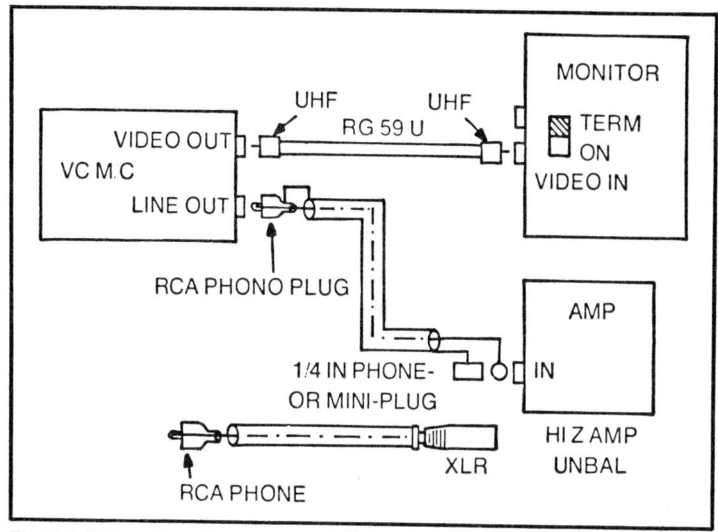

Fig. 5-2. TV monitor with separate audio.

3. A preamplifier is used. This can have several inputs, and they can be switch selectable between functions or inputs such as a tuner, a tape deck, a record player, an auxiliary, or a mike. A playback characteristic switch also may be included, labeled **nab, ccir, riaa**, etc.

The **line out** of the machine should be used, and the tape or auxiliary inputs are the best connections to make. **Nab** is probably the best characteristic to use. Figures 5-4, 5-5, and 5-6 illustrate these connections.

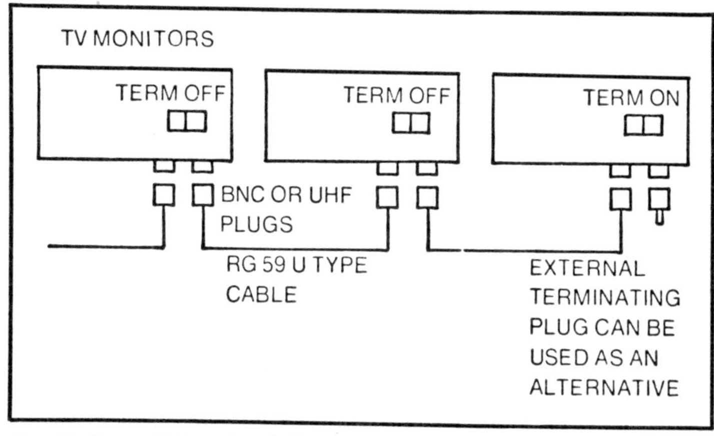

Fig. 5-3. Several TV monitors in "loop through."

A Small Studio System

The **video out** of the machine is connected to one of the **video in** positions on the video switcher or mixer in the studio control room. The problem that will be encountered here is that the video out of the machine is a fully composit video signal, and the inputs to the mixer require a noncomposite signal. Sync is usually fed in separately from the video inputs. To mix the cassette signal with the studio sync by plugging into a noncomposite input is likely to cause trouble and is a connection which should not be made.

Most of the smaller video mixers are made with one or two composite inputs so that they can accommodate VTRs, and one of these should be used. Often these will be associated with one or two select buttons on the control panel which are separated from the main controls used for switching the noncomposite camera signals.

It is impossible to fade in and out of the cassette signal, and any cuts to and from it will cause a vertical frame roll. The only way of avoiding this is with an editing machine which has the studio video output fed to the video input of the machine. This acts as a stabilizing signal and will allow cuts and fades to and from the cassette machine without introducing frame roll.

The audio most likely will be controlled by a small audio mixer. Most of the time these are likely to have standard broadcast 600Ω, 0 level inputs and mic level inputs. Some models are provided with high level, high impedance inputs. Preferably the **line out** of the cassette machine should be connected to a high level input.

If only mic level inputs are available the line out will have to be attenuated or "padded down" as in Fig. 5-7. A further complication is that the cassette machine outputs are unbalanced and the mixer inputs may be balanced.

If both channels are required to be heard on the program, then both line outs must be connected to different inputs of the mixer, or alternatively the monitor out can be used instead. Figure 5-8 shows a common audio mixer, and Fig. 5-9 shows the wiring of the input connections which use the XLR-type plug.

Stereo Playback

Because the two audio channels are quite independent of each other, it is possible to record and playback the audio in stereo. For playback of stereo sound, the **line out** sockets must be used because they are the only independent outputs. Most stereo hi-fi

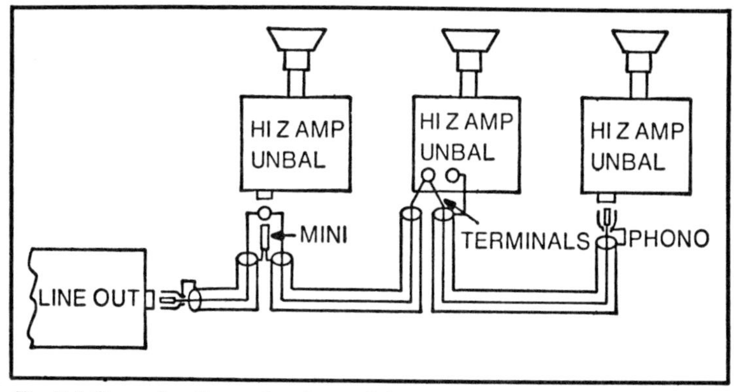

Fig. 5-4. Several audio monitors in parallel.

amplifiers and other general purpose amplifiers are high level, high impedance inputs, as are most audio recorders, so no difficulties should be experienced with the connections. It is also possible to use the **phones** output for this mode, as shown later.

Separate Audio Systems

This mode has several uses, such as feeding two different language commentaries to separate audio systems or feeding Channel 2 into a program and Channel 1 into a small control room monitor for listening to postrecorded cues. The manner of connecting the machine will be dependent upon exactly what is required and what is available but in general will be no different from anything already described.

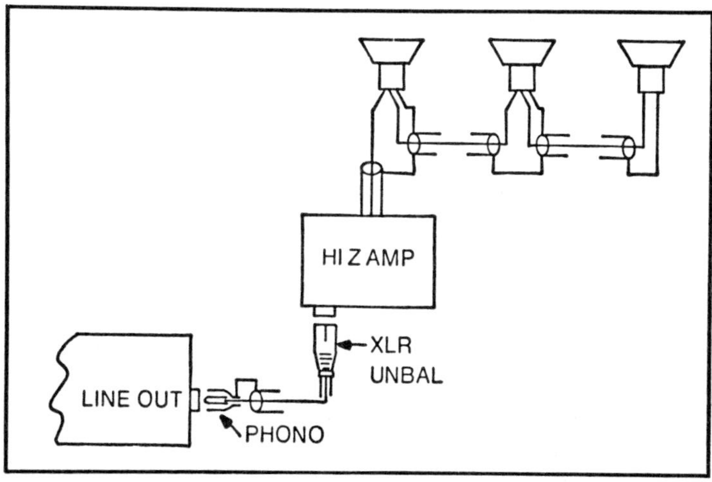

Fig. 5-5. Audio monitor with several speakers in parallel.

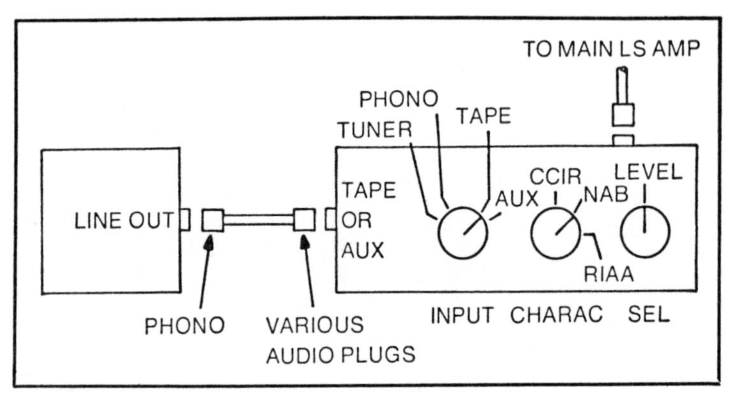

Fig. 5-6. Audio preamp input.

Video and Audio Distribution Systems

Large buildings and building complexes such as apartments, offices, and schools often distribute the video and audio over separate lines, and for this distribution amplifiers are used in each of them. Video distribution amplifiers (VDAs) often have a terminating or a loop through input and about four outputs, and the gain is set to be unity.

For audio, a bridging or high input impedance amplifier is used. The gain can be set to anything convenient for the installation, and again up to four outputs are available.

The **video out** of the machine is used, and the **line out** is used for the audio. Figure 5-10 shows a typical setup.

Copying Onto Another Videocassette or VTR

The outputs and inputs of all the cassette machines are compatible with most of the inputs found on the common VTRs. The audio **line out** is connected to the **aux** or **line in** of the other machine, and the video connections are obvious.

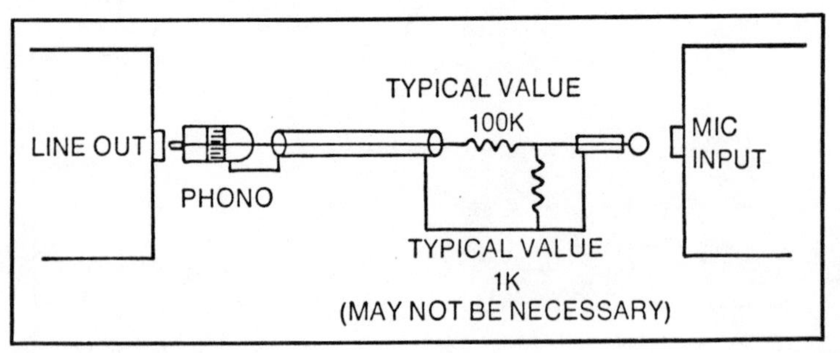

Fig. 5-7. Attenuator or "pad."

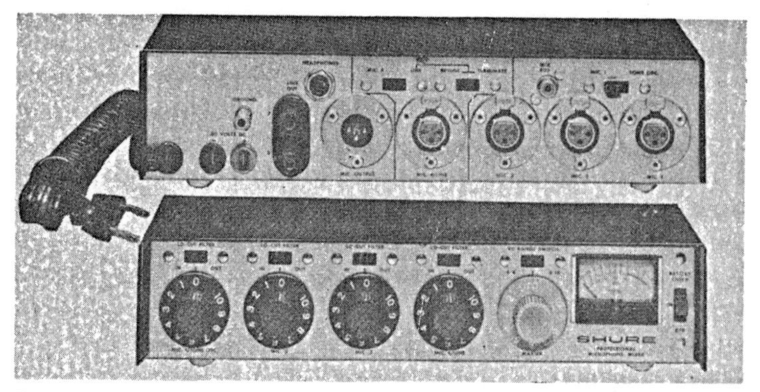

Fig. 5-8. Front and rear of Shure mixer (courtesy of Shure).

With machines built to the EIAJ standards, any combination is possible, and no diagram is needed to explain these. With the various 1 in. machines the connections and plugs will be different, and each must be tried on its own merits.

Monitor Output

In all of the previous situations the **line out** has been recommended for use. This usually produces the required result because the audio has been recorded onto one channel only, and the correct connection involves finding the correct channel.

It is possible however, to use the **monitor** output in all of the above examples. The audio output at this point is selected by a switch on the front panel, but the actual details of what the audio is varies slightly from model to model. The advantage of this output is that it can combine the audio from both channels or can switch from

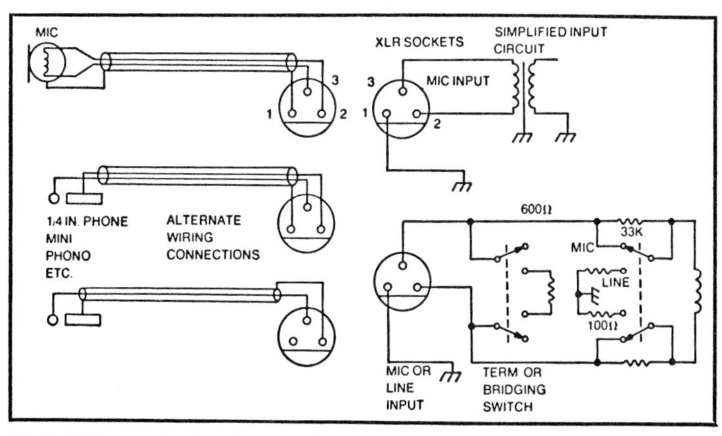

Fig. 5-9. Shure mixer input connections.

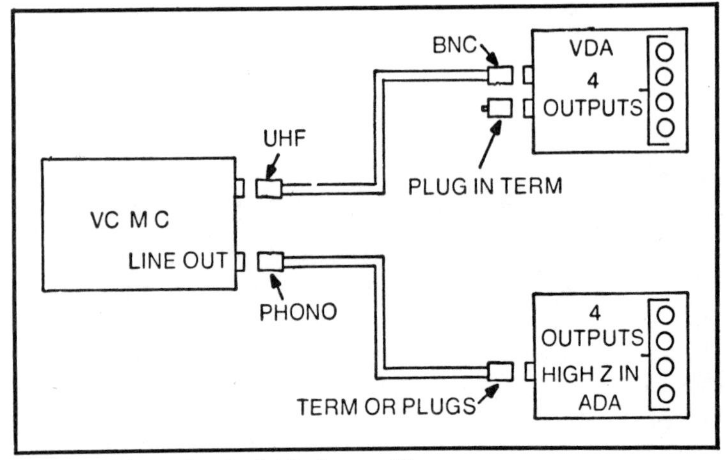

Fig. 5-10. Video and audio distribution amplifiers.

one channel to another. This is useful in academic examinations which require alternative questions to the same video or which have answers on the second track for student self-checking.

Phones Output

This output is useful for many purposes. The most obvious is to plug a set of headphones into the socket and listen to the program in privacy and without disturbing other persons in the vicinity.

A stereo audio plug is used on the phones, and the audio heard is controlled by the audio select switch. However, the audio fed to the phones is different from that fed to the monitor output and the RF modulator: in position 1, Channel 1 is fed to both ears: in position 2, Channel 2 is fed to both ears; and in mix. Channel 1 is fed to the *left* ear and Channel 2 to the *right* ear.

Examples of this facility are to listen to the program on phones while it is faded down at the mixer. The mixer then can be faded up at the correct time so that only the desired audio is copied or broadcast. Cues can be placed on Channel 1, which can instruct when to fade in and out the program audio on Channel 2.

Because 8 or 16Ω phones are specified, this output can be used to drive many other inputs. However, it must not be used with a **mic** input because the level is much too high. The two output levels available are selected by the **phones** switch, position 2 is the loudest. When this output is used as a source to feed an input, the **line** or **aux** input must be used. Figure 5-11 shows typical connections.

INPUTS AND RECORDING CONNECTIONS

Recording on a videocassette is required at many times and in many different places, so both video and audio inputs are made available on many models.

When recording, the connections to the machines must be made in addition to those required in playback. So in general the most difficult aspect of recording is the connections, and as with playback it is the audio which is the most complicated.

Video Inputs

Two video inputs are provided to the record electronics on the machines. The simplest for the operator is that from the internal TV tuner. This requires no connections other than the antenna at the rear of the machine, where a normal TV or CATV antenna or lead is connected. The input select switch is set to the **int** or **TV** position, and this switches the video and audio to the input of the machine.

The other video input is a **video in** socket which enables various external sources to be connected. In this mode the select switch is set to the **ext** position, which routes the video and audio inputs to the record amplifiers.

The external video input must be a fully composite video signal, because no separate sync or drive inputs are provided on these machines. Most outputs from other nonbroadcast equipment are fully composite, so these connections present no difficulties. The **video in** is a terminating input and no loop through is possible, so the cassette machine must be the last in line with any multiple connections. See Fig. 5-12.

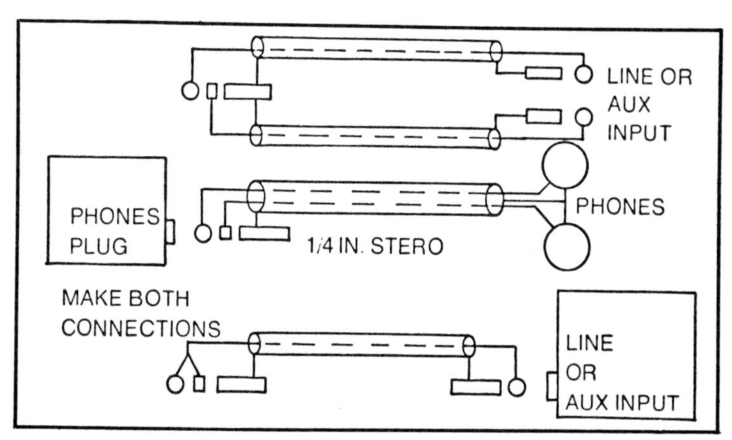

Fig. 5-11. Phones output.

The video levels from external sources can be between 0.5V to 2.0V. The video record amplifiers are automatic gain controlled and thus the correct level of signal will be put onto the tape. The **agc** also works for the **off air** signals and, provided they are not too weak or too strong, good recordings will be obtained.

Video Connections in Record

In general there are several common sources which will be used to supply the input to the machine.

Camera. The output of the camera should be plugged directly into the **video in** socket. This must be a fully composite signal with a stable sync signal, and most nonbroadcast cameras with interlaced sync are satisfactory for use.

A portable camera cannot be used because the cassette machines do not have horizontal and vertical drive outputs and have no provision for accepting an adaptor box. With a correct adaptor box, some portables may be usable with the cassette machines.

TV set video output. This mode can be used to record **off air** programs if an internal tuner is not provided in the cassette machine. A connection is made between the video output of the TV set and the video input of the machine.

Copying from another VTR or cassette machine. This is the same as any other VTR copying procedure. The video output of the playback VTR is connected to the **video in** of the cassette machine.

Output of a studio. The line out of the studio is connected to the video input of the cassette machine. Ensure that the studio output from the mixer is fully composite.

Audio Inputs

In the **record** mode, as in playback, the audio connections are more complicated than the video. Audio from the internal TV tuner is recorded on the cassette if the input select switch is in the **tv** or **int** position. This requires no connections other than the antenna. In this mode the audio is recorded onto Channel 2 only, and the input of Channel 1 is grounded, so it is not possible to add a commentary to a program while it is being recorded—this must be added later.

With the select switch in the **ext** position the inputs of the audio channels are from the input sockets at the rear of the machine. Two types of inputs are provided, a **mic** and a **line** or **aux**. One is provided for each channel, and they are identical as described below.

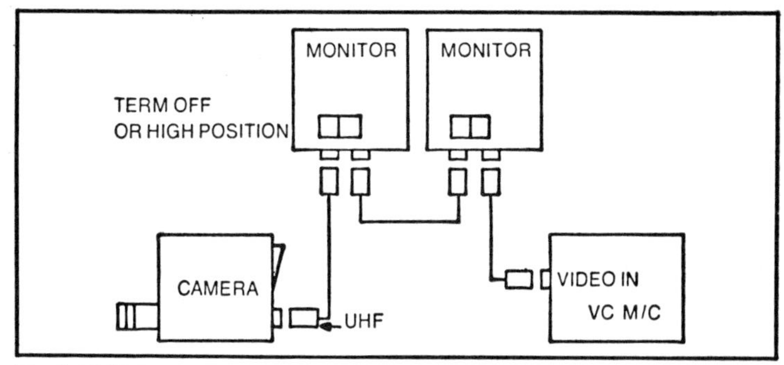

Fig. 5-12. Video in on a cassette machine.

Mic in left dub input. This is a normal microphone input to the *left* channel or Channel 1. It accepts a level range of −70 dB up to −30 dB, and the impedance is nominally 10 kΩ unbalanced. An RCA phono-type plug is used. This channel should be used for dubbing extra speech into a recording.

Mic in right input. This is identical to the above except it does not work in the audio dub mode.

Most microphones will work these inputs, but it is advisable to try a microphone before committing oneself to it.

Aux in left dub input. This input is for use with the outputs of most other items of audio equipment. It accepts a level range from − 22 dB up to +10 dB, and the impedance is nominally 100K unbalanced. This channel should be used for audio dubs.

Aux in right input. This is exactly the same as the input to Channel 1, but it does not work in the audio dub mode.

A miniplug is used for these inputs so that confusion with the **mic** input will not arise.

In the normal record mode using an external source, it is possible to record onto both channels simultaneously, because both are active and the previous audio is erased from both by the main erase head. In the **audio dub** mode only Channel 1 is active, and the Channel 2 audio remains untouched—being neither wiped nor recorded upon. The input to Channel 2 is grounded in this mode. The selection of which mode and which input to use is obviously dependent upon the situation and is shown in the following examples.

Examples of Input Connections

In many respects these examples are identical to those illustrating the output connections, and the same general remarks apply here.

Recording from a Microphone. The microphone is simply plugged into the **mic input**. Either channel can be used in the normal record mode, but only Channel 1 if audio dubbing is envisaged. Most microphones will work, especially those from the same manufacturer, but it is advisable to check with a test recording first because both the levels and the impedance may be unsuitable.

Recording from a Studio Mixer. Small mixers are found in studios, auditoriums, etc. and will be encountered frequently. Most have a **line out** or a **high level** output, which will be around 0 level. Both balanced and unbalanced outputs will be found on different models. These should be connected to the **line in** or **aux in** of the cassette machine.

Occasionally a mixer will have a low level output. This may be low enough in level to feed into the **mic** input, but it may be too high. It also will probably be too low for the **line input**. In this case it is best to attenuate it or "pad" it down to a slightly lower level.

Not all mixers have a level meter or a master level control, so sometimes the exact level out will be in doubt. Again, a test recording is advised in these circumstances. The audio output connections for the Shure mixers are shown in Fig. 5-13.

Using the Line Out of Other Equipment. This should be connected directly to the **line in** or **aux in** of the cassette machine. In most cases the impedances and levels will be suitably matched and no difficulties should be experienced. The main problems will be matching the plugs and sockets and converting from a balanced to an unbalanced line.

Most audio equipment does not have a level meter, and so again the levels may be in doubt and a test recording should be made first.

Stereo Audio Recording. Because stereo sound is a proven consumer item, and because some TV programs have been broadcast in stereo in conjunction with FM radio stations, the ability to record in stereo onto a cassette is considered an advantage. All of the previously described cases can be performed simultaneously on both channels. Ideally the outputs from each piece of playback equipment should be the same, but this may not be the case. The dual outputs of a stereo record player amplifier will be identical but the outputs, of a TV set and an FM tuner may be different. The setups in Figs. 5-14 and 5-15 indicate what might be expected.

The Output of Another VTR. This is identical to a normal VTR copying setup. The **line out** of the playback machine is connected to the **line in** of the cassette machine.

Audio Output of a TV Monitor Set. This is a **line out** in both level and impedance and should be connected directly to the **line in** or **aux in** of the cassette machine. The input select switch on the cassette machine must be in the **ext** positions. The **TV vtr** switch on the TV set must be in the **TV** position, otherwise it will not feed the received audio signal to the output. This connection is effectively the same as that used for the 8-pin, which is discussed next.

8-PIN CONNECTIONS

The 8-pin is a special connector originally designed as a quick convenience connector to be used between a VTR and a TV monitor or "jeeped" TV set. A jeeped TV set was the forerunner of the TV monitor. It had the video and audio brought to an output plug, and video and audio input was possible. It usually required a small attachment, known as a "jeep box'" to be attached to effect the connections to the modified internal circuitry.

With the 8-pin plugs and cable a program can be recorded **off air** and then played back into the same TV set with no plug or cable

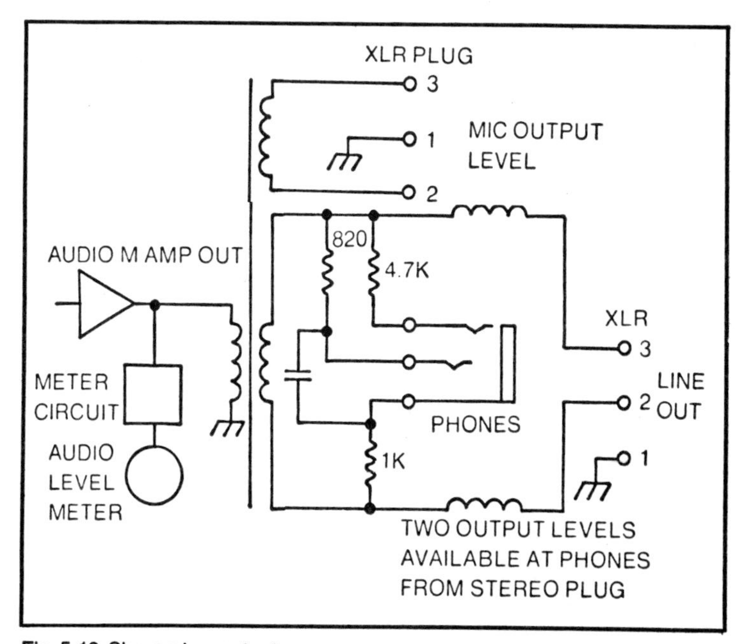

Fig. 5-13. Shure mixer outputs.

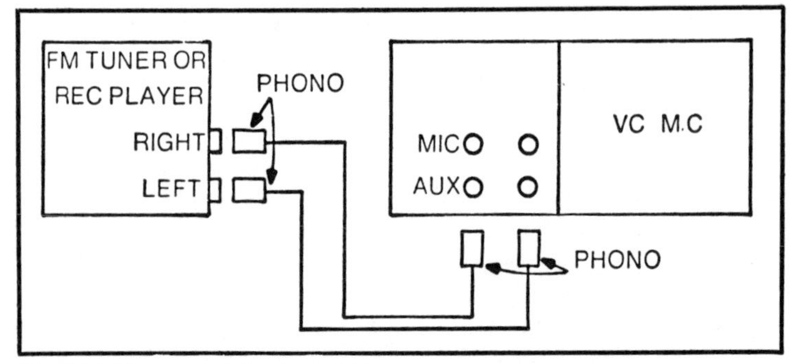

Fig. 5-14. Stereo recording connections.

changes. With a color TV and color VTR or cassette machine the recorded program is in color. With monochrome equipment a monochrome picture is recorded and played back.

To effect these connections a special multicable is used, with four individual shielded cables enclosed in one sheath. A male 8-pin plug is mounted on each end, and a female 8-pin socket is mounted on both the VTR and the TV set. The cable can be used either way round but the plug will fit into the socket in only one way. This one connection at each end completes the audio and video connections for all record and playback purposes.

Figure 5-16 shows a TV set with an 8-pin and other plugs mounted at its rear and Fig. 5-17 shows the designations of the pins on a TV monitor and a VTR. In this, note the terminals and the functions associated with them, because this explains why direct interconnection between VTRs and cassette machines is not possible with the cable.

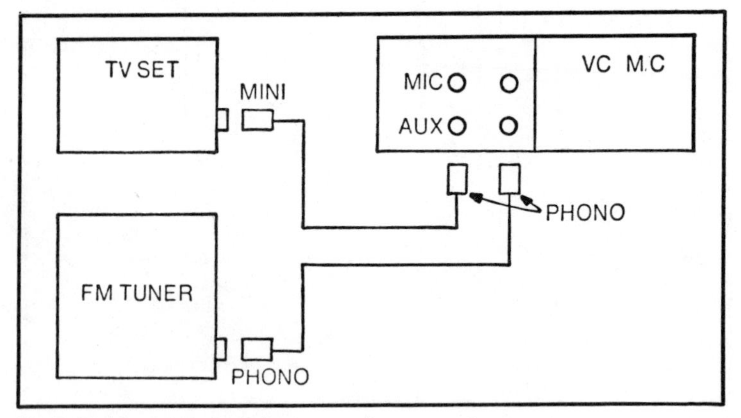

Fig. 5-15. Alternate output stereo recording connections.

Recording OFF AIR from a TV Monitor

This is the easiest use of this connector and is the main reason why it was designed. One plug is inserted into the socket of the VTR or cassette machine and the other into the TV set. The TV-VTR switch on the TV set is set to the TV position, and the desired program is selected and tuned in the normal manner. The input select switch on the cassette machine is set to **ext**, and the recording mode is initiated in the normal manner. Figure 5-18 shows the setup in pictorial fashion. Another TV monitor can be plugged into the machine outputs and the program can be viewed in the E-E mode if desired.

Immediate Playback of OFF AIR Recording

If the tape is rewound and put into the playback mode, it can be seen and heard by switching the TV set to the VTR position. No replugging is necessary. Obviously, with this connection any other prerecorded tape can also be played back.

These two simple examples cover the full use for which the 8-pin cable was designed. It is not intended for interconnections between VTRs, cassette machines or any other purpose. However, with a few simple modifications its use can be extended.

VTR Copying With The 8-Pin

The reason the 8-pin cannot be used to interconnect two VTRs or cassette machines is that it would parallel like functions instead of interconnecting complementary functions. Reference to Fig. 5-17 will make this clear. There are two ways of alleviating this dilemma: to use the TV set in an unusual manner and to rewire the cable within the plug.

Copies using a TV set. The connections for this mode are shown in Fig. 5-19. The normal 8-pin connection is used between the cassette machine and the TV set, but the machine to be used for recording is connected as shown in Fig. 5-19. Note which sockets have been designated for these connections; they do not correspond with their normal functions, and this is definitely a "trick" hookup. It is understood by referring to Fig. 5-20. which shows diagrammatically the outputs of the cassette machine and the inputs to the TV set. The played back tape can be viewed on the monitor in the TV set. The played back tape can be viewed on the monitor in the normal manner, but it must be in the VTR position, and the 75Ω termination switch at the video **ext in** must be set to **off**. The audio levels and impedances are satisfactory for this connection.

It is likely that this mode may never be used, because other ways exist of copying, but it does come in useful when lack of other cables is encountered and an 8-pin is available.

Direct VTR Interconnections with the 8-pin. For this it is necessary to modify the internal wiring of the plug at *one end* only. This modification is shown in Fig. 5-21. The modified plug should be painted red and should be used as the input connector to the recording machine only. It will not work when used for any other purpose.

The 8-Pin Splitter Box

These can be very useful as a central connector for several interconnection purposes. The one shown here provides simple copying facilities from a cassette to another cassette or VTR and allows monitoring with an 8-pin and a TV set or an ordinary monitor. It is one of many such boxes which can be easily built. The only precautions are to ensure the video is neither unterminated nor doubled terminated and that the audio inputs are correct. See Fig. 5-22.

RF SYSTEM

The RF system can be considered to be a complete subsystem within the main videocassette machine. A typical block diagram is shown in Fig. 5-23 and it consists of three main sections: the TV tuner, the RF modulator, and the antenna switching.

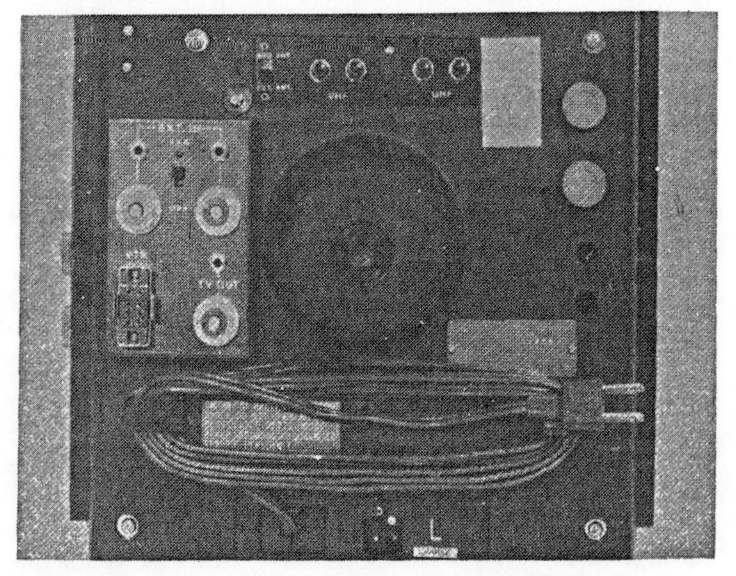

Fig. 5-16A. Rear of TV set with 8-pin connector (courtesy of Sony)

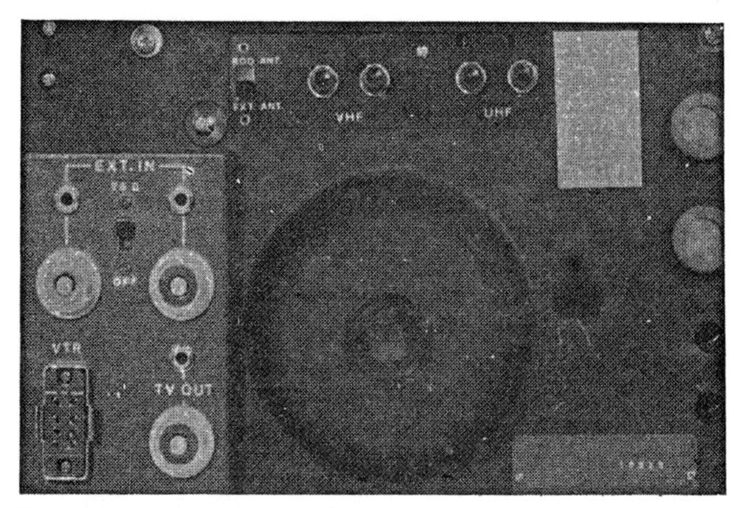

Fig. 5-16B. Video input panel (courtesy of Sony).

In many ways the RF system is different from the remainder of the cassette machine, and it was approached by the designers with an attitude completely different from that used when designing the previous helical VTRs. There are four main reasons why the RF system is best treated separately from the rest of the machine:

● Both the audio and the video are carried simultaneously in the same RF cable, modulated onto a standard TV channel frequency, and do not require separate cables or attention. Note that this is a shielded 75Ω cable and not the domestic 300 twin lead.

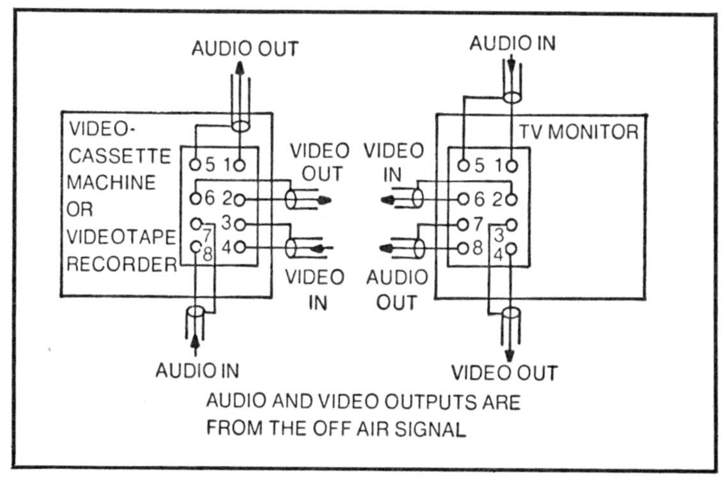

Fig. 5-17. Pin designations on VTR and TV.

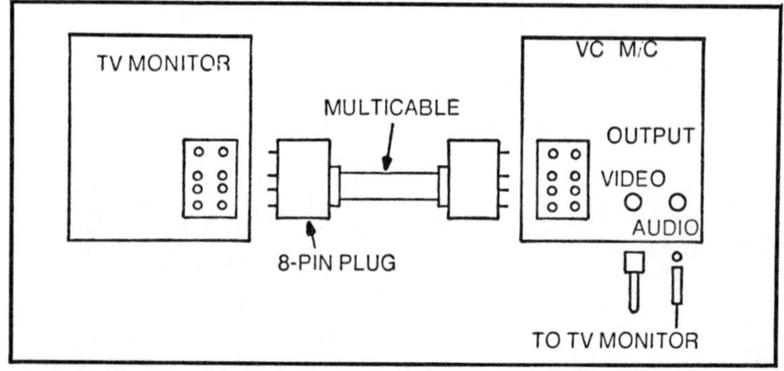

Fig. 5-18. Off air recording with 8-pin.

●The RF levels out of the machine are governed by FCC rules, are set at the time of manufacture, and are not controllable by the operator.

●The distribution of an RF signal through an MATV or a CATV system is very different from the normal audio and video connections.

●The antenna loop-through connections in the machines are different from any other connections or switching.

Each of the three sections are covered separately, followed by examples of RF use.

TV Tuner

To win acceptance as a consumer item the ability to record TV programs OFF AIR and then play them back into a normal domestic TV set was considered vital. The manner in which this has been

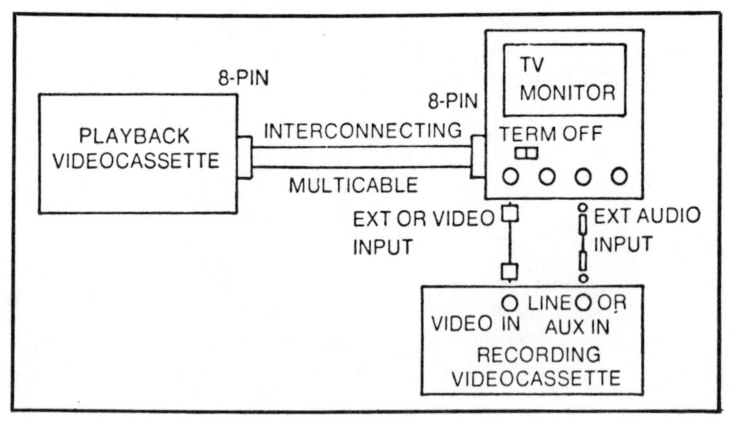

Fig. 5-19. Copies with a TV set.

done in the cassette machines makes their inclusion into a domestic setup very easy and does not introduce any difficulties.

To do this a standard TV tuner has been incorporated into many of the machines. This is a standard type of commercial TV tuner which is found in domestic TV sets and contains both the VHF and the UHF channels. Two sets of antenna terminals are on the rear of the machine, and the TV antenna or the CATV cable are connected to them just as with a normal TV set. On the front panel of the machine are the two usual channel selector dials, each with its fine tuning ring, and they are used just as a TV set. An extra provision is a tuning indicator, a meter whose pin reaches a peak reading at optimum tuning.

If the desired channel is turned to and the input select switch is set to the TV or INT position, then the TV program will be recorded onto the cassette and can be viewed and heard on a monitor or TV set connected to any of the output sockets.

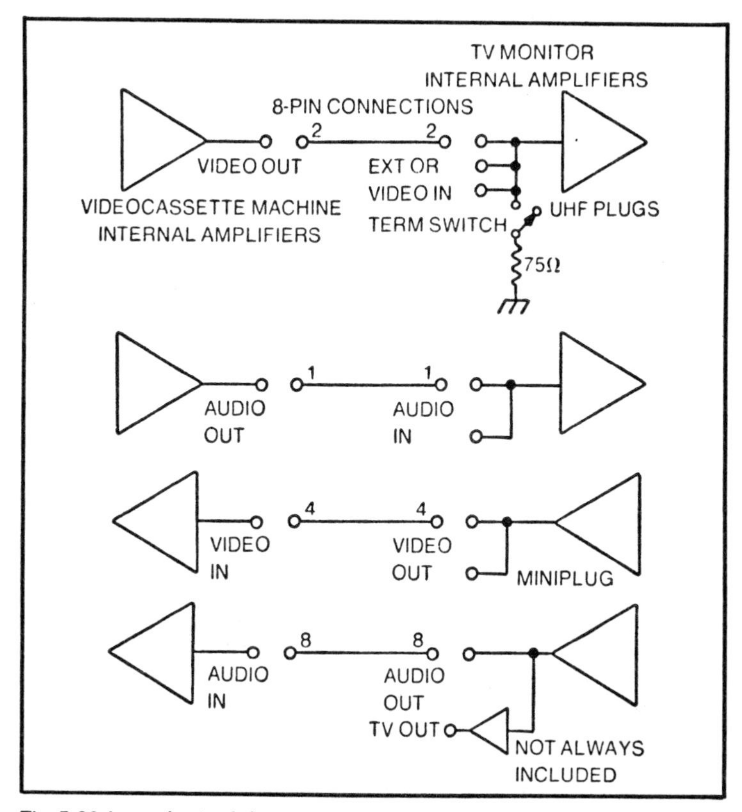

Fig. 5-20. Ins and outs of plugs.

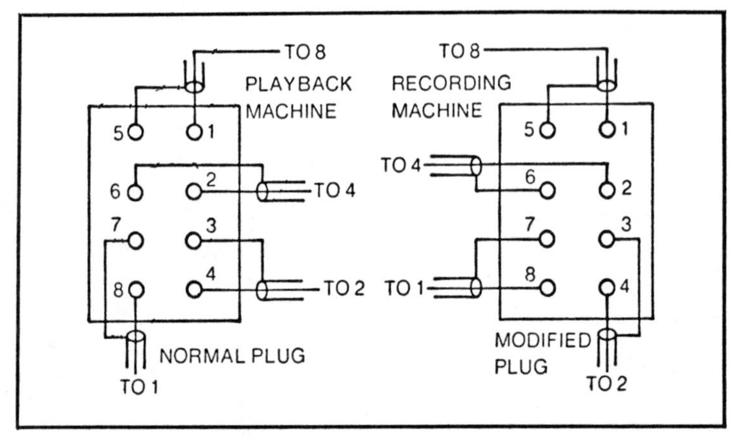

Fig. 5-21. 8-pin modifications.

All machines with a tuner can record **off air** programs in color and then play them back immediately into the same TV set in color. This is possible without any changes of cables and plugs and even without taking the cassette out of the machine; it merely needs to be rewound. In many cases it is not even necessary to change the channel on the TV set.

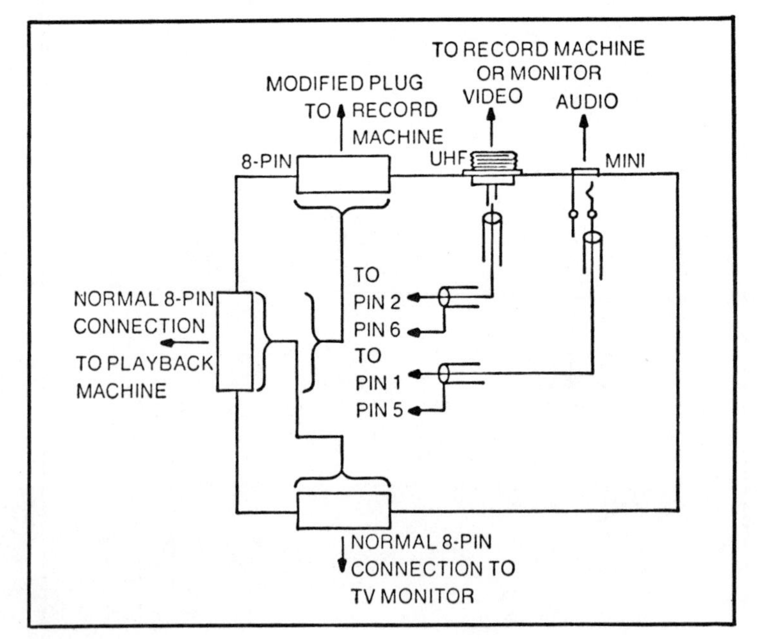

Fig. 5-22. Splitter box.

The RF Modulator

This is a small TV transmitter of very limited power. Its purpose is to allow a cassette to be played back over a normal unmodified domestic TV set. Unlike other VTRs, it is permanently installed and is not removable by unplugging and pulling out from a recessed compartment.

Any input to the machine in the record or standby modes is fed to the input of the RF modulator and appears at its output. If a tape is being played back, then the tape video and audio are fed to the modulator. The audio is selected with the audio select switch on the front channel.

The output of the modulator is switch selectable between two standard TV channels; Channels 3 and 4 comprise one selection, while other machines will use Channels 5 and 6. The output appears at an F plug on the rear of the machine; this is a 75Ω output which is connected through a matching transformer to the VHF antenna terminals of the TV set. The TV set is tuned to the channel selected by the switch on the back of the machine. This output is labeled either RF or VHF out, and the two terms are used interchangeably in most literature when referring to this output.

When the machine is set up to be used with a MATV or a CATV system, or even a domestic antenna, direct connection is not always possible to the TV set. (This is covered in an example in this chapter.)

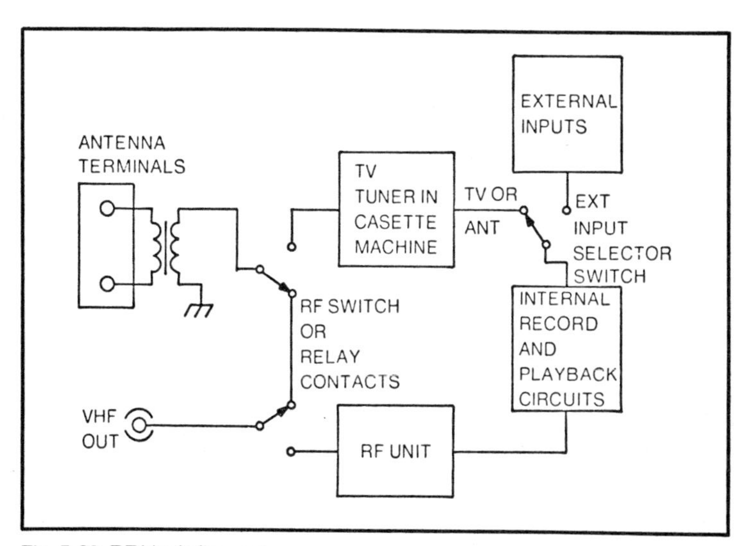

Fig. 5-23. RF block diagram.

As far as the user is concerned the modulator is a nonserviceable item, and if it gives trouble it should be replaced with a new one. Its output is governed by some strict FCC rules, and this warning is given by many manufacturers. See Fig. 5-24.

The audio and video inputs to the modulator are permanent connections inside the machine and cannot be switched out. In both cases the inputs are high impedance and act as bridging inputs to the main video and audio outputs. This ensures that if the unit is removed no interruption of other normal functions will occur.

Antenna Switching

When the machine is powered up for recording or playback, the output of the video and audio circuits are fed to the RF modulator inputs, and the external antennas are fed to the tuner inputs. When depowered, the antenna inputs are switched from the tuner inputs to the RF output socket. The inputs to the modulator remain untouched. This enables the antenna to be routed to the antenna input terminals of a TV set, which is permanently connected to the RF output of the cassette machine.

The actual switching arrangements vary from model to model, and details should be checked with the operator's manual. The diagrams in Figs. 5-25, 5-26 and 5-27 show several examples. The main difference to be observed is whether it is possible to record one program OFF AIR while viewing another. This is not possible on some machines but is presented as a useful facility with others. Note also that the UHF antenna does not "loop through" most machines; an external UHF splitter must be used.

Examples of RF Use

When using a video cassette machine in RF installations there are certain items of equipment which will be encountered that are not found elsewhere. A brief description of the most common of these is given before the examples of use are covered.

CAUTION

Connection between the machine RF OUT terminal and the antenna terminals of a TV receiver should be made only as shown in these instructions. Failure to do so may result in operation that violates the regulations of the Federal Communication Commission regarding the use and operation of RF devices. Never connect the output of a machine to a receiving antenna or make simultaneous (parallel) antenna and machine connections to the antenna terminals of a TV receiver without the use of an approved splitter.

Fig. 5-24. FCC warning.

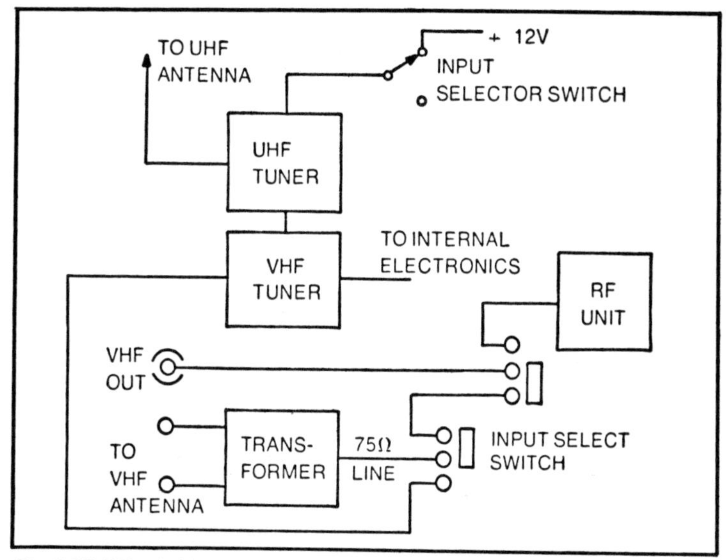

Fig. 5-25. RF switching system.

The Antenna Transformer. Almost all domestic TV sets are made with 300Ω antenna input terminals only. Because most RF distribution equipment, such as MATV or a CATV system, uses 75Ω equipment and cable, a transformer is required. To refrain from its use will cause inferior pictures due to the mismatch of the line and their terminals, and it could contravene FCC

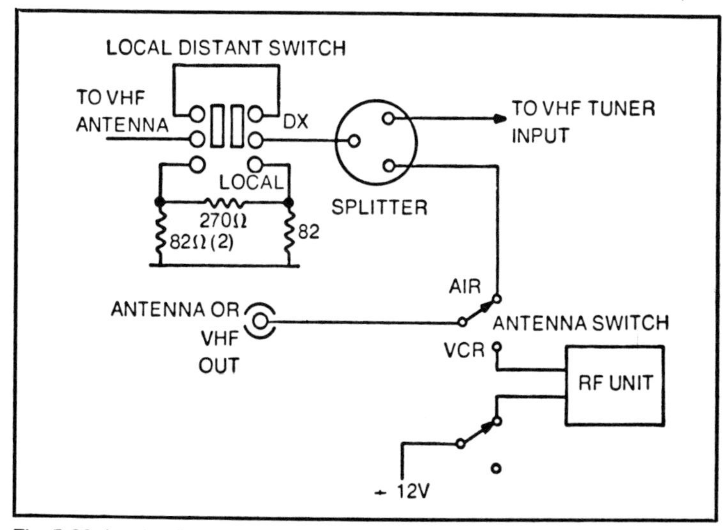

Fig. 5-26. Another RF switching system.

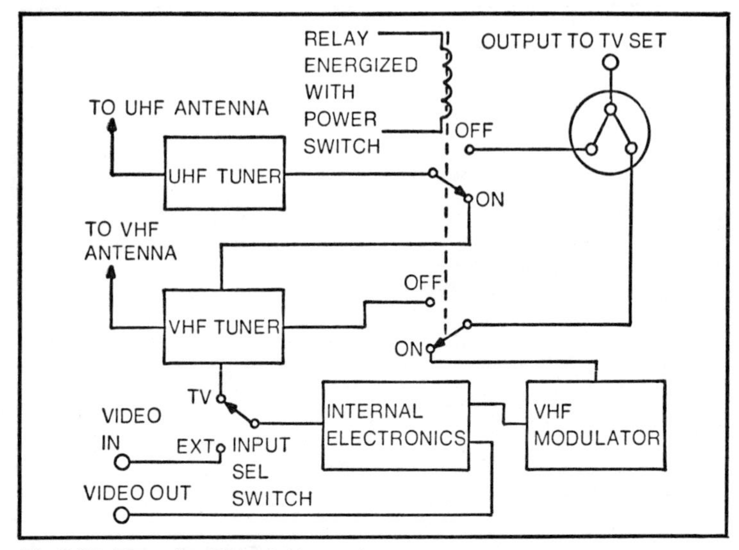

Fig. 5-27. Alternative RF switching system.

regulations concerning RF devices. A photo and the connections were shown in Fig. 2-2.

The Splitter. If an RF cable or antenna is to be fed to two different inputs, these inputs cannot be paralleled in the same manner as most hi-fi audio connections. The RF signal must be properly "split" so that it can feed the two inputs. To do this a "splitter" is used. Figure 5-28 is a photograph of a two-way

Fig. 5-28. Two-way splitter (courtesy of Jerrold).

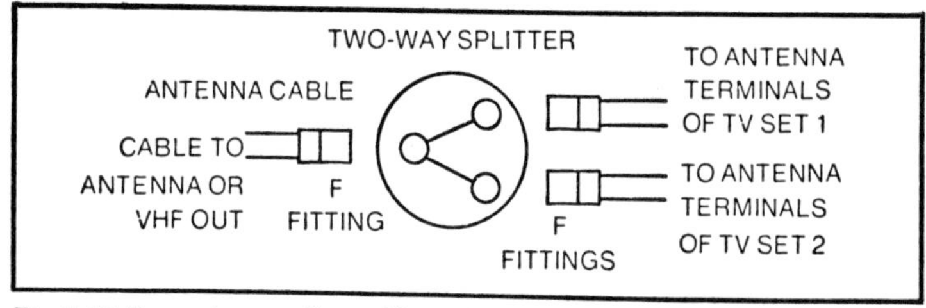

Fig. 5-29. Two set connections with a splitter.

splitter, and Fig. 5-29 shows a simple two-set connection. Splitters are made for both 75Ω and 300Ω lines and must not be intermixed. The signal is dropped about 3 db at each output with a two-way splitter. Up to four-way splitters are manufactured, but these have correspondingly larger drops in signal. VHF splitters and UHF splitters are different, and each usually will not pass the other signals; this should be checked before use. When connected in reverse, splitters will combine signals. Any unused outputs must be terminated with a 75Ω terminating plug.

RF Attentuators. An attenuator is a device which reduces the level or strength of a signal. RF attenuators are properly shielded and terminated and will reduce the level by a given amount. They can be either fixed or switchable. The amount of attenuation introduced is marked in decibels.

Field Strength Meter. The FSM is used to determine the level of RF signal received on an antenna or in a cable. It is connected to the antenna or the cable and then turned on. Basically it is a TV tuner with a calibrated meter on the output. The tuning dial is adjusted to get a peak reading on the meter, which indicates the strength of that particular channel in decibels. They are used extensively in the MATV and CATV industries.

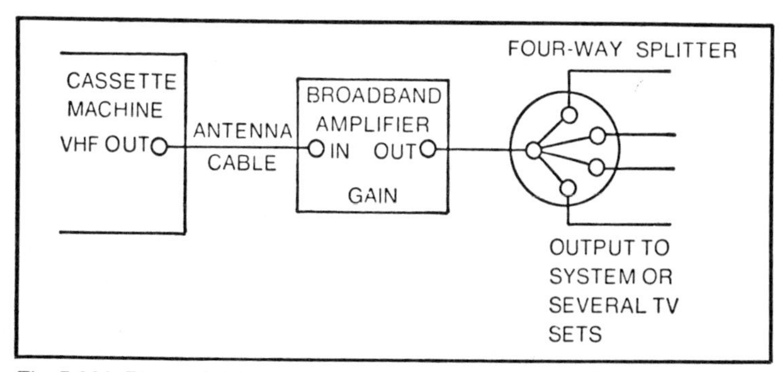

Fig. 5-30A. Playback into several TV's with a broadband amplifier.

RF Bandpass Filters. These are devices which will accept or reject a certain part of the RF frequency spectrum. They are usually used to prevent interference between channels on a cable system. They are either preset to one channel or are tunable.

RF Amplifiers. Two types of RF amplifiers are made: broadband and narrow band.

- **Broadband.** These are made to amplify a wide range of signals an equal amount. Usually they are made to cover the entire 13 TV channels and the FM radio band. Often they cover Channels 2 to 6 (only) and 7 to 13 only. A wide variety of controls are often found with these which should never be touched without the proper RF alignment equipment. A loop-through input is usually provided with a termination switch, and up to four outputs are common. The gain can be preset for unity or something higher. These are usually found in the head end of MATV systems and as distribution amplifiers in MATV and CATV systems.
- **Narrowband.** These are also called "strip" amplifiers, and they will pass one channel only. They are used in many MATV and CATV head ends where separate antennas pointing at stations of different signal strengths are found. Care must be used in setting them up, because in the rejection of any adjacent channels they may reduce the level of the audio and color signals. Usually only one input and two outputs are provided.

The reason for the preference for the use of one above the other is not often clear and is a matter of local conditions and the personal choice of the designer of the system.

Playback into a TV SET Antenna. The RF or VHF output of the cassette machine is connected to the VHF antenna terminals of the TV set, and the set is tuned to the same channel selected by the channel select switch on the rear of the machine. Because the machine has a 75Ω output a transformer must be used to match the 300Ω antenna terminals of the TV set. Most such transformers have an F connector on one side and antenna-type leads on the other, so a cable with two F connectors is required for this connection. This cassette is played back in the normal manner, and the audio is selected by the audio select switch on the front of the machine.

With this simple setup no interference with other TV sets should result, and a satisfactory color picture and sound can be attained on the TV set.

The level out of the RF output is such that it can only satisfactorily feed one TV set. Two TV sets can be connected,

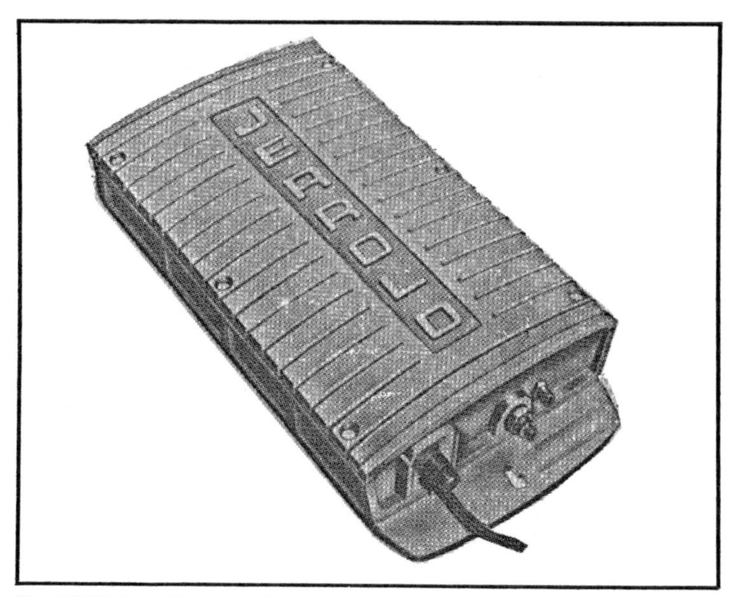

Fig. 5-30B. Broadband amplifier (courtesy of Jerrold).

provided a splitter is used, but this is about the maximum that will produce a good picture. To feed more sets than this, some RF amplification will be required, as the next example. Figure 5-30A demonstrates these connections. (From now on, the plugs are assumed and not shown.) Figure 5-30 B is a photo of a typical broadband amplifier.

Playback into an RF Distribution System. An MATV or a CATV system are examples of RF distribution systems. In

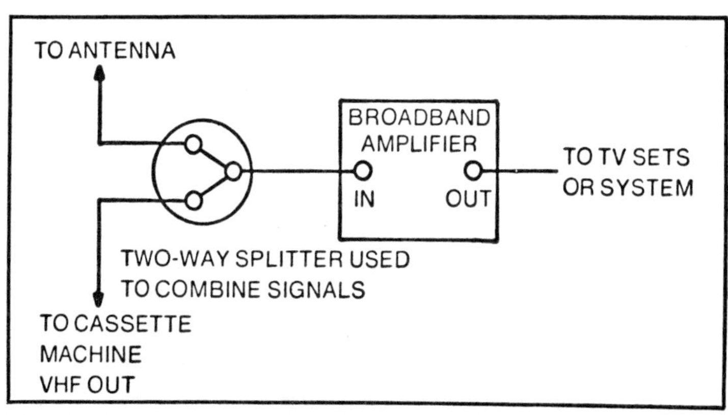

Fig. 5-31. Playback into TV system with a broadband amplifier.

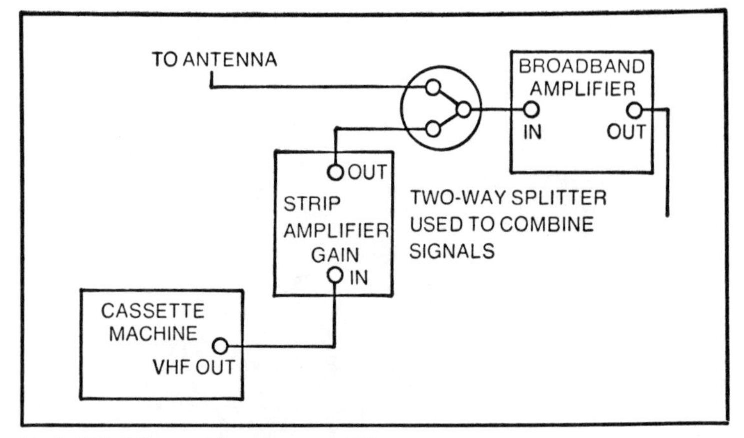

Fig. 5-32A. Strip amplifier with a cassette.

principle, this situation is no different from the first example. The output of the machine is directly connected to the input of either a broadband or narrowband RF amplifier; a transformer is not required because these have 75Ω inputs. The outputs of these amplifiers are either already connected into a system or must now be connected into it. The following situations are typical of what can be found in practice.

An existing system using a broadband amplifier. This is shown in Fig. 5-31. A splitter is introduced at the input of the amplifier, and the connections are made as in Fig. 5-31. The level of the cassette RF is measured at the output of the broadband amplifier, and it should be exactly the same as the other channels. If not, it should be attenuated or amplified with a strip amplifier, as shown in Fig. 5-32A. (Fig. 5-32B shows a strip amp.)

A field strength meter, FSM, can be used at one of the broadband outputs, or it may be necessary to use a splitter, as shown in Fig. 5-33. Note that this will reduce the level in this output leg by about 3 dB.

A temporary setup with a broadband amplifier. This is typical of the situation which must be faced at such events as

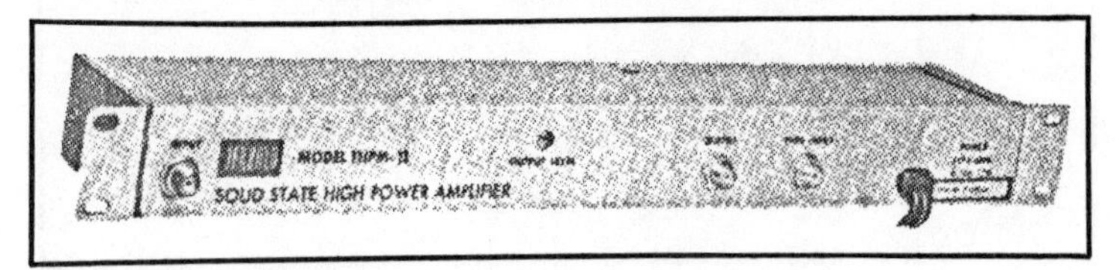

Fig. 5-32B. Strip amplifier (courtesy of Jerrold).

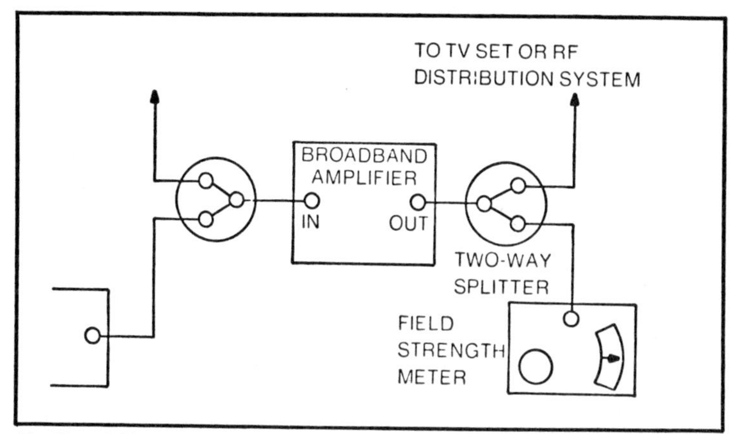

Fig. 5-33. FSM used to measure levels.

exhibitions and public demonstrations. Two diagrams explaining the setup are shown: Fig. 5-34 has only the cassette machine as the input, and Fig. 5-31 can be used when other TV channels must be included with the cassette.

The gain of the amplifier is set so that the input to the TV sets will be about 0 level. With just the cassette machine this is easy to achieve, but with the other TV channels either a strip amplifier or an attenuator may be needed. The main point to remember when setting up such a system is to have the TV sets the same "distance" from the output of the amplifier so that they all receive the same RF level.

An existing system using strip amplifiers. This is shown in Fig. 5-35. Note that each amplifier has an individual input, and the outputs are connected in a "loop-through" mode. One end is terminated, and the other feeds the system. The strip amplifier

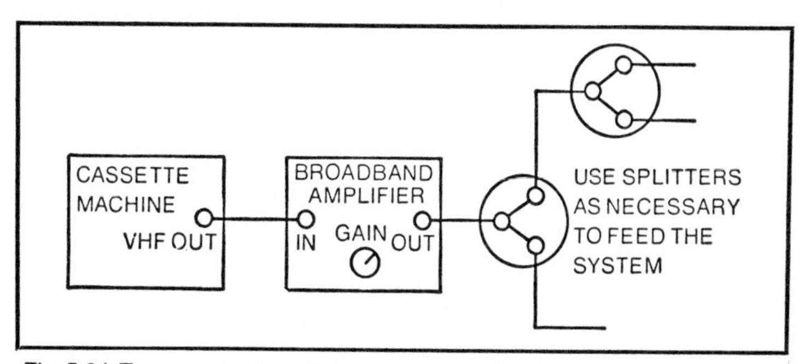

Fig. 5-34. Temporary setup with broadband amplifier.

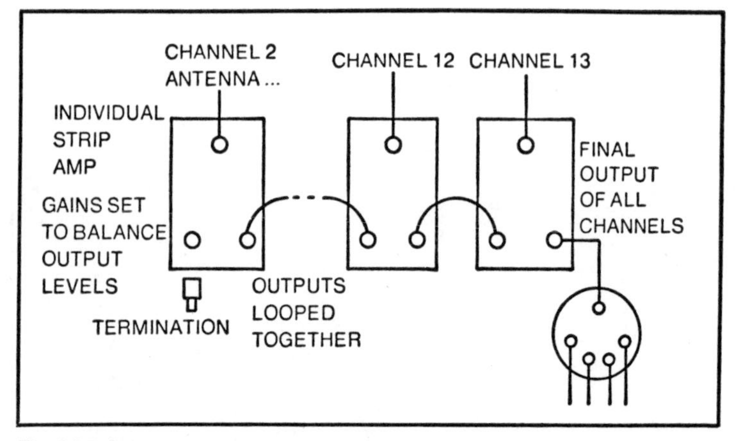

Fig. 5-35. Strip amplifier set up with antennas.

with the cassette input is just added to the existing system, and the gain control on the amplifier is adjusted so that its output matches the others. An easy way to measure the gain of level of each channel is to introduce a splitter as shown in Fig. 5-36 and to connect the FSM to the extra output. This will give a level drop of about 3 dB in the main system, but this will not cause any trouble at the antenna terminals if the level is sufficient in the first place.

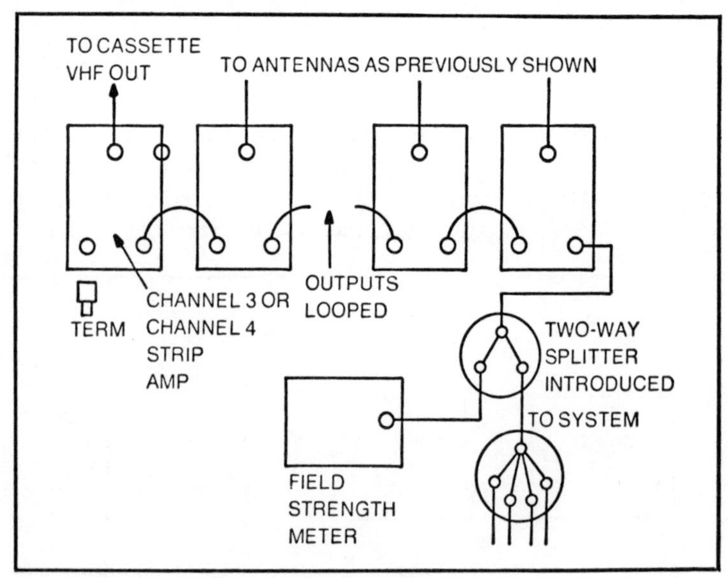

Fig. 5-36. Strip amplifier set up with VC added.

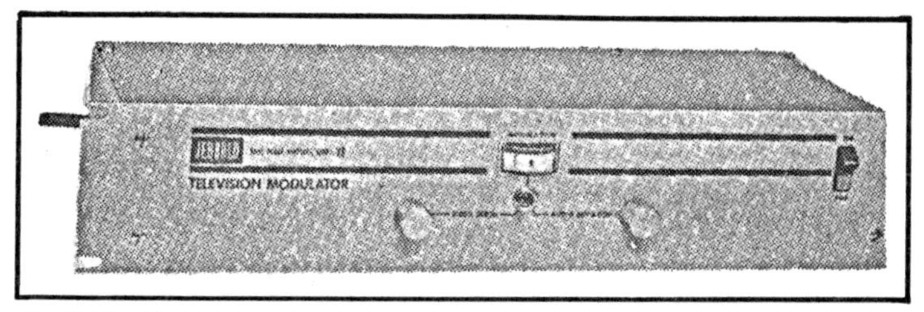

Fig. 5-37. TV modulator (courtesy of Jerrold).

Playback into an RF or TV Modulator. In many MATV and CATV systems provision is made for the introduction of program material from a local source, such as a camera or a VTR, which does not have an RF output. Sometimes, even if an RF output is available, it is necessary or desirable to closely control the level of the output before it is added to the system. In both these cases the video and audio outputs are fed into a "modulator." This is a small TV transmitter, the output of which can be set to a desired level and then fed into a system. The audio and video input levels can be controlled and so can the amount they are allowed to modulate the RF carrier. The advantage of this system is that the channel can be chosen by the system designer, and all the necessary levels can be set to give an optimum picture and sound within the system. A meter is often provided on the modulator to indicate the video and audio levels, and the output RF level is set with an FSM. Figure 5-37 shows a commonly used modulator. They work on one preselected channel only, and the channel cannot be changed; if another channel is desired, then another modulator is necessary.

A common mistake here is to select the same channel on the TV set as selected on the Betamax or Umatic tuner. This is wrong.

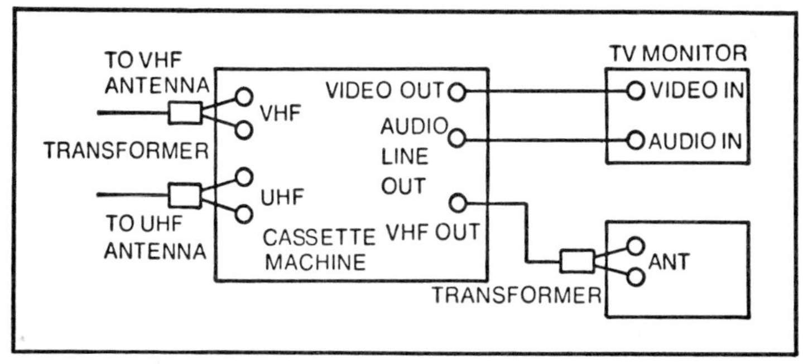

Fig. 5-38. Antenna record with direct connections to VC.

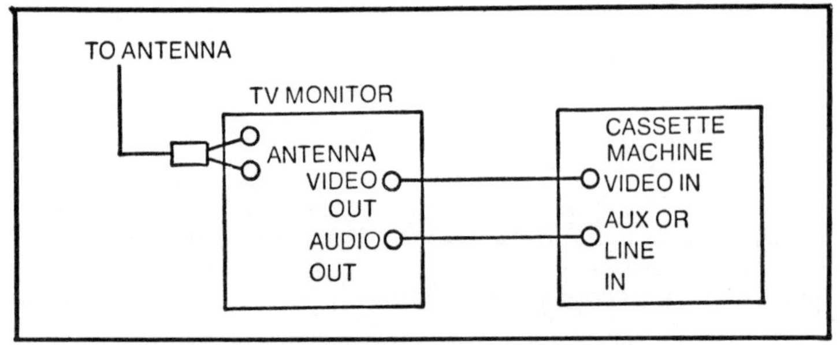

Fig. 5-39. Antenna record with connection to TV and then to VC.

The TV set tuner is set to Channel 3 or Channel 4—depending on which is selected by the switch at the rear of the cassette machine, or which RF unit has been supplied for your area.

The actual station received off-air is selected ONLY by the Betamax or U-matic tuner.

Recording TV Programs. There are three basic ways of recording TV programs as shown below.

● **Direct antenna connection to the cassette machine.** The VHF and the UHF antenna are connected to the inputs to the machine, and the RF or VHF output of the machine is connected to the TV antenna terminals. (Alternatively the video out and the audio **line out** can be connected to the inputs of a monitor.)

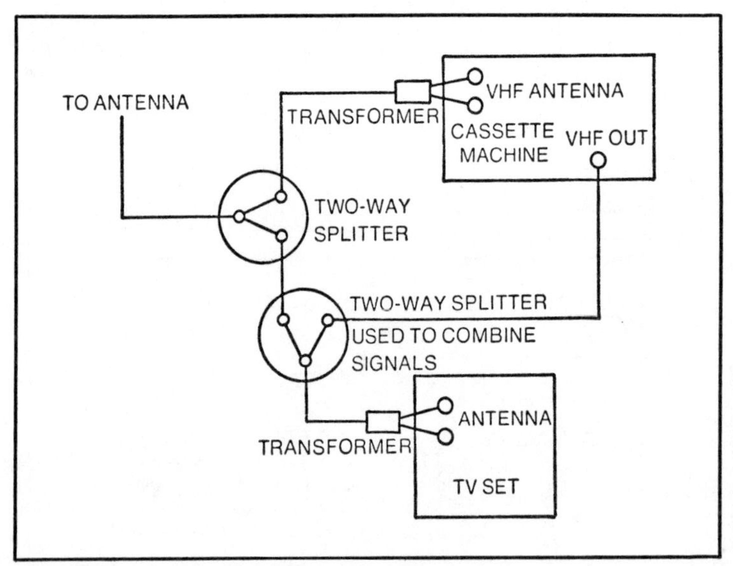

Fig. 5-40. Record and view separate channels.

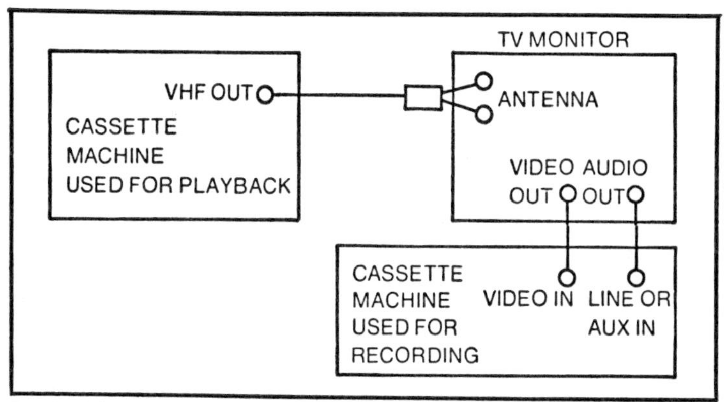

Fig. 5-41. RF copying.

The TV set is tuned to the channel selected by the cassette machine RF switch, and the required TV channel is tuned in the normal manner using the tuner on the cassette machine. The fine tuning ring is adjusted to obtain a peak reading on the tuning meter, and the picture is observed on the TV set or monitor. Often it is best to offtune the fine tuning ring so that the tuning meter is slightly away from its peak. This avoids interference beats between the audio and the color and removes any "herringbone" or "moire" pattern from the screen. Once optimum tuning has been achieved recording can begin. Figure 5-38 shows this setup.

● **Antenna to TV set, 8-pin to the cassette machine.** The antenna is connected to the TV set in the normal manner, and the TV is tuned in the normal manner. An 8-pin cable is then connected between the set and the cassette machine as described previously.

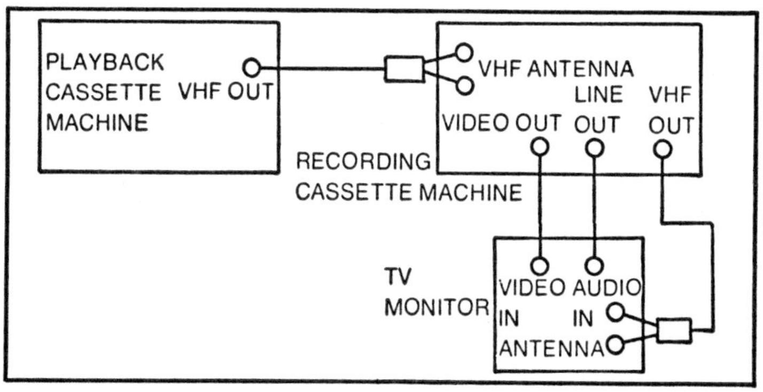

Fig. 5-42. RF copying.

● **Antenna to TV set, with separate connections to the cassette machine.** This is the same as above, but separate audio and video connections are made between the TV set and the machine. See Fig. 5-39.

Recording From a CATV System. This is exactly as described above, but with two minor exceptions. First, all CATV systems are VHF only and a UHF connection is not required. Any UHF stations which are carried have been converted down to an unused VHF channel. Second, the CATV line will be 75Ω, and so the transformer with it must be retained when changing from the TV set to the cassette machine.

Do not connect any cassette outputs to the incoming CATV line.

OFF AIR Recording of One Program While Viewing Another. This may not be possible with all models, so the operator's manual should be checked first. A typical setup for this is shown in Fig. 5-40. Note first that the antenna connections are made to both the cassette machine and to the TV set, and the antenna connection to the TV set is combined through a splitter with the RF output of the cassette machine.

The TV is set to the channel required to view the output of the cassette machine. The machine is then tuned into the desired program for recording while viewing the TV set. When recording has started the TV set is retuned to the second channel one desires to watch. After recording, playback can occur as previously described.

Note here also that the cassette machine tuner and the TV tuner are on different stations. Again, it is wrong to have both tuners on the same station.

Use of RF Connections for Copying. A normal RF playback into the antenna terminals of a TV set provides a method of copying a tape. The cassette is played back into the TV set via the RF output and antenna terminals, as described, and the audio and video outputs of the TV monitor are connected to the input of another cassette machine or VTR. The TV set is tuned just as normal, and the recording is conducted in a normal manner. If the set is tuned correctly this can provide satisfactory copies of a tape. An alternative method is to connect the RF output of one cassette machine to the antenna terminals of the second and then to use the internal TV tuner to receive the signal. Simultaneous viewing can be made at the output of the recording machine. Both systems are shown in Figs. 5-41 and 5-42.

6
Cuing and Editing

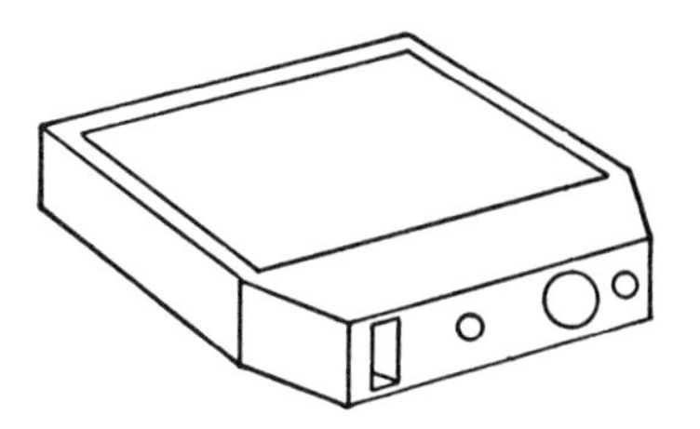

Cuing and editing are among the most important operations undertaken with a videotape recorder, and they present the greatest difficulties when using videocassette machines. At first sight these seem easy procedures, merely requiring the selection of some point in the program to be used as a start or as a place to change over to new video information. Although the production decisions about these points are fairly easy to make, implementing them is not so easy, especially when some or all of the material is on prerecorded tapes.

The operations required to put together a complete program on a tape require considerable operational and technical skill, and the open-reel machines are ideally suited to such operations. With the videocassettes the situation is very different. These are basically post production machines which were designed as enclosed tape systems to be used mainly for playbacks in nonproduction areas. They are at their operational worst when used as advanced production machines in the areas of editing and program assembly. However, many establishments have gone into video at the cassette level and are finding that copying as well as cued playbacks and editing are required; and they are experiencing the operational difficulties associated with these modes.

This chapter is devoted to explaining the nature of these difficulties and presenting procedures to circumvent them.

Cuing is discussed first because it is essential to both editing and copying, with practical examples used as illustrations of the situations found in practice. Editing is covered next, with the mechanical details of the machines which are pertinent to understanding editing described first, and again typical situations are used to explain the operational principles and to show how cuing is effected in practice.

After this, the problem of editing with the cassette machines is covered, concentrating on the operational procedures and the limitations imposed by the nature of the cassette system and showing how to work within these limitations.

The chapter concludes with a brief introduction to the specialized equipment becoming available to facilitate cassette editing operations.

CUING

The word "cue" comes from the theater, where its meaning is well known. With the advent of radio and TV the cue was still used in the same way, but it took on a wider meaning because it was now possible to take cues from prerecorded material.

Two approaches to cuing are required in the TV studio, where many programs are built by recording one section at a time. The first approach concerns the studio artist, who begins an action or speech by taking a cue from the end of the previously recorded section. The second is where the technical staff use the cues to perform the necessary functions such as starting the machines or pushing the record and edit buttons. It is in these technical aspects of cuing that the difficulties arise. These involve setting up the machines and starting them at precise times so that the audio and video will be ready for use at a later time. In many cases this is not easy to accomplish, and much thought, planning, and rehearsal often are required to get the correct timing of the tapes. This is especially true when editing is required, and it becomes especially difficult with the cassette machines.

There are five basic situations which exist in practice, each differing in complexity and difficulty. The first of three are relatively easy to achieve; the main problems lie with the last two, which form many of the situations found in studios today.

The operational difficulties are based on one simple fact: Although an artist can wait until a cue and then begin his action or speech, this cannot be realized with a videotape machine. It takes a definite amount of time for a machine to run up to speed and to stabilize its picture, and this run-up time must be considered when using a tape machine. This produces several operational problems and difficulties of a technical nature.

Simple Playback for Viewing Purposes

This is the simplest of the modes in which cuing must be considered. The exact start time of the program on the tape is not important. What is required is that the machine runs up and stabilizes in as short a time as possible so that there is little delay or waiting in the start after command to begin has been given. This means the tape must be set up at some time prior to the start of the video and then started on cue.

If the starting point for viewing is at some point within the program material on the tape rather than at the beginning, it can create some difficulty in setting up the tape prior to playback, but it still does not present a critical situation.

Studio Recording

When a program is being assembled in sections with a live artist in the studio, the tape is played up to the point where the artist must begin the new section. He is cued, and simultaneously the edit button is pressed. Provided the artist is aware the tape is running, he can be ready for the cue, which can be by hand or taken from watching a monitor.

Both of the above modes are easy with open reel and cassette machines.

Playback of a Tape for ON AIR Transmission

In this case the program on the tape must begin at an exact moment in time, and so the tape must be set up much more accurately. The tape must be left stationary at some point so that when it is started on a cue, the run-up time to the beginning of the video is known exactly. In this way the program will begin at an exact moment. The problem here is to choose the point on the tape at which it will rest until required, and this involves knowing how long it takes for the machine to start and then stabilize.

With practice this operation can be perfected with both open reel and cassette machines.

Playback for Copying

In this mode both the record and the playback machines must be allowed to come up to speed before the copying can start, so both tapes must be started some time prior to the beginning of the playback video. However, the actual start time is not important, in contrast to the above case.

If a straight copy of a whole tape is required, this presents no problems. Both tapes are just started and, provided enough run-up time has been allowed, a good copy will result. If only a section is to be copied and this starts in the middle of the playback tape, then some difficulty can be experienced in setting up the playback tape and in choosing the right moment to initiate the record mode. Often this cannot be done with just two machines, and a studio switcher-fader is required.

Playback for Editing

This is similar to the playback for an **on air** case, because the actual start time of the video is now very important. It requires much more critical operation than the previous situations, and this is where most of the operational troubles lie.

This situation is not easy to achieve, and it requires careful thought and planning in its execution. It reaches its most difficult stage with a program being assembled from prerecorded sections of tape and is typical of many copying and editing situations in studios today.

Cuing is very important in all VTR operations, and the user of cassette machines soon finds the need for these modes. These techniques must be thoroughly mastered if full use of studio and cassette machines is to be enjoyed. With cassette machines certain operational difficulties exist which are not found with the other machines, and these make the operations much more complex, uncertain, and time-consuming. The actual cuing techniques are closely allied to the editing operations, where they are required most often; and this is where they are covered.

PRINCIPLES OF VIDEOTAPE EDITING

Videotape editing falls into three main modes, **assembly** and **insert** with **audio dub**, a subsidiary mode provided so that audio can be added to make a cohesive program from a series of assembled video sections. These three modes are described briefly in this section, followed by a discussion on editing machines. Then the operational aspects of editing are described.

Editing Modes

- **Assembly editing.** This is recording a section of video onto the end of video already on the main program tape. When this new section is put onto the tape, the material in front of it is never returned to, except in playback. Once a new section has been added, further sections can be similarly added to make a complete program. On playback of the completed tape, it should look like a smooth program from a studio.

To perform assembly edits the tape is played until the "cut-in" point is reached, and then the **edit** button is pressed. This changes the machine over from the **playback** mode to the **record** mode, and from that point a normal recording takes place. At the end of the section the machine is stopped just as with a normal recording. On playing back the tape the cut-in should look like a normal camera cut, and the end should be followed by a section of blank tape with noise.

Figure 6-1 is a representation of an assembly edit on tape, where it can be seen that the noise is due to the erased section of tape which was not yet recorded upon when the stop button was used.

● **Insert editing.** This is the most difficult mode of editing, but it is becoming more available on machines.

Insert editing is the placing of a section of new video into a program, and at the end of the new material the original program is returned to. This introduces several operational difficulties and requires certain interior differences in the machines. Because the original material on the tape will be returned to, the tape must at all times remain in perfect step with the incoming video (see the section on servos), and this necessitates the internal electronic changes. The most important operational difference is that a new control track is not recorded onto the tape; the original is preserved, and the erase heads are energized differently to allow a smooth "cutout" and a return to the original material on the tape. Figure 6-2 shows how an insert is arranged on the tape.

● **Audio dub.** This is the provision of adding audio to the video already on the tape. It is covered fully in a later section.

Editing Machines

Before the operational aspects of editing are covered, it is important to discuss three points about the machines which must be understood. These are mechanical and electronic considerations which have a great affect on the operations.

● **Servo lockup.** All videotape machines have servo circuits to control the rotating heads and the capstan, and these are explained in more detail in Chapter 10. Their operational importance is that they must be given time to stabilize before the machine can either record or play back a video picture.

In the record mode the servos must "lock up" to the sync pulses of the incoming video signal, and this must be a completely stable signal. In a copying or editing situation this incoming signal is the playback video from the playback machine, and thus the

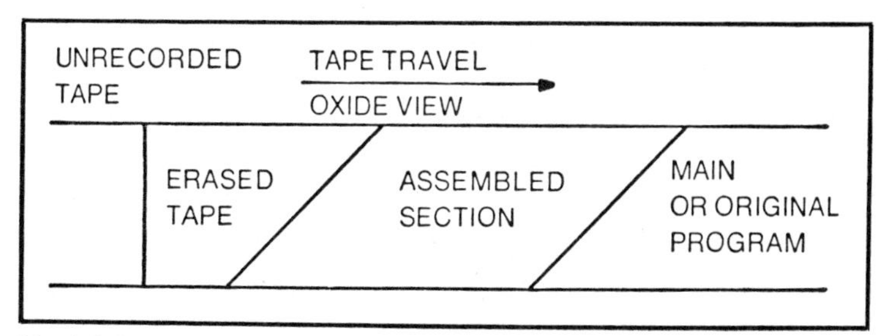

Fig. 6-1. Assembly edit on tape.

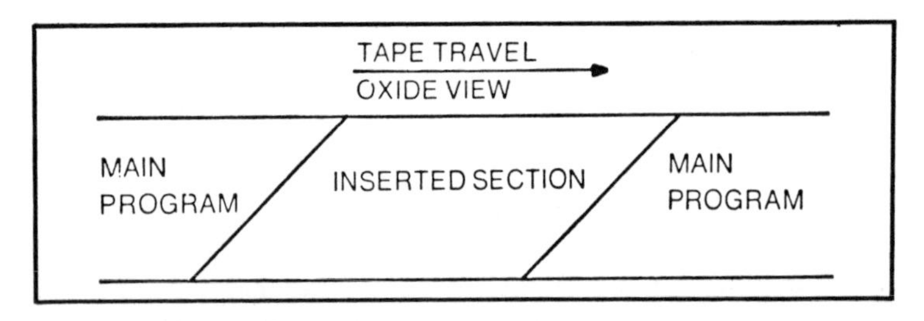

Fig. 6-2. Insert edit on tape.

playback machine must be fully up to speed and stabilized before its signal can be used and copied.

Hence in an edit or copying session all machines must be allowed to run up and stabilize before they can be satisfactorily used, and this means that a "run-up" time is required at the beginning of each tape. If insufficient time is allowed for the machine to stabilize, then although a picture may be seen on a monitor, it will probably not be stable enough for the servos of the record machine to lock onto, and on playback the opening of the tape will not be stable and picture tearing and rolling will be seen. Further copying of this tape will only make matters worse. At least 10 seconds should be allowed for this run-up time with any VTR, and this is especially true with the cassette machines, where the auto threading must be considered.

An important technical and operational point is that both the record and playback machines can stabilize on a black signal. This has no video, but it does have sync pulses. This enables both machines to lock up and become stable without any picture actually being viewed on the screen. This introduces certain operational difficulties, but it solves several technical problems.

● **Audio offset.** Figure 6-3 shows the layout of the heads in a cassette machine, and Fig. 6-4 shows the positions of time coincident points on a tape. The offset of the erase and audio from the video can be easily seen.

The operational problem this produces concerns the section of tape marked AE. If these represent the points where new video is edited onto the tape, then the new audio begins at point A and the old audio is erased from point E. The section AE is recorded over without being erased and so could contain a double audio. The only way to avoid this is to have about 10 seconds of silence at the end of every section put onto the tape. Some machines obviate this with the audio turning on at a slightly different time than the control track and the video.

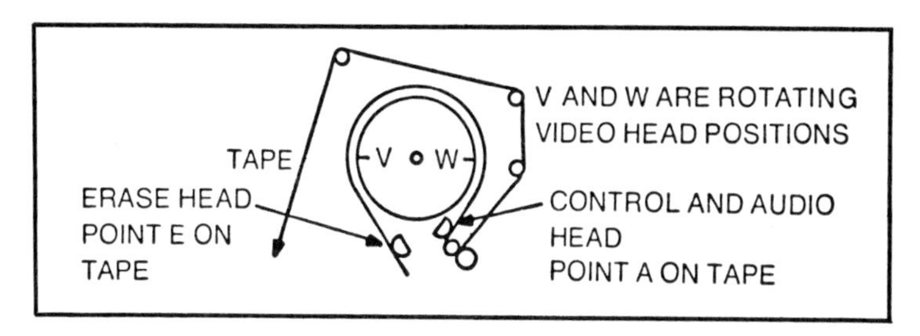

Fig. 6-3. Head layout.

● **The erase heads.** The positions and use of the erase heads differ, depending upon the editing mode used.

In an assemble edit the full erase head wipes the tape clean and the rotating heads record new video tracks, as shown in Fig. 6-5. The section of the tape *VWEF* is not erased and is recorded over twice by the video heads. In many cases the video current is sufficient to impress the new video over the old with no problems, and in some machines extra video current is applied for about 3 seconds. At the end of the assembly section the stop button cuts all current to all the heads, thus ending the recording mode.

This could leave a section of tape in AE which has no video, no control track, and no audio. To prevent this from causing problems when the next section is added, the audio and control tracks can be timed to start and stop after a slight delay. Another successful technique is to reposition the audio erase heads in the same block or head as the audio heads. The various models tend to differ in what has been done, and each must be examined separately. If any

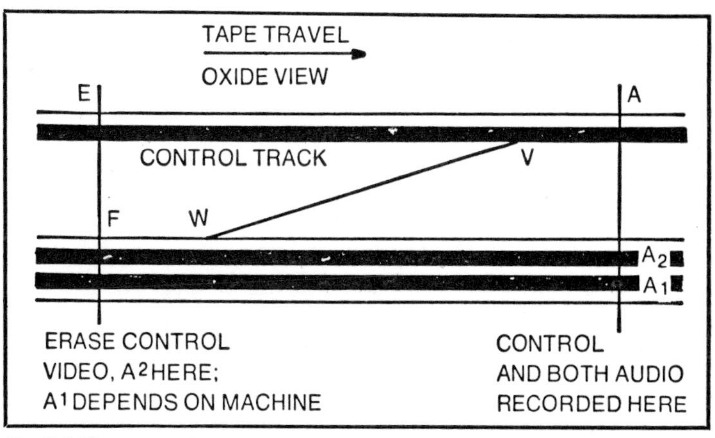

Fig. 6-4. Time coincident points on tape.

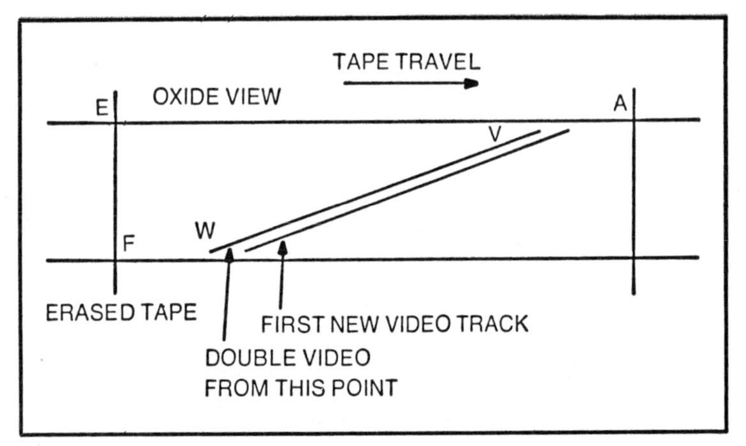

Fig. 6-5. New video tracks.

doubt exists, then a few experiments should be tried, and 10 seconds of silent black should always be used in productions.

With insert editing the situation is a little different. The cut in has to be just as definite as with the assembly mode, but the cut out also must be just as clean. To effect this the machines that have the insert mode have a "flying-erase head." This is similar to the rotating video heads, and it is mounted in the drum just ahead of the record heads, and it is mounted in the drum just ahead of the record heads. At the cut-in point it is fed erase current, and it wipes only the video track which is about to be recorded upon. At the cut-out point it is shut off, and the next video tracks are not touched. The main erase head is not used in this mode, and so the control track is untouched. A separate erase head is still required for both audio tracks.

Operationally the insert mode is very attractive, but it has one subtlety which can cause serious problems. In the insert mode a new control track is not recorded. So if it is used, then there must already by a control track on the tape. This means that the tape must have something already recorded on it—even if it is only silent black. Often, to obviate difficulties, it is worthwhile to record silent black along the whole length of the tape; this lays down a control track which will keep the servos locked up and will prevent serious editing problems and picture tearing and mistracking.

OPERATIONAL PROCEDURES IN EDITING

A serious problem is encountered when there is no signal on the playback tape prior to the start of the required program material or if there is no signal too soon after the end of the material. This means that there is no sync signal or control track for the machine servos to lock up to, and this can ruin any attempt at copying or editing. For this reason the following operational procedure is advised in all cases.

All video that is intended for further use of any kind should have at least 30 seconds of silent black at both the beginning and the end.

This ensures a stable sync signal and control track and also obviates the problems due to audio offset and erase head positions.

Assemble Editing

This is a simple procedure which can be very easily learned. The program tape is merely played up to the point where the new material is required and the EDIT button is pushed. The only operational problem is ensuring that the tapes are in synchronization so that the new material begins at the correct time in the program. There are three main sources of providing new material, as described in the following sections.

● **Studio.** This is a most common situation, and almost all programs from a TV studio are put together in this way. The opening of the program is recorded onto the tape, and the action continues until a convenient stopping point or until a mistake occurs. The tape is then rewound to a convenient point and replayed. At the decided upon edit point the **edit** button is pressed and the studio cast continues its action. A simple setup of equipment is shown in Fig. 6-6. If possible the artists in the studio should be allowed to see a monitor so that they can judge their start point. In all cases the studio talent should be instructed to remain silent after any stop and also to keep quiet until they are cued to commence. In this way audio problems will be avoided at the edit point.

● **Another videotape machine.** Assembling prerecorded sections from different tapes into a complete program is a regular undertaking in a TV studio and is something all operators should be able to do. The main problem now encountered is ensuring the video from the playback tape starts at the correct time, and this is best performed with the following procedure and the equipment shown in Fig. 6-7.

The point on the playback tape which represents the start of the program material to be added to the main program is determined, and this is labeled Cue 1. A point about 10 seconds before this is determined and is labeled Cue 2. The entry point for the new video is found on the main program, and this is called Cue 3. It is now necessary to determine Cue 4; this is the point in the main program where the main tape will be left stationary, and this point is about 8 seconds before Cue 3. The main tape is set up at Cue 4 and left.

The playback tape is now started from a point about 30 seconds before Cue 1. When Cue 2 appears the main program tape is started. This tape now locks up to the stable video incoming from the playback tape, and if Cue 4 has been correctly chosen, then Cues 1 and 3 will be coincident in time. If they are, then the tapes are reset, the procedure is repeated, and the edit is made.

If this time coincidence does not occur, then a further trial is made with a new Cue 4. The operational difficulty is choosing the cue points so that the machines have time to run up and stabilize and then be in exact synchronism so that the cutover can be made.

To successfully achieve this timing the run-up times of the machines must be known, and it must be possible to reset the tapes with minor adjustments if the first run is incorrect. It usually requires a few practice runs to get the synchronization right, and this means some trial-and-error adjusting of the cue points. This

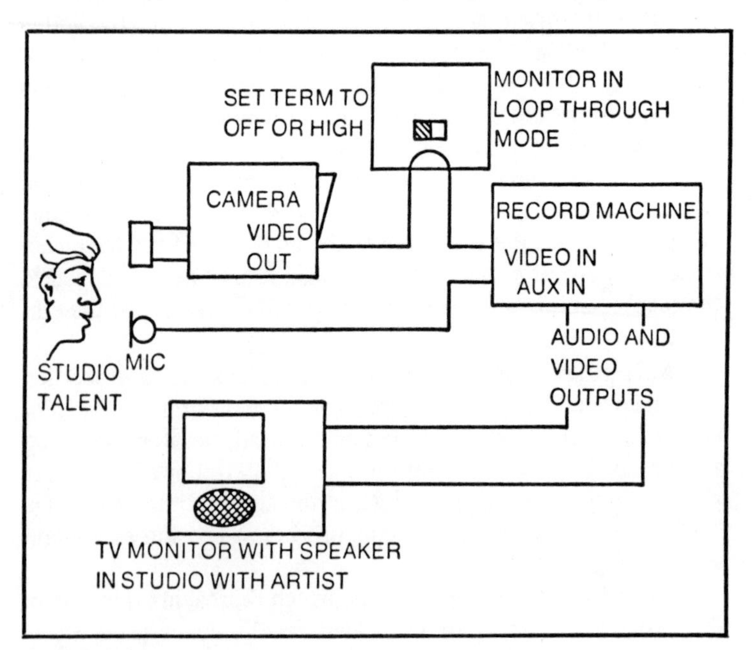

Fig. 6-6. Studio editing setup.

adjusting often can be done with the aid of the tape counter and by adjusting the reels by hand.

For many purposes this method produces good, reliable edits and copies of program tapes.

Figure 6-8 illustrates the cues on the tapes.

A variation of this simple procedure is to reverse the tapes and the cues so that the main program tape is run from a point 30 seconds back and the playback tape is left stationary. This can work well, but it often does not give time for the record machine servos to adjust and stabilize to the incoming playback video, hence longer run-up times are needed—usually about 15 to 20 seconds.

● **A film chain.** This is midway between the previous two cases. The camera is sending out a silent black signal to which the VTR machine can lock up, but the action has not started, so it resembles a studio setup. It is akin to the VTR copy in that a run-up time for the film is required and must be cued correctly. Operationally it is similar to the above case, but a much smaller run-up time can be tolerated with the film chain than with a playback tape.

Insert Editing

Insert editing is more difficult to accomplish than assembly editing and, as much as possible, is a mode to be avoided, especially with cassettes.

● **Studio.** To have an artist or performer in a studio record an insert into a tape, the person must know exactly what is required

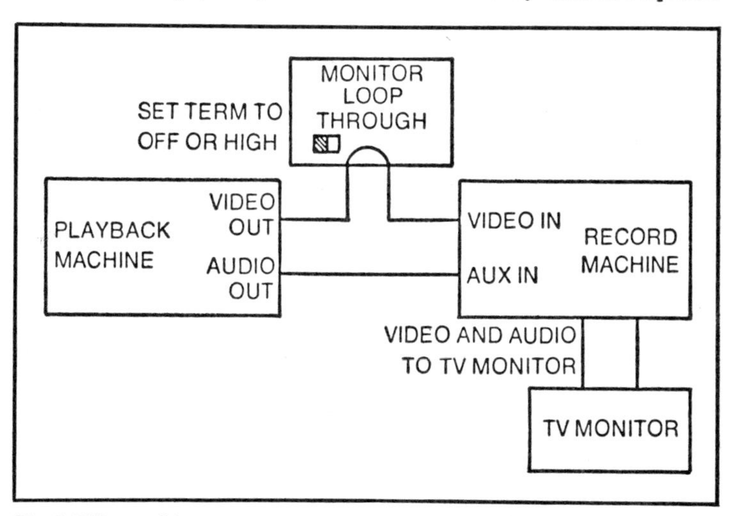

Fig. 6-7. Tape editing.

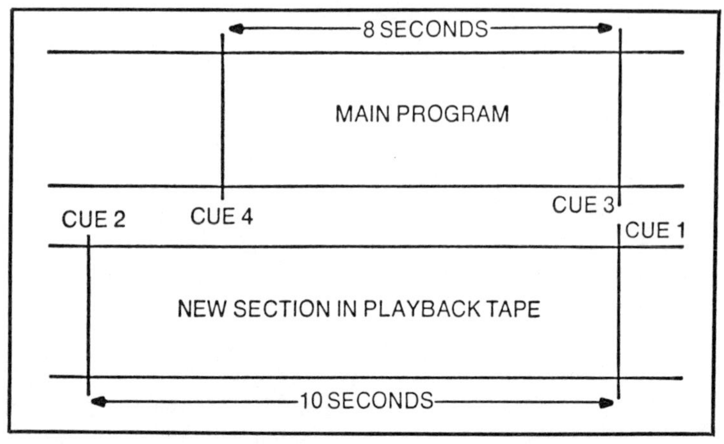

Fig. 6-8. Edit cues position on tape.

and must rehearse it several times to ensure getting it correct. A mistake in the middle of a section is not serious, but an overrun can ruin the whole recording. As with the assemble mode, the tape is played up to the point where the artist is to start, and the cut-in or edit button is pressed. The artist must now time the action, preferrably with a stopwatch, and must finish at the correct time so that the cut-out button can be pressed before the resumption of the original required video.

Due to the tight cuing difficulties, it is often better to insert a section longer than that actually required, picking the in and out cues at convenient places where the least critical timing is needed.

● **Inserts from another tape.** To insert a section which has been prerecorded onto another tape presents the most difficult and critical operation possible with videotapes, and as far as possible this mode should be avoided. Operationally it is similar to using tape sections in assembly, but now each insert must be of a certain length. Because an overrun cannnot be tolerated and the tape insert cannot vary in speed like a live artist can, the start point and the cues must be very accurate. The cut-out point also must be initiated just as accurately.

Should any doubt exist about whether the insert will be successful, then it should not be attempted, because once a good section of tape is ruined it must be completely redone.

● **A film chain.** This is essentially the same as the previous case. In general all editing situations can be improved operationally by using a studio switcher-fader instead of going directly from one machine to the other.

Audio Dubbing

This is the ability to record audio only onto the tape while the video and servo circuits remain in the playback mode. This facility is useful in a variety of cases: a noisy environment may prevent a good audio track being recorded in the first place, a second language may be required on a finished tape, and separate video sections edited together seldom have a cohesive audio track. Audio dub is a continually used facility with VTRs.

Audio Heads and Tracks. Before the operation can be fully understood, the effects of the erase and record heads on the two audio tracks must be known. Figure 6-9 shows how the tracks and heads are set up, with the audio track No. 1 erase shown.

An important point to remember is that the **record** button will cause the erasure of both audio tracks, whereas the **audio dub** button will affect only Channel 1 audio and leave Channel 2 and the video untouched. Audio which has been recorded onto Channel 2 cannot be separately erased.

Figure 6-10 shows the alternative audio and erase head positions for audio dub in both an editing and a nonediting machine. In both cases each audio track obviously has its own record-play head, but only Channel 1 has its own individual erase head. Channel 2 shares the main erase head with the control and video tracks. When the **record** mode is used, both erase heads are activated, and all the information on the tape is erased and newly recorded. In the **audio dub** mode, only the Channel 1 erase head is activated; so Channel 2, the control track, and the video remain untouched.

The main problem in the **audio dub** mode with the nonediting machine is the offset of the erase and audio heads. This length of tape amounts to about 3½ seconds in time, and it can end up with double audio. With the edit machine this is not a problem due to the proximity of the erase head to the audio head.

Audio Dub Operations. Audio dubbing can be done from any source, and either the **mic** or the **aux** inputs can be used separately or together. The **input select** switch must be in the **external** position. The cassette should be inserted into the machine and played up to the point where the audio is to be added. As the required point approaches, the **safety** button should be held down and the **audio dub** button pressed exactly on cue. The audio dub mode is ended by using the **stop** key.

If the new audio is to be added from the start of the tape, then the **audio dub** button can be held down and the **play** button pressed, or the tape can be run out by using the **play** button and

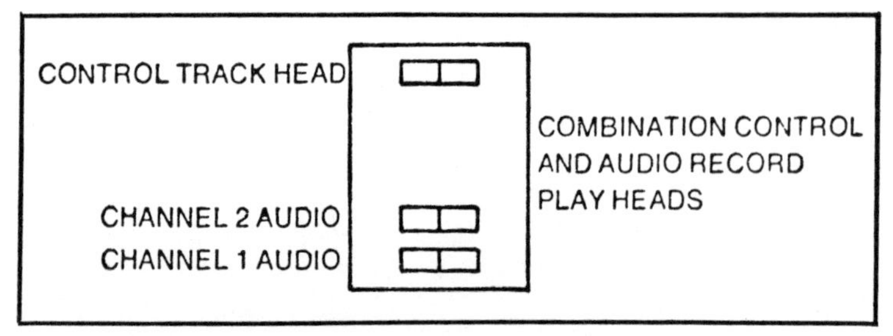

Fig. 6-9. Audio erase head position.

then the **pause** button. If the **audio dub** button is now pressed, this mode will be initiated when the **pause** is released.

At all times care must be exercised when using this mode. It is very easy to clip the beginning and end of the audio, and it also is easy to get a period of double audio. If audio dubbing is envisaged, then at least 10 seconds of silence should precede and follow every section of video recorded into the tape. If a microphone is used for the audio source to be dubbed, then it must not be placed too close to the speaker, which may be playing back the tape for cues. Because Channel 2 audio is not erased in this mode, sections of it can be copied onto Channel 1 if required.

CUING AND EDITING WITH VIDEOCASSETTES

The previously described principles of editing and copying cuing are fairly simple to achieve with the open reel machines and are common operations in any studio. With the spread of the cassette machines these capabilities of editing and cuing have

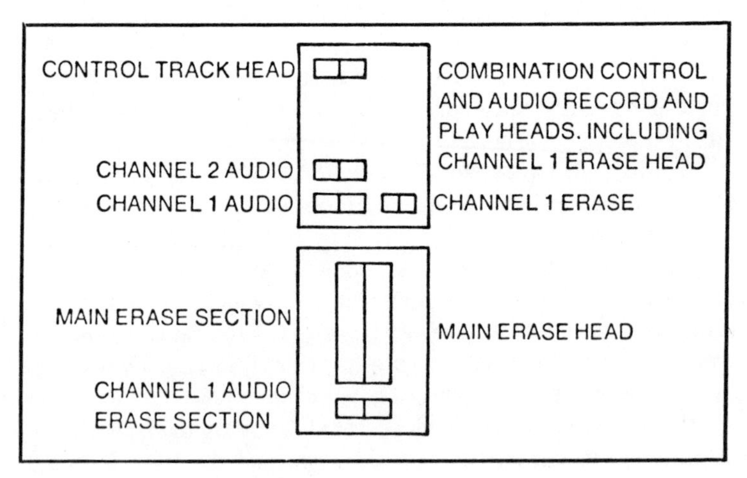

Fig. 6-10. Audio erase head position.

remained just as important, and several machines now available are manufactured with either an assemble or insert capability.

However, due to their design there are certain inherent operational disadvantages which make these two modes very difficult to achieve. There are three main reasons why this is so:

1. When the **stop** key is used the tape retracts back into the cassette and rewinds for a few seconds, so the cue point is lost. Because this short rewind is not the same for different times in the tape, it is impossible to rely upon the tape to be cued.

2. The tape reels are not accessible for minor adjustments by hand, so the cue points cannot be refined or adjusted.

3. There is no video or audio during the run-up time. These are muted until the tape is up to speed.

If some lack of precision in the actual start of the added material can be tolerated, then this inexact setup method can be used. It is absolutely useless for exact cuing, such as for a commercial to be played **on air**, or to provide the next video section in an assembly edit.

Video cassettes are not made for this type of work, and this must be realized by the user. If the cassette machines are required to perform these types of operations, then certain extra operational facilities must be built in or the machine must be modified to accept them.

The most often required modifications to the cassette machines are as follows:

- **A pause** control—This stops both the tape transport and the rotating heads without retracting the tape back into the cassette. This is an easy modification to incorporate into machines which do not have it as an original feature. With the one-reel cartridge machines, the **stop** control does not cause tape retraction and is in effect a pause control.
- **A still frame** control—In this the tape transport is stopped but the heads are allowed to rotate, thus producing a still frame picture on the screen. This is the most desirable condition, and it is found on many open reel machines. Due to the peculiar mechanical construction, it is very difficult to incorporate in the cassette machines and virtually impossible to effect as a modification.
- **Video and audio suppress until the program start point**—All cassette machines have an audio and video mute during runup as part of their normal circuits, but there is no provision for keeping these from the outputs until required. It is easy to modify

the audio for this but impossible to modify the video because this also would suppress the much needed sync signals. To effect this it is best to use a studio switcher-fader.

- **Demute the audio for cuing**—This would allow prerecorded cues on the unused track to be used, especially in the silent black period.

Of these modifications the most essential is the **pause** control. With this, most of the editing and cuing operations can be achieved, although care is still needed in its use. The pause control provides the ability to stop and start the machine on an exact cue. Once this has been done, the cue point should be checked in a trial run to see if it is correct. If it is not, then the only way to adjust it is to play back the tape and stop it at some other point and try that. It is impossible to wind one of the reels backward or forward a few turns by hand as with the open reel machines.

Accurate cuing points can be set up with the cassette machines, but often the procedure is time-consuming and tedious.

To make the cuing process with the cassette easier, certain procedures should be adopted at all times.

- About 30 seconds of silent black should be on the front of every section of video. This is especially important if these sections have no video prior to the program entry point.
- About 30 seconds of silent black also should be at the end of every section.
- Recording countdown cues on the second audio track is useful in establishing when the opening video point is about to arrive. A point is chosen about 30 seconds ahead of the start, and the audio dub mode is used. Seconds are just counted off, starting at 1 and continuing until after the video arrives. It does not matter which number coincides with the video—it is just remembered and used as the cue. These can be heard over phones on playback.
- Using the counter to assess the position of the tape and the runup time is useful, but it is not always accurate. It can be used provided it is checked for each place in the tape. Some machines are provided with very accurate resettable counters, which can be used quite successfully.

Umatic Editing Systems

An editing system consists of two machines connected to and controlled by a remote editing panel. With such a set-up it is possible to perform video assemble and insert edits on a normal cassette with any combination of video and audio tracks. All editing systems use the Type-2 tape transport, as only this type has the necessary stability.

The earliest of these systems was the Sony RM-400 Control panel with two VO-2850 machines. As this is a very easy and very popular system its operation is worth covering briefly. Figure 6-11 shows a block diagram of the system, and Fig. 6-12 is a photograph of the items.

The RM 400 control panel is divided into three sections. The two outside sections are control panels which will completely control each of the cassette machines, and the central panel controls and performs the editing. In operation each machine is treated independently of the other. First the counter memory is cleared by pressing the **record** button on the **slave** machine, and this sets up an E-E picture. It is now possible to set up the **master** tape. Note the terminology used: the **master** tape is the **playback** tape, and the **slave** tape is the tape onto which the final program is recorded. The **master** tape is played up to the point where the video is required to be inserted onto or added to the main program, and then the **pause** button is pressed. This sets the counter to 000. The **pre roll** button is now used and this rewinds the tape for about

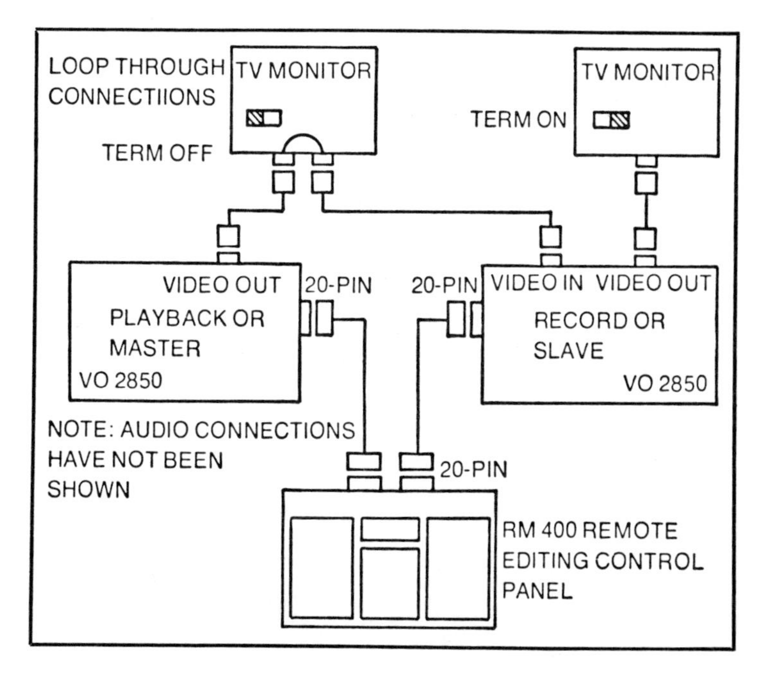

Fig. 6-11. V02850 setup.

10 seconds and then plays the tape until it is 5 seconds before the 000 point. The tape is then parked at this point.

The **slave** or recording tape is now treated in the same way, using the controls for this machine, and it also is left parked 5 seconds before the program cut-in point. Both **standby** lights are now on, indicating that the machines are ready for the edit. Both of these setup points can be found easily by using the **search** button, which rolls the tape at one-fifth its normal speed.

If an **assembly** edit is required, then the **assemble** button in the center panel is pressed and the wider **start** button is used to commence the function. The machines now both roll simultaneously, and the control circuits in the RM400 make the edit at the correct time.

Either machine can be displayed on the counter, but it is best is display the **master** or playback tape because its control track is counted on the playback; it is impossible to count the control track of the other machine because it is recording. This procedure aids in resetting the tapes if the edit was not properly made.

The **insert** mode is conducted similarly, but the slave machine is monitored on the counter instead. In the insert mode, the **cutout** button can be used to end the inserted video whenever desired, and it is possible to cut back in with another insert and cut out again as many times as desired.

By using two monitors the edit points can be viewed accurately and the video from both machines can be seen. Note that the video and audio connections must be made because normally they are not routed through the remote control panels.

The counter will indicate times from 59.9 seconds before to 59.9 seconds after the edit point, and this allows ample time for

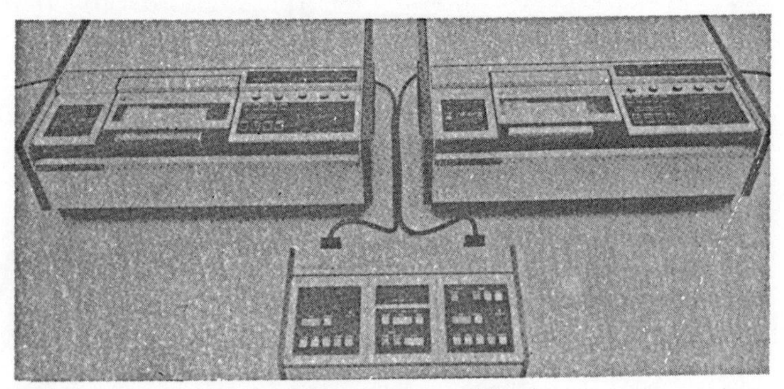

Fig. 6-12. Two V02850's and RM400 (courtesy of Sony).

adjustments of the edit point. The edits are made to within an accuracy of 0.2 seconds. It connects to the RM400 panel via a 15-foot length of cable which has a 20-pin connector on the end. This cable also carries the power to the RM400.

The 400 weighs 7.7 lb, is approximately 16 × 8 × 2 in. and consumes 6.5 W of power. It is built entirely of ICs, and the display is of LEDs.

A single VO2850 can be used as a recording machine to assemble a program from a studio camera or from another type of tape machine, and a typical setup is illustrated in Fig. 6-13. To enable only one operator to control both the camera and the cassette machine, an RM410 remote control panel can be used. This has the same layout and functions as the deck and will control all the functions of tape motion and editing. This connects to the machine with a 15-foot cable and uses the same 20-pin plug as before. It weighs only 3 lb 10 oz and is about 6 × 6 × 1½ in.

Although this first system is adequate for most industrial and educational purposes, the broadcast industry requires much more accurate editing facilities. To accomodate the broadcast industry the BVU-200 series of machines was developed, and several editor control panels have been introduced to work with these and other machines. The Sony BVE-500 series is an example of editors that are designed to work only with the BVU-200 series of machines. Figure 6-14 is a photograph of one of these. This unit contains all the remote controls for the record and playback machines, plus a panel which allows edit points to be selected, displayed and memorized. It will then control both machines and will preview the edit, then perform the edit and afterwords will review the edit.

An example of an editor made to work with a variety of machines is the Convergence Corporation ECS-103, shown in Fig. 6-15. This editor is typical of those which can perform the most complex operations. The desired operations, such as cut-in, cut-out, etc. can be programmed and memorized, and the whole sequence of operations controlled by a microprocessor. Often a TV monitor will be used as a display for the in and out times.

Some editor panels use push buttons to control the tape motion, and speeds ranging from 1/20th to twice normal in reverse and forward are quite common. Other editors use a joystick or rotary control knob to give continuous speed variation or incremental speed jumps.

The facilities available on modern editors vary considerably, and care should be taken when using these with the various models

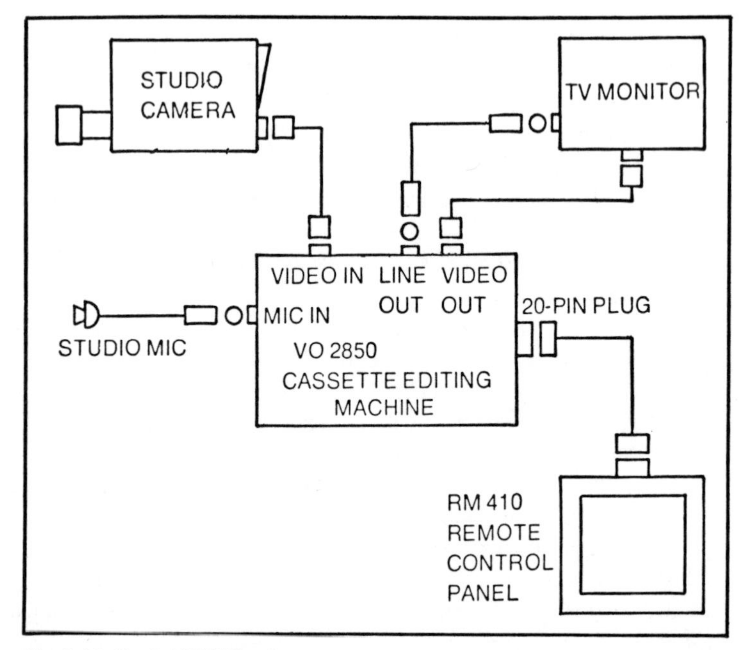

Fig. 6-13. Single V02850 setup.

of cassette machines. Many editors require that the machine be sent to the manufacturer for modification. The actual details on performance and operation are also best obtained directly from the manufacturer because their capabilities tend to differ considerably. A further complication is that some different models of cassette machines require slightly different modifications.

In general, the editors are for the more expensive Umatic machines, although the industrial Betamax models can also be remote controlled. The main use for editors is in TV news, production houses, and industrial and educational studios. They are not generally considered a consumer item for home use at this time.

The internal circuitry and construction of these editors is quite complex and is a subject in itself.

The Digital Audio Units

One of the recent innovations in TV has been the digitizing of both the audio and the video signal, and performing signal correction and processing on the digitized version. Much work has also gone into recording these digital signals on tape.

The technical limitations and drawbacks of the conventional analogue tape machine are well known, and it has been shown that these disadvantages can be overcome by using a digitized audio signal on tape. Although several high quality professional digital machines have been produced (BBC, 3M, Mitsubishi) these are very expensive and are currently beyond the means of the average consumer.

An interesting application of this idea is the two digital audio units or adaptors, produced by Sony, which permit a digitized stereo audio signal to be recorded on a standard unmodified VTR. Two models have been produced. One is for consumer use with a Betamax or similar VTR, and the other is for professional studio use with a Umatic or other broadcast machine.

These two adaptors digitize the audio signal and then convert it to a 'psuedo video' signal which can be recorded on a normal VTR. Both are low cost items which plug into the VTR, and thus leave the VTR still available for recording TV programs. They also make it possible to use existing TV facilities for transmitting high quality stereo audio over a normal TV channel.

Both units can record and playback the original audio signal with no errors. Both also eliminate the narrow dynamic range, limited frequency response and poor signal-to-noise ratio of the standard audio machine. Wow and flutter are so low they cannot be measured and distortion and print through are completely absent. The size and cost are both such that they can be considered for serious consumer use as well as in the professional studio.

Fig. 6-14. Sony videotape editor.

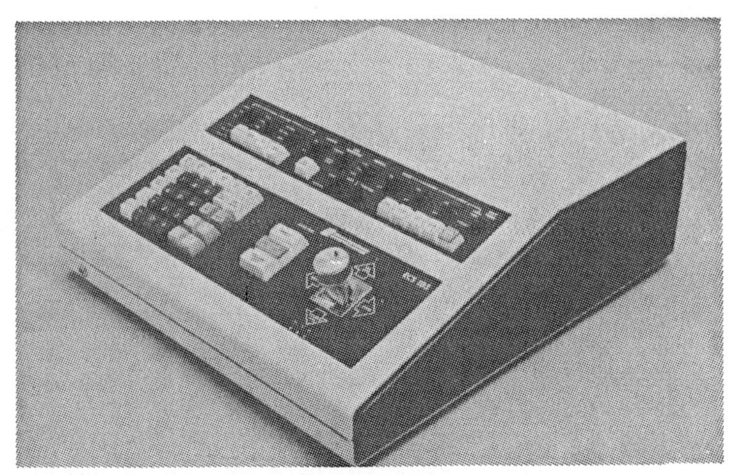

Fig. 6-15. Convergence corp. videotape editor.

Introduction to A-D Conversion. An analog signal is one which is continuously varying and has an infinite number of levels between its highest and lowest levels. The information is contained in the different voltage levels, and in the manner in which the signal changes from one level to another. The familiar audio electrical signal is a typical analog signal.

Analog signals have several advantages and disadvantages. Typical disadvantages are their low bandwidth and the ability to assume any level. Typical disadvantages are that the dynamic range is often limited, distortion is present, and the signal to noise ratio is often not as good as desired.

Digital signals, such as those found in computers and much industrial control equipment, have only two levels - high and low - or ON and OFF. One of the main advantages of digital over analog signals is they can be transmitted and amplified many times without the introduction of severe distortion, as the distortion can be easily removed see Fig. 6-16. The main disadvantage is the much wider bandwidth.

Table 6-1 compares typical specifications for professional studio analog and digital audio recorders.

Audio signals are now digitized and transmitted in several professional applications. For example, many long distance telephone calls are carried digitally, and the BBC in Britain is using 'sound in syncs' for interstudio distribution of program audio. Also, several professional digital audio recorders have been announced (BBC, 3M, Matsushita).

A/D Conversion. To convert an analog signal to digital it is sampled by a series of pulses, as in Fig. 6-17. This produces a 'pulse amplitude modulated' or PAM waveform. The amplitude of each PAM pulse is now compared to a series of reference voltage levels in a 'quantiser' (which is several comparator circuits). The bits in this output represent the amplitude of each separate sample pulse in the PAM. These bits are produced simultaneously and can be recorded on a separate channel on a tape. In this example only 4 levels are used and only 4 bits are produced. As all of these bits cannot be transmitted over a line or radio link simultaneously, they are fed into a serial-to-parallel (s-p) converter to produce a serial output that can be transmitted.

If a signal is to be quantized into many levels instead of just a few, then a different a/d method is used. This performs the conversion much faster and its output is usually in the form of binary counting. Binary counting keeps the number of bits in the digital word to a minimum for the number of levels used. The large number of levels make it difficult to record the bits on tape in separate tracks, so usually the serial output pulse train is recorded onto the tape.

The number of bits in the pulse train is related to the number of levels by the equation:

$$\text{levels} = 2^N$$

Where N = number of bits.

An analog signal must be sampled at more than twice its highest frequency if it is to be reproduced accurately when reconstructed later. This is called the Nyquist condition. So the number of samples transmitted over a line is equal to the number of samples taken in the original, times the number of bits used in the digital code:

$$\text{transmitted frequency} = \text{No. of samples} \times \text{Number of bits}$$

A high quality audio signal must be sampled at about 30-40 Khz and needs about 13-16 bits to be faithfully represented. So the resulting digital pulse train will have a frequency given by:

$$\begin{aligned}\text{frequency} &= \text{sample rate} \times \text{number of bits}\\ &= 40\text{ khz} \times 16\\ &= 480\text{ khz}\end{aligned}$$

The digital signal has a much wider bandwidth than the original analog audio signal. Recording and playing back such signals on tape can be achieved, but it does produce some problems not found in the simple audio machine.

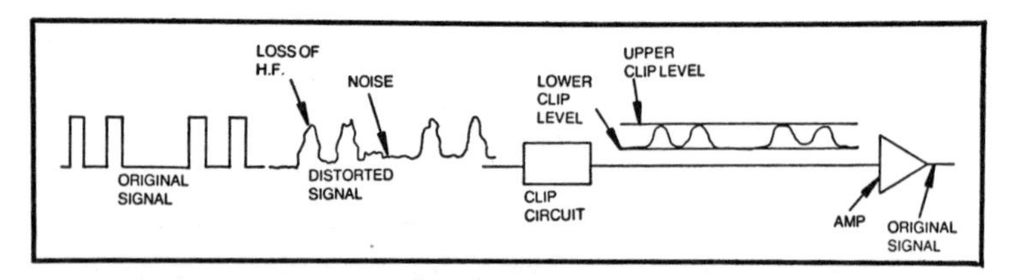

Fig. 6-16. Improving a distorted digital signal.

D/A Conversion. The process of converting the digital train of bits back into an audio signal or analog signal is very easy. The serial pulse train is converted back to a parallel signal so that all the bits appear simultaneously for each sample time. These are now all applied to a resistive ladder, as in Fig. 6-18. The output of the ladder is a step signal which takes on an analog value equal to the quantized value of the original samples. A low pass filter removes the high frequencies in the edges of the steps and produces a copy of the original signal.

Digital Signal on Tape. To put a bandwidth of 500 KHz or more onto a tape two things are required: a high writing speed and a small head gap. To use a head gap of 0.5 mil would require a tape or writing speed of 500 ips. If a normal audio type machine were used, then very large reels would be needed to hold the large amount of tape required, and even then very little program time could be put on one reel.

One way around this problem is to record the digital signal from the quantizer on different channels, with a separate bit from each sample on a separate channel. This reduces the speed and conserves tape, and is the idea on which the recent digital studio machines are based (however, the actual method used is a variation of this idea). Another way to obtain the high writing speed for the digital signal is to use a rotating head, as in a VTR. In fact, a VTR can be used.

Table 6-1. Comparison of Analog and Digital Audio Recorder Specifications.

	Analog	Digital
Frequency Response	20 Hz - 16 Khz + or − 3 db	DC - 20 Khz = or − 1 db
S/N ratio	55 db	80 db
Wow and Flutter	0.05 to 0.1%	below the bounds of measurement
Overall Distortion	0.5 - 2%	0.03%
Dynamic Range	65 db	90 db

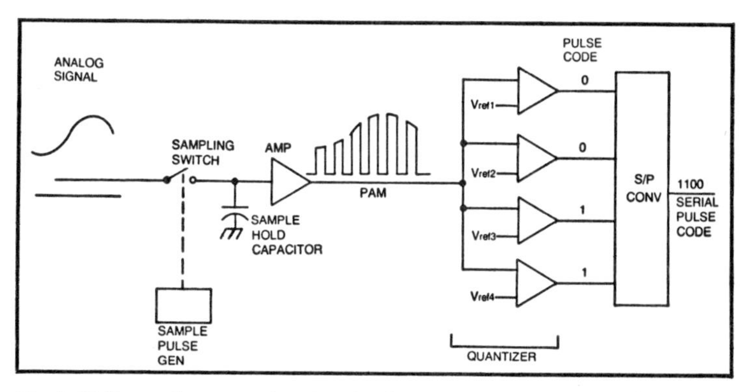

Fig. 6-17. Converting an analog signal to digital.

Using a VTR. A VTR can be used to record the serial train of digital pulses which represent the audio signal. It has the required head gap, writing speed, and economy of tape. However, in a VTR the heads cannot be allowed to rotate freely. They must be servo controlled and referenced to the vertical sync interval of the TV signal. When recording a digital audio signal on a VTR, the vertical and horizontal sync pulses must also be supplied. The digital audio must be placed in groups of bits between the horizontal sync pulses, with no pulses in the vertical interval. For this reason the term 'pseudo video signal' is used. See Fig. 6-19. To arrange the signal in this manner it is first fed into a digital memory and then read out in a suitable manner.

For several reasons a simple digital code is not recorded onto the tape. Several special codes are used, such as biphase, delta modulation, NRZ, etc.

NRZ is simple to produce and playback, requires very simple circuitry, and it is non self-clocking. With a VTR a non self-clocking

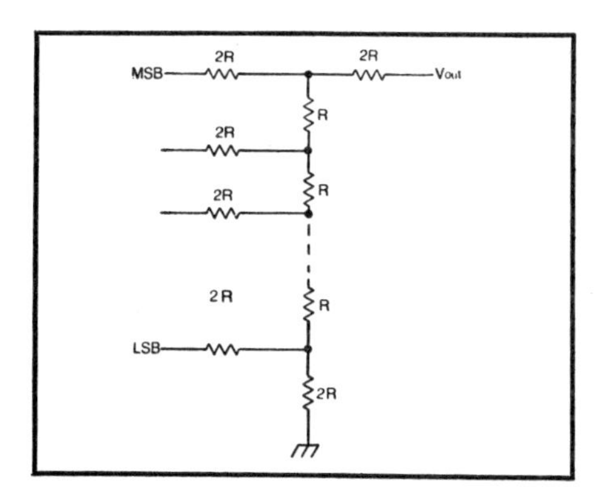

Fig. 6-18. Resistive ladder.

code can be used as the sync pulse and can provide the clocking. So NRZ is used in the VTR type of digital audio machines. The NRZ (non return to zero) digital code modulates the normal FM signal which is put onto the tape.

Drop-Outs and Data Configuration. One of the main problems with video tape is drop-outs. A very small drop out is noticable on the small video tracks and so a compensator must be used. In video the drop out compensator (DOC) replaces the signal loss with a repeat of the previous horizontal line.

In digital audio a drop out results in missing bits and the errors introduced into the signal are quite unacceptable since the effects are very noticable to the ear. Replacing the lost information with a repeat of the previous section of the signal does not work, and so a video type DOC cannot be used. Several very interesting ideas have been introduced to overcome this problem.

To ensure that the lost signal can be regenerated four basic ideas can be used:

- Repeat the signal in another part of the tape. This wastes tapes and there is no guarantee that another drop out will not lose the repeated information.
- An error detecting or even an error correcting code can be added to the signal.
- Adjacent digital words in the original signal are 'scrambled' when put onto the tape. That is normally adjacent words are not recorded next to each other. This works because a small amount of information can be regenerated by the method used in Item 4.
- Regenerate lost information by interpolation. A lost bit or word can be interpolated from the two which are normally adjacent

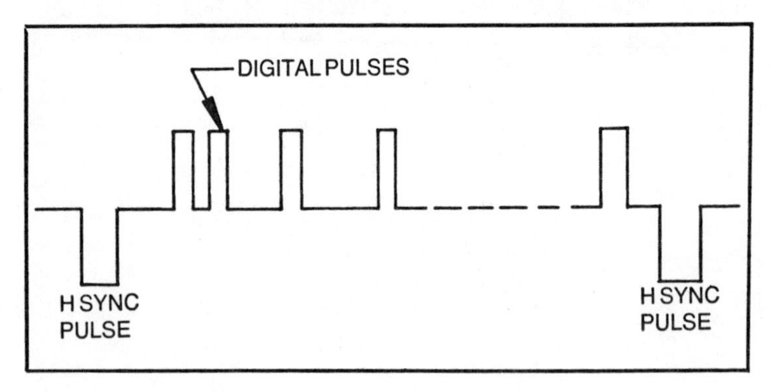

Fig. 6-19. Digital pulses in an active video line.

to it. Hence, the reason for scrambling the signal on tape. It prevents adjacent words being lost in the same drop out.

A combination of the second, third and fourth ideas are used.

When the digital words appear at the output of the A/D converter they could immediately be put onto the tape. Instead, they are put onto the tape in a different order so that drop out compensation can be achieved. The data words from both stereo audio channels are first fed into a memory and are then put onto the tape in an 'interleave' system, where normally adjacent or successive words are placed on different horizontal lines. As most drop outs are less than 1/6 H lines in length this allows any missing information to be regenerated by interpolation from its two normally adjacent words, which are placed elsewhere on the tape and are not lost in the same drop out.

In the simpler consumer units a check code is used to ascertain if information is missing. In the professional units a more complex interleave is used, with a full error correcting code. This is a special code and not one of the standard computer codes.

Listening tests have shown that this is a satisfactory method. However, if more than one successive word is lost, interpolating more than 2 words or extrapolating several words does not work, as the listening tests have also shown.

The Interleaving System. To use a data interleave system the words must be put onto the tape in a well defined manner. To make the circuits easy a whole number of words must be placed in one horizontal line. To do this the NRZ code must have a definite relationship to the H sync pulses, and so must the frequency of the samples in the original analog signal. This means a careful choice of master clock frequency, sampling frequency in the A/D converter, the number of bits per word, and the number of words per horizontal line.

When the information is put onto the tape the vertical interval is included in a period of about 17.5 lines which contain no audio data. This allows the vertical sync pulse and the head switching points to be clear of interference and also allows edit codes, etc. to be placed in the normal vertical interval.

This system of interleave is necessary as the drop outs are very noticable and must be covered up. Deciding on how many words per line, number of bits per word and how to interleave etc. was a complex procedure since several conflicting requirements had to be met. The amount of data put onto the tape is determined by the overall characteristics of the tape system, such as writing speed, frequency response, wow and flutter, etc. This leads to two slightly different systems for the Betamax type machines and the Umatics. The consumer or Betamax type system is the simplest and easiest, and is covered next.

The Consumer Digital Audio Adaptor. The high frequency and other characteristics of the Betamax type machines limit the quality of the audio which can be digitized and recorded onto the tape. However, it does provide an audio quality considerably better than the best analog audio machines on the consumer market.

In this, the 'B' system, the digitized audio signal has 13 bits per stereo channel, which are combined into a single 26 bit word. Three words are placed in each horizontal line and each group of 3 is followed at the end of the line by a 16 bit check word. This is called a 'cyclic redundancy code' (CRC) and is used to determine where any errors may lie which occur in that line. The total number of bits in each line is 94.

These words are put onto 245 lines. so a total of 735 words are used in each field. Figure 6-20 shows the interleave system; the numbers in each block represent the digital words to be recorded in that field. Each field is divided into 8 interleave blocks; the first 7 contains 92 words and the 8th contains 91 words. The CRC is added at the end of each horizontal line to detect the drop outs and this allows a drop outs to be corrected by interpolation over a range of 46 words, or 15 H lines.

Figure 6-21 is a simplified block diagram of the recording system. The two stereo audio channels have separate input amplifiers, equalization, and sample and hold circuits to produce the PAM. The PAM output is fed through an electronic switch and the A/D converter is time shared by both channels. The A/D output is fed into the memory circuit, which is a static RAM of 8K bits. The bits are read out of the memory in the interleave order and the CRC bits are then added. These now pass into the video output amplifier where normal video sync is added and the final 'pseudo video' signal is fed to the input of the VTR, where it is FM modulated and put onto the tape.

The electronic switch, A/D converter, memory read-in and read-out, and video sync are all generated in the sync and pulse generator circuit, which is controlled by the master clock.

Figure 6-22 is a simplified block diagram of the playback circuit. The video signal from the VTR is amplifier, fed through a sync separator, the CRC detector, and then into the main memory.

The memory has two main outputs. One feeds the DA converter where the bits are converted back into the audio signal. The other path feeds the interpolation circuit, which forms the missing audio caused by the drop outs. The complete signal is now fed to a switch which separates the two stereo channels, and LPFs feed each signal to separate output amplifiers. A master sync circuit controls all the playback operations, and this is controlled by the playback sync signal from the tape. The master clock has a frequency of 7.04896 Mhz. Dividing this by 160 produces the sampling frequency of 44.056 Khz; and dividing the master frequency by 448 gives the horizontal frequency (15,734 Hz). The transmission rate of the bits is 1.762238 Mbits/sec.

The Umatic or 'A' System. This is for use with the professional machines, and is much more complex than the simple consumer system. The main difference is that the higher frequency response and greater tape stability of the Umatics allows a 16 bit digitized audio signal to be used, and allows the bit density on tape to be doubled. A much more complex interleave is used and a full error correcting code is employed instead of the linear interpolation of the simpler B method. However, the same sampling frequency is used, and the 13 bit system can be used if desired.

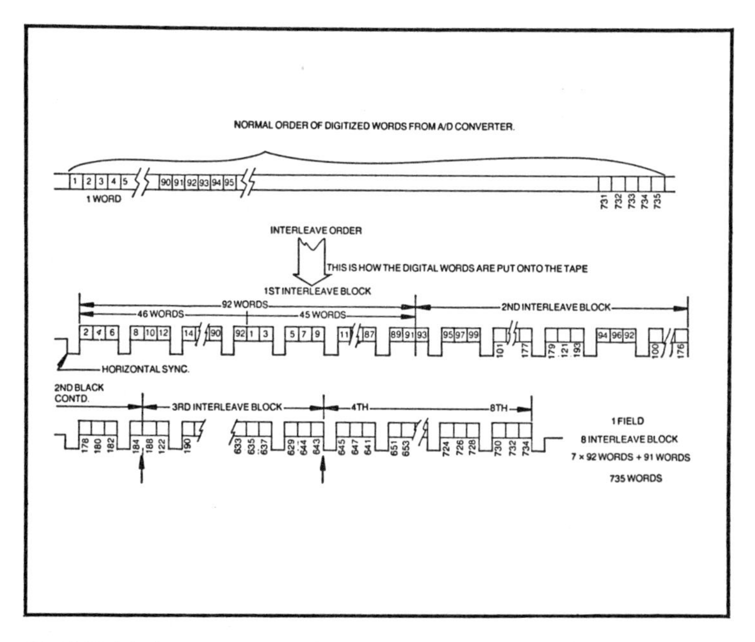

Fig. 6-20. Interleave system of PCM-1.

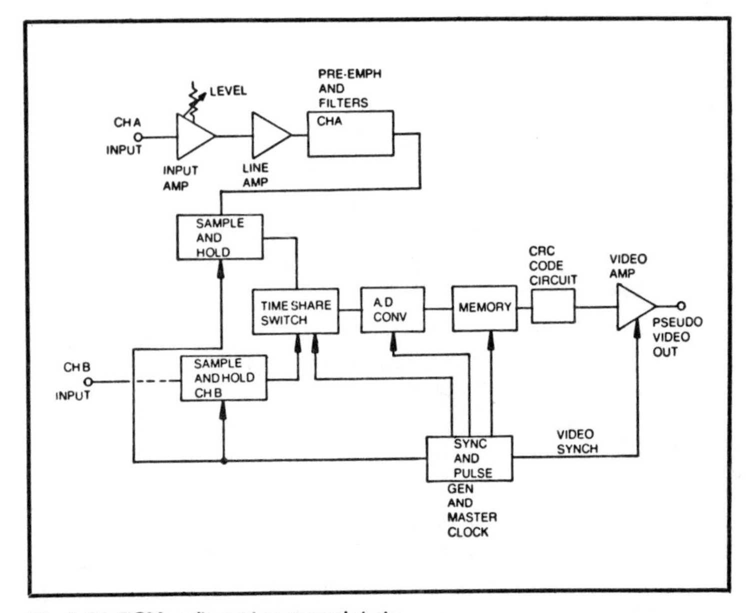

Fig. 6-21. PCM audio system record chain.

Figure 6-23 shows the tape interleave system. The block diagram and circuitry are similar in principle to the B system.

CONCLUSION

Table 6-2 shows the main characteristics and specifications of each of the adaptors. These excellent figures are due to the fact

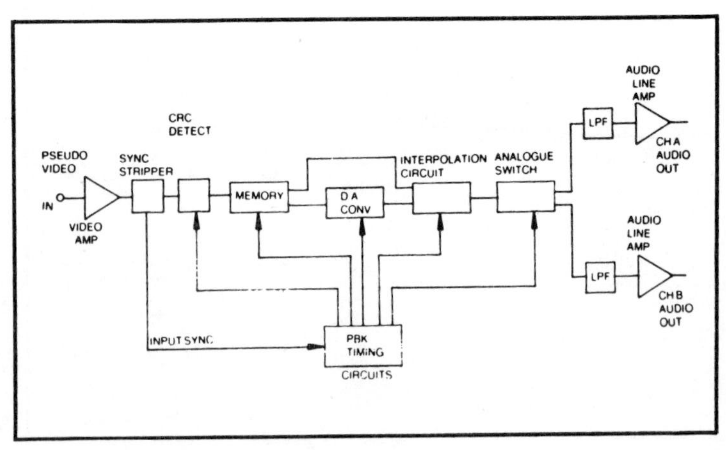

Fig. 6-22. PCM audio system playback chain.

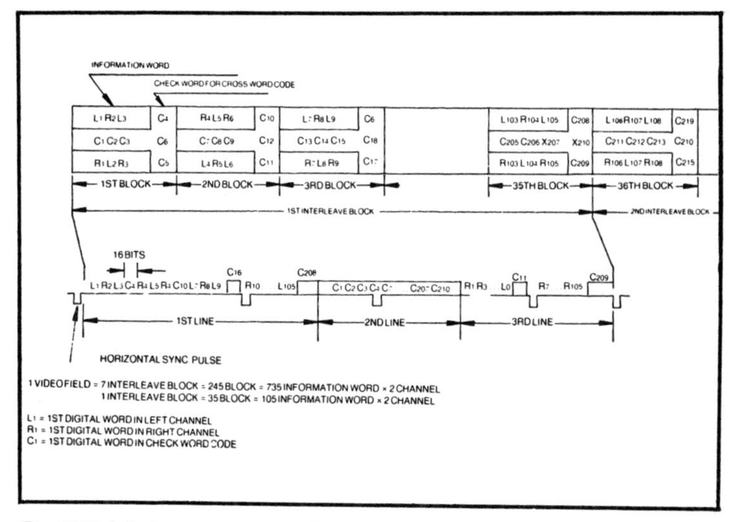

Fig. 6-23. Interleave system of PCM-1600.

Table 6-2. Comparison of Consumer and Professional Digital Adaptors.

	Consumer Unit PCM-1	Professional Unit PCM-1600
Number of channels	2	2
Sample Frequency	44.056 Khz	
Transmission rate	1.726 Mbits/sec	3.5795 Mbits/sec
Quantization	13 bits	16 bits
Dynamic range	85 db	95 db
Harmonic Distortion	0.03%	0.03%
Wow and Flutter	Below measurement	
Frequency response	2 Hz - 20 Khz +or-1 db	20 Hz - 20 Khz +0.5 db - 1 db
Drop out method	Interleave and CRC code	Interleave and error correction code.
S/N ratio (approx)	85 db	95 db

that the performance of digital tape systems is not limited by either the tape or the mechanics of the decks.

Both adaptors permit digital to digital dubbing of tapes, as the VTRs can be synchronized by normal video methods. Electronic editing is possible with unmodified editor panels, using SMPTE time code or control track pulses.

These two units are very new and are not yet established in the marketplace, but they do show a new and interesting direction in the application of digital audio and the use of VTRs.

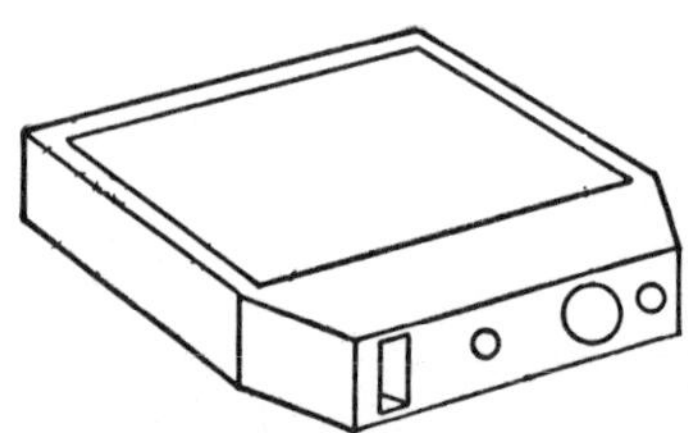

7 Cassette Limitations and Ancillary Equipment

The gradual growth of video in industry and education has been made possible by the production of equipment designed specifically for this type of television. Much of this equipment is of excellent quality and admirably fulfills the tasks for which it was designed. But it is not broadcasting equipment, and it should not be used as such.

As an illustration of the difference, consider a typical small family car used for going to work and general purpose transportation. No one would seriously consider entering it in a grand prix race. And conversely no one would drive a grand prix racer to work every morning. Both items may be an excellent piece of machinery, but each has its own use.

A similar situation exists with video equipment and the user should realize that the videocassette machines are not broadcasting equipment. If they are used as such, then some serious problems will arise which will often confuse the users as well as impair the quality of the recording or playback.

The purpose of this chapter is to explain why these machines are not broadcast equipment and what troubles are likely to occur through this type of misuse.

CASSETTE LIMITATIONS

Inherent in the design of the videocassette system are certain limitations which do not allow the machines to fully meet FCC specifications with respect to sync stability, noise, and color signal requirements. The FCC has adopted certain rules for equipment for very practical reasons, and much of the cost and complexity is incurred in meeting these criteria.

Outside of broadcasting, a lower level of quality can be tolerated, and this allows simpler equipment to be built and used with an attendant reduction in cost while still producing pictures good enough for closed-circuit work.

When nonbroadcast equipment is used with the broadcast items, or the same level of performance is demanded from it, certain problems arise which are difficult to either obviate or explain away. These can be grouped into five main categories: sync problems, time base errors, color problems, signal-to-noise ratio, and bandwidth limitations.

The first three are somewhat related and are capable of being corrected to some degree. The last two are independent of the first three but are related to each other to some extent.

Sync Problems

The FCC has set certain specifications regarding the stability of the television signal which can be broadcast over the air, and they have been formulated for the purpose of ensuring that all TV stations produce the same type of signal and that all equipment can receive and work reliably with this signal—even under the worst conditions. The standards are based upon practical experience with making good television signals.

The main concern is that the sync signals contain as little "slop" or as few errors in their timing and level as possible, because it is these which govern the stability of the picture when it is viewed on a TV set or monitor. For broadcasting, these specifications are very stringent and are set out in a document published by the Electronic Industries Association and known as RS170.

In an industrial or educational setup, a lesser quality of sync signal can be tolerated, and often that conforms to the specification known as RS330. This is basically the same as RS170 but has some subtle differences. For most closed circuit operations it will produce good pictures, but it is not good enough for broadcasting. The main characteristic of these two signals is that the lines on the screen are all carefully interlaced and of equal length, and they all start evenly on the left-hand edge of the screen. A high quality picture with no annoying jitter or crawling is seen. Usually both these sync signals are produced in a separate sync generator, and it is used to drive the entire station or studio.

In the small, inexpensive industrial and surveillance cameras a very crude type of sync signal known as "random interlace" is produced within the camera itself. Although this makes the camera a self-contained item, this type of sync is absolutely useless for serious closed circuit TV and is out of the question for broadcasting.

The sync signal has a profound affect on all videotape recorders. The stability of the recording and the servos in the machine are directly related to the stability of the incoming sync signal; and the more stable the sync, the more stable the machine and the recording. In playback, the servos rely on the playback sync from the tape to control both the machine and the tape. If the sync is not stable, then a stable picture will not be obtained.

If a nonbroadcast type of sync is used when recording, then the timing errors it may contain will be recorded onto the tape, and they can be so great that the playback stability of the sync will make it impossible to reliably copy or view the picture. This becomes especially true with the helical type of VTRs, where the inherent stability of the machine will add to the timing errors of the incoming sync. In fact, if the random interlace type of sync is used, it is possible that the recording will be so unstable that it will never play back satisfactorily.

Errors due to a bad sync signal are likely to occur in addition to a class of machine errors known as *time base errors*. It should be noted that many broadcasting machines will not work without a correct broadcast type of sync; their circuits will just shut down.

Time-Base Stability

Time-base stability is a measure of the stability of the playback picture from the tape and is a most important figure when it comes to transmission over long distances, ccpying, editing, and especially the play back of color. It is one of the most critical specifications to be found in broadcasting VTRs, and typical figures are about 50 nsec (nanoseconds) for a monochrome picture and about 6 nsec for color. It is a figure significantly absent from the details published about helical VTRs and cassette machines.

To understand what this means, consider a horizontal line of a picture. It takes a total time of 63.5 μsec (microseconds), with the visible portion of the line using about 53 μsec and the sync lasting about 10.5 μsec. A variation of 2 μsec in the playback timing easily can be seen as sideways jitter on a monitor and is unacceptable. Much less than 1 μsec is required for acceptibility.

One of the biggest problems in the early development of videotape recorders was getting a sufficiently good tape transport and head scanning which could produce a rigid sync signal on playback. In fact most of the cost and complication of the modern broadcast VTR is in the mechanics required. The modern broadcast VTR is good enough to play back a monochrome tape to meet FCC sync specs with a minimum of horizontal jitter and thus produce an acceptable picture. However, the slight variations in the sync timing which do exist will completely prevent the playback of a color picture.

These timing errors in the playback are caused by tape stretch due to handling, temperature and humidity changes, small mechanical imperfections in the machine, and the gradual wearing of the heads.

Further instabilities also remain due to the limiting factors of the mechanical parts of the head and capstan servos. These timing errors appear in both the recording and the playback and are in addition to any errors which may already be in the incoming signal. The following list of problems are all caused or exacerbated by time base errors and cannot be ignored in a quality TV system:

- The time-base errors and instabilities will compound with successive generations of tape copies, so multiple copies will soon become unplayable as the errors will build up to the point where the sync timing will be so far off that no VTR servo or monitor will be able to lock to it.
- Time-base errors cause color problems long before they cause sync problems. It is important to know that no VTR will play back color directly; all need extensive electronic correction circuits.
- Broadcast equipment is made with very tight sync circuits and will not work with anything less then perfect FCC sync. This is deliberate because it is essential to know about trouble in the studio long before it can affect the viewer. This produces problems when attempting to copy onto broadcast VTRs from the nonbroadcast machines because their playback sync often is not stable enough.
- A signal which must travel a great distance over a cable suffers degradation. Often this is in the form of rounding the sync pulses and attenuation. This will give sync problems in the receiving equipment. An unstable sync source will only add to such problems.

In broadcast VTRs much of the circuitry in the playback electronics is for the sole purpose of correcting the sync signal and the time-base errors as they are played back from the tape, and much of the remaining circuitry is for the correction of the color signal.

In the nonbroadcast equipment much of this corrective circuitry has been omitted in the interests of size and economy, and much was not considered necessary in the light of the machine's intended use. Hence time-base errors are an ever-present and serious drawback to nonbroadcast VTRs—and the videocassette machines must fall into this category.

Although they contain corrective circuits for the color, in general the success of these machines lays in their use with the type of monitors which have a fast recovery time constant in the horizontal circuits. These allow the monitor to quickly lock up to any changes in the sync timing, and thus the visual effects of these errors are very neatly masked. Hence, use of the cassette machines with a monitor recommended by the manufacturer will always produce a very good, stable picture on that monitor.

It must be clearly understood that broadcast equipment will do the exact opposite of this—it will expose time base errors very quickly—and for this reason the interface of cassettes with broadcast equipment is often unsatisfactory to the point of being impossible.

Color Problems

The FCC has very tight specifications concerning the quality of the color signal, and like the sync signal these are adhered to by all broadcasting equipment.

Two main methods have been adopted for the recording and playback of the color signal, and are known as the "phased" and "nonphased" methods. All broadcasting equipment uses the phased-color system due to its superiority. However, it is not an easy system to use where large time base errors are found, and this is one of the reasons why the helical-type machines generally have adopted the nonphased color method, which is also known as the "heterodyne" or "converted subcarrier" method. Although this system is simpler and is satisfactory for nonbroadcast work and does allow fairly good copies to be made, it can and often does produce problems when attempting to integrate with, or copy onto, broadcasting equipment.

Although color editing is possible with this system, the success of the edits rely more upon the use of vertical interval switching and the fast recovery monitors than upon electronic correction.

There are other color problems associated with the differential phase and differential gain of the cassette machines. The difficulties are caused by timing in a system containing other equipment, and a lack of sync inputs. These are too complex to be covered briefly, but the user should be aware that other and more subtle problems do exist and that they are very real.

Signal-to-Noise Ratio

This is a measure of the amount of "snow" or noise which can be seen in a picture:

An S/N ratio of 40 dB means that no noise or snow can be seen by the naked eye.

An S/N ratio of 35 dB still allows an excellent picture, with traces of noise discernible by a trained eye.

An S/N ratio of 30 dB means the noise is becoming noticeable but is not objectionable.

An S/N ratio of 20 dB means the noise is easily noticeable, and it corresponds to fringe area TV reception.

If the S/N ratio gets as bad as 10 dB, then the picture is unusable.

Most broadcast equipment will claim an S/N ratio better than 45 dB, thus guaranteeing that no noise is visible in the picture. With nonbroadcast equipment, claims vary between 35 and 40 dB, and this can be achieved when the equipment is properly set up, but in many cases of general use anything between about 25 and 35 dB will be found.

Common causes of noise are inadequate light in the studio and improper adjustment of cameras or other equipment at the time of recording. Noise is further introduced by copying, passing the signal through equipment, and transmission over long lines or over the air. Because noise cannot be eliminated by subsequent processing it is essential that it be eliminated at its source. Noise is a high-frequency phenomenon, and it can be removed to some extent by lowering the high-frequency response of the system, but this reduces the quality of the overall picture because the ability to resolve fine detail is impaired. This approach can be tolerated more with a color picture than with a monochrome picture because the color hides much of the fine detail in a picture—but there is a limit which is soon met.

Although noise-free pictures can be seen played back from a videocassette machine, in general their S/N ratio is borderline for broadcasting.

Bandwidth

This refers to the overall frequency response of the system, which is a measure of the fine detail which can be seen in the picture. The greater the bandwidth, the more fine detail possible. It is often expressed in "lines" rather than as a frequency, with 350 to 400 lines being typical for a monochrome picture and about 250 to 300 for color. These are usually quoted under the title of "horizontal resolution."

Due to the lower carrier frequency used in the recording process than that used in broadcasting machines, most videocassette machines do not have as great a bandwidth, and often the resulting picture is barely good enough for broadcast use.

Also, the heterodyne color system restricts the color bandwidth to less than that required by FCC specifications. Although this is often not noticeable to the eye, it has placed a restriction on the overall quality of the signal which will become apparent after much copying or transmission over long lines, where attempts will be made to correct for signal deficiencies.

ANCILLARY EQUIPMENT

In an attempt to overcome and correct the problems just discussed and to guarantee pictures which will meet FCC specifications, several items of equipment have been produced to be used in conjunction with helical VTRs and videocassette machines.

The problems caused by noise and bandwidth limitations must be treated at their source or, in other words, the original equipment must be good enough not to exhibit these problems, because subsequent correction is not possible. The problem of nonphased color must be handled by the playback circuits of the cassette machine, and no further correction is usually possible.

With regard to the sync and time base errors, some external correction is possible, and most corrective equipment concentrates on these areas. Three main items of equipment are used: processing amplifiers, dropout compensators, and time-base correctors.

Processing Amplifiers

Usually called "proc amps" these are one of the most important items of ancillary equipment in any TV system. Their main use is to add a good sync signal to a composite TV signal which has been passed over a long video or RF line and has suffered attenuation and distortion. By adding a stable regenerated sync signal to the video and amplifying the video to a usable level, a substandard signal can be restored to near perfection.

A further problem is that the sync signal obviously must be continuous. Any breaks would upset the picture on the screen and cause it to break up or have other instabilities. All cassette machines and many helical VTRs use two heads, and a switch from one to the other must be made at some time in the picture. This switching causes minor disturbances in the sync signal and is responsible for the "flagging" seen near the top or bottom of the replayed picture. By adding a continuous sync signal this can be eliminated.

Dropout Compensators

All tapes suffer from dropouts. These are momentary losses of the signal from the tape and are caused by faults in the tape oxide, dirt on the tape heads, or scratches in the tape caused by lack of care or long use. These losses of signal appear as flashes on the screen; they are irregular in size, frequency of occurrence, and place on the screen. In excess they are annoying to the viewer.

They can be covered up quite easily to give a clean picture which is free from any dropouts. The manner in which this is done is to sense when a dropout of signal from the tape has occurred and then to replace the horizontal line of video with a repeat of the previous line which was free of dropouts. The techniques for accomplishing this will not be covered here, but the same method is used in all cases. Many cassette machines contain a small "DO Comp" as an internal circuit, but in many cases it must be provided as a separate external item. Often they are manufactured in the same chassis as the processing amplifiers and form part of a complete video correction package.

Time-Base Corrector

The time-base corrector is a device which will accept a played back tape with all of its time-base errors and then remove all those errors and present a stable picture with a sigid sync signal, thus making it usable in a broadcasting system.

Time-base correctors fall into two types, analog and digital. Both do basically the same thing, which is to store the TV signal, a line at a time, and then march it out at a later time. In this way the marching out of the signal can be made independent of the time it went into the memory and thus can be made free of the timing errors. A computer-type digital memory or a controlled delay line are the most used memories for the signal. To help in this timing correction, a signal may be fed from the corrector to the video input of the tape machine, thus controlling the servos in the playback mode to aid in the correction.

The exact mode of use and the connection must be checked with the instruction manual because they differ from model to model. (A complete discussion of how they work is beyond the scope of this book and will not be attempted.) They all rely upon integrated circuits and new electronic techniques, and they are very complex and expensive. It would appear that they are very successful in correcting the pictures from the helical VTRs and are making cassettes and portable tapes usable in broadcasting.

Cameras

Several cameras are produced for use with the professional and consumer videocassette machines. Since the manufacturers literature describes these quite adequately, only the most important points will now be covered. In general, any camera from any manufacturer will work with the machines from any other manufacturer.

Two main types of cameras are produced for use with cassette machines - color and monochrome. All color cameras are either one or two tube, and have the full broadcast RS-170 sync signal with color burst. These present no problems when used with any VTR. The monochrome cameras present a different situation. The camera must have its own internal sync generator, and here care is required. Many small monochrome cameras do not have the RS-170 sync signal. They have either the 2:1 interlaced 'industrial' sync signal (RS-330), or they have a random interlaced signal. Both of these should be avoided with cassette machines.

The professional Umatics will not handle the non broadcast sync very well, as they have several identification circuits which look for a stable vertical pulse and equalizing pulses, and the absence of these can produce erratic operation.

The consumer machines must have a stable and regular horizontal sync pulse in order to prevent interference due to the lack of guard bands, and for the ½H FM shift to work. If a camera with less than perfect sync is used then visible interference patterns will be seen on the screen, and they cannot be removed.

In order to reduce the size and cost of the cameras the electronic circuits are almost exclusively ICs, and many of the controls considered essential in a studio are missing. The quality of the picture depends upon the engineering of the camera, and whether vidicons, saticons, or other tubes are used, and this is reflected in the price. For example, some models are used in many broadcast stations for use with the Umatics. Whereas the small monochrome cameras are sold solely for use as a home entertainment camera.

All cameras plug into the video input of the cassette machine and are complete in themselves. However, several use a 10-pin conector, which supplies audio, vertical drive and control pulses to enable the machine to be stopped and started by the trigger on the camera. These connections should be checked when trying to mix cameras and cassette machines from different manufacturers. For example; pulling the trigger on one camera supplies a ground to the deck, but another camera supplies 9v on the same connector pin.

Also missing on many small monochrome cameras is an electronic viewfinder, but the professional color cameras all have a small 1 inch electronic viewfinder. Other typical provisions are a microphone mounted in the camera, and a **mic jack** on the camera for a hand held microphone.

Figure 7-1 and Fig. 7-2 show two different types of portable cameras, both with the portable Betamax unit SLO-340.

Timer Units

Many machines have a timer unit built in, while others have to use an accessory timer. The built in types usually have a digital LED readout driven by standard TTL or MOS ICs, and these simple circuits are covered in the service manuals.

The accessory units are usually mechanically driven. They contain power outlets at the rear which are powered and de-powered by the amount of time set on the timer. The machines must first be placed in the required mode, such as record, and then the power is removed by using the timer setting switch. The timer then supplies power at the set time, and the cassette machine switches itself off at the end of the tape. The timer will automatically cut power about 4 hours after it turns on.

A word of warning here. In effect, these timers are the same as the average 'coffee pot' timer found in the home. But the power required by a cassette machine is usually much higher than that of a coffee pot,and using a domestic coffee pot timer instead of a proper timer will probably burn it out. This is extremely dangerous, especially in an unattended home. So, only use the proper timer recommended by the manufacturer.

FURTHER VIDEO PROBLEMS

Other factors also influence the quality of the TV picture in nonbroadcast situations, and these are not usually given the same consideration as in broadcasting. These are as follows: degradations due to multiple copies, losses and sync problems due to long lines, environmental changes, and bad tape handling. The last two items are covered elsewhere in this book, but the first two will be discussed.

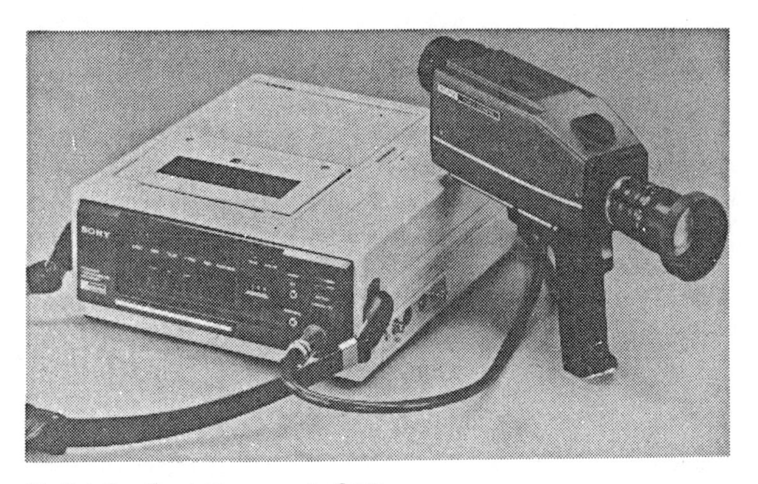

Fig. 7-1. Small portable camera by Sony.

Multiple Copy Degradations

There are three main errors which can be introduced during a copying process. These are noise, color distortion, and accumulated time-base errors. Little can be done about the noise and color errors other than to use good equipment and to treat it properly. The time-base errors can be corrected to some extent.

Any time-base errors introduced into a copy will remain there. Although a time-base corrector will remove the errors introduced by the playback at the time it is used, it will not remove recorded-in errors. For this reason any tape intended for multiple copies must be made on as high a quality master machine as possible, and a time-base corrector should be used every time it is copied. If a copy of the copy is to be made, then time-base correction is essential.

The effects of cumulative time-base errors in multiple copies can be offset and hidden for a long time by the use of the fast-recovery monitors, but eventually a generation will be reached where the errors are too great for even these. Once a serious time-base error has been allowed to be recorded onto the tape, the tape can never be corrected to the point of complete stability.

Long-Line Problems

Two types of line are used to distribute video signals, the video line and the RF line. Both, when used properly, will allow video pictures to be sent over very large distances with unperceived loss of quality. The TV networks and many CATV systems are testimony to this. On the other hand, an improperly treated line very quickly will produce a rounding of the sync signal and a loss in picture quality and detail and will introduce noise.

If the originating signal is a tape playback, then any errors in the playback signal will be added to the line losses, and the received picture could easily be unusable and even uncorrectable.

To obviate these problems, a time base corrector should be used at the sending end, intervening amplifiers should be used along the line, and a proc amp should be used at the receiving end.

Fig. 7-2. Larger, but still portable Sony camera.

If the received signal is to be recorded, then a high quality machine should be used whenever possible.

CONCLUSION

As with most other items of equipment, this type of videotape machine has both advantages and disadvantages. It should be understood that they were developed to be used as nonbroadcasting items and that this is their correct use.

A broadcast VTR such as the RCA TR70 or the Ampex AVR2 are about the size of a small car and cost around $100,000; and so it is senseless to compare the smaller cassette machines with these and expect the same performance. If the same level of performance were possible, no one would continue to make or buy the large machines. The cassette machines do a more than adequate job in closed circuit television, but by themselves they fall short of several broadcasting requirements, and they should not be compared to the broadcasting machines. Unfortunately, a comparison is too often made by persons who are not in command of the knowledge or facts, and costly mistakes in equipment purchases often result.

Although most users will never need to use the cassette machines in a broadcasting situation, there are many persons close to broadcasting who do use the machines and often wonder why they are not so readily accepted. It is true that they record and play back beautiful color and are as reliable as anyone could wish them to be, but for the reasons outlined in this chapter they do fall short of broadcast specifications.

At this time the potential user of cassettes should realize that for nonbroadcasting work—such as in an office or a school—the cassettes are perfectly good machines and will do anything which is required of them. The limitations explained in this chapter will only come to the fore in two major situations: when they are interfaced with broadcasting equipment and when high quality multiple generation copies are required.

8
The Umatics Type Videocassette Mechanisms

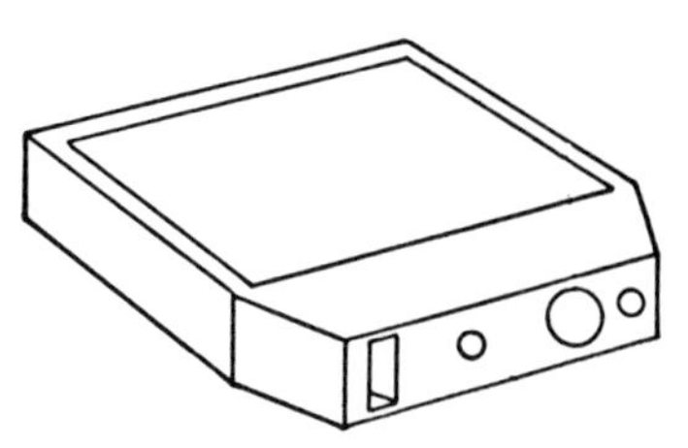

The Umatics were the first videocassette machines to be introduced into the US. They quickly established themselves in industrial and educational studios. Although the term 'Umatic' has become the most common and accepted generic name for these machines it is really a Sony trade name. Strictly speaking, the term 'U-type' machines should be used.

These are now made by several manufacturers, and although there are differences between models, all produce tapes which are interchangeable. The earliest machines had the 'type 1' mechanism. This was the first automatic video tape threading mechanism, and it established the reliability of these machines. Later models have the 'type 2' threading and transport, which made the machines stable enough to be used for editing and limited broadcast purposes.

This chapter will include a brief description of the Umatic format and cassette, the type 1 mechanism, the type 2 mechanisms, and a brief introduction to the broadcasting machines. The full mechanical details are not covered, because this would be impossible in the light of the complexity of the machines and the differences from model to model. This chapter is suitable as an introduction for the user who wishes to understand these machines and do simple maintenance. It is also for the person who wishes to become a service technician and perform full maintenance and alignment. It is not intended to supplant the service manuals for any model, which should always be consulted prior to any maintenance.

TAPE TRACKS AND FORMAT

As with other helical machines, the video tracks slant across the tape, but in this format they do not extend from one edge to the other. Two audio tracks are placed on the lower edge of the tape and the control track is placed on the upper edge. This is shown in Fig. 8-1 which includes the important dimensions.

In this format provision has been made for an auxiliary track. This is for recording editing codes or other information which will be useful for timed stops, playbacks, etc., or for any other desired production information. At this time the heads, electronics, inputs and outputs are provided only on the broadcast machines; and the SMPTE time code is most often used on this track. The location of the track is shown in the diagram. It is placed so that it runs through the part of the video track where the vertical interval is recorded, because this results in minimum interference with the video playback.

Table 8-1 gives the most important dimensions of the Umatic format.

THE CASSETTE

A cassette is shown in Fig. 8-2. It is 8 ¾ × 5½ × 1 ¼ in. in size, its weight is 1 ¼ lb, and it can hold a maximum of 60 minutes of tape, with 10 minutes the usual minimum. The tape is ¾ in. wide and has a high density chromium dioxide or cobalt doped ferric oxide for its active coating.

The cassette contains two open-sided reels on which the tape is wound, with the oxide facing out. These reels are overlapped to conserve space, as in Fig. 8-3. The ends of the tape are attached to the reels with a clear plastic leader, which is used for sensing the end of the tape. The cassette is shaped so that it will fit into the machine in one way only, thus making improper insertion impossible; and it will accept a specially shaped label which aids in identification and orientation.

The tape crosses an opening along the front edge of the cassette, where it is accessible for pulling out and threading into

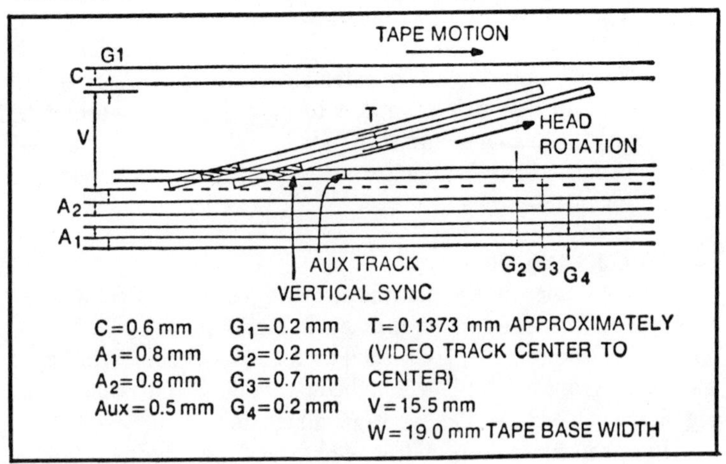

Fig. 8-1. Tape tracks for Umatic system.

Table 8-1. Dimensions of the Umatic Format.

Tape width: 19.00 ± 0.03 mm
Tape speed: 95.3 mm ± 0.2 mm/second
Video track angle: 4° 54′ 49.1″ with the tape still
Video track width: 85 microns
Head to tape, or scanning speed: 404.3 in./second
Drum diameter: 110 mm
Control track: 29.97 Hz pulses
Chroma signal: 688.374 kHz
Typical machine dimensions: 20 × 20 × 8 in. approximately or 25 × 20 × 8 in. approximately
Typical weight: 48 lb or 65 lb
(These depend upon whether the machine is for recording or playback only).
Temperature Range: 41 to 104°F in operation.
Operating position: horizontal only
Resolution: 320 ± 10 lines
250 ± 10 lines

the machine. This opening is covered by a spring loaded flap or door, which is usually marked with instructions to keep fingers off because it is for protecting the tape. On insertion into the machine the spring loaded catch on the door is pushed away, thus allowing the internal mechanism of the machine to flip the door up and expose the tape for threading.

To prevent erasure of a prerecorded tape, a small plastic cap—usually red—is removed from a hole in the underside of the cassette. When the cassette is inserted into the machine a small pin on a microswitch rises into this hole and disables the record circuits.

THE TYPE 1 MECHANISM

The Type 1 mechanism was the first to be introduced with the original Umatics. Since a complete description is given in the service manuals it will not be given here. However, it is instructive to remove the top of the machine and observe the threading and unthreading procedure.

To load the cassette into the machine all that is necessary is to push it into the opening at the front and let the machine do the rest. The cassette is shaped so that it will only go in the correct way, and when pushed fully into the compartment it drops to the operating position. This now opens a protective door or flap on the cassette and places it so that a threading pin seats behind the tape in the triangular notch on the underside of the cassette, as shown in Fig. 8-2. This pin is attached to a pivoted arm on a large threading ring. This is the center of the mechanical construction and everything is linked to it in one way or another.

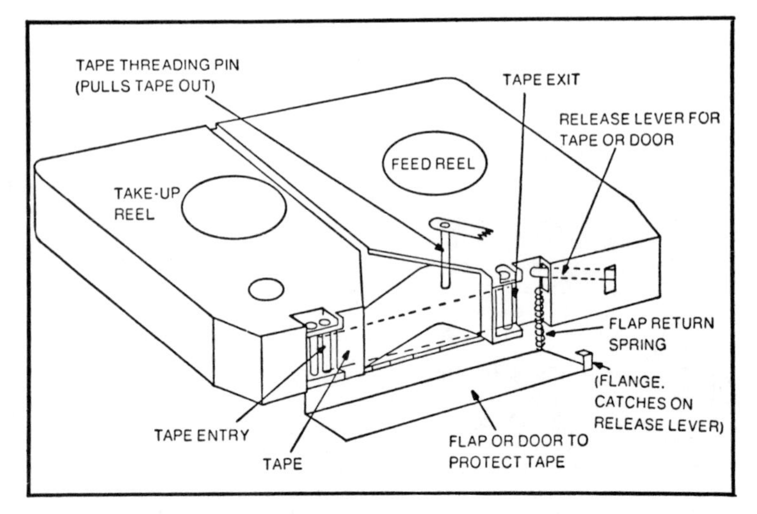

Fig. 8-2. Cassette with flap or door open.

When the **play** button is pressed the ring is rim driven by a DC motor, and it rotates clockwise when viewed from the top. This causes the arm and pin to pull a loop of tape out of the cassette and around the threading path, as shown in Fig. 8-4. The threading ring is mounted on a slant so that it actually pulls the tape in a downward spiral, and to aid in this difficult procedure the tape guides are mounted around the ring. This threading cannot be actuated without the cassette in place, because the dropping of the cassette into position pushes down the arm of a microswitch which allows

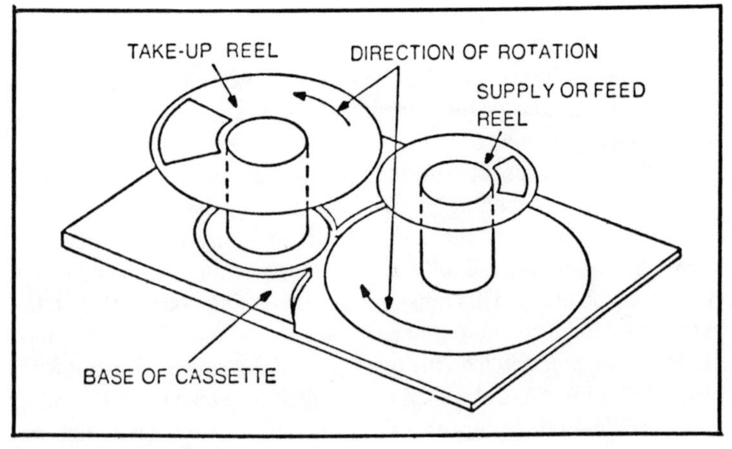

Fig. 8-3. Overlap of reels in cassette.

power to be supplied to the control circuits. This is shown in Fig. 8-5.

An alternative method of extracting the tape is shown in Fig. 8-6. Here the pull-out arm is not on the ring but is mounted separately. This pulls the tape out to the position shown in the dotted lines and allows the guides on the tape ring to guide the tape around the drum. This is somewhat easier mechanically for the tape and for servicing the machine. Other methods also are in use for tape extraction and threading, but these two are the most common.

The pressure roller is mounted on the threading rim, and it follows the tape loop around the complete threading path as the ring revolves. Just before the tape is fully wrapped around the drum a notch in the ring closes a microswitch. This powers the main motor, and the heads and capstan begin to revolve. As the threading is completed the tape is wrapped around a revolving head drum and is placed very close to a capstan which is already up to speed. Just before it completes its travel the ring depowers the threading motor by actuating another microswitch. The tape is now fully threaded and lays against the heads and guide ready to be transported.

At this point a complex mechanical action closes the pressure roller, and the tape begins to traverse its path. At about the same time the electronics are powered and the selected function commences.

This whole procedure takes about 5 seconds, and a picture appears on the screen about 6 seconds after the play button is pressed.

During this threading procedure an orange or yellow warning light on the front of the machine is illuminated, and nothing must be touched while this is on. When the threading is complete this goes out, and then the other functions may be selected.

When playing or recording, the tape is pulled from the supply reel in the cassette and around its path at 3 ¾ in./second and is pulled back into the takeup side of the cassette, which is puck driven by the main motor. The heads rotate at 1800 rpm and give an effective head-to-tape speed of about 404 in./second, which is suitable for satisfactory color recording and playback.

At the end of the program or at a time when the STOP button is pressed, the tape is unthreaded and retracted back into the cassette. The unthreading is the exact reverse of the threading process. The pressure roller releases, and the ring begins a

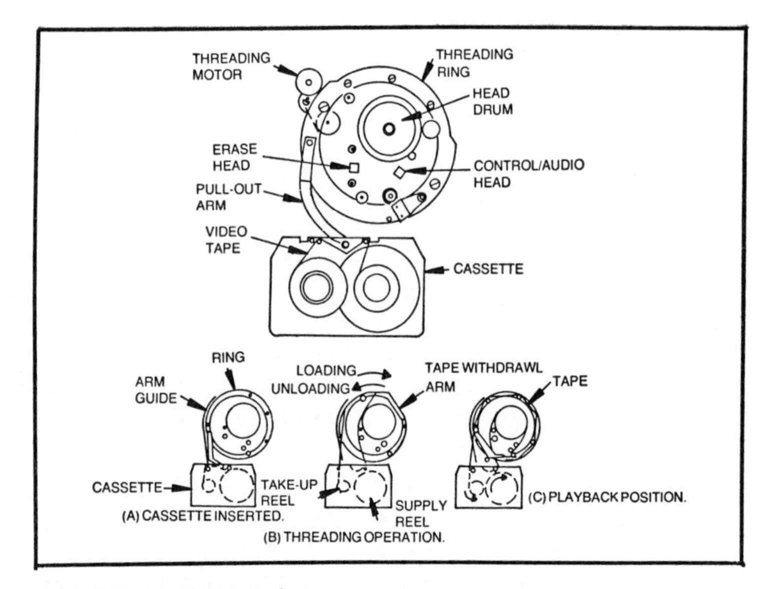

Fig. 8-4. Umatic tape threading.

counterclockwise movement as it is rim driven by the DC motor. The tape is pulled into the cassette by the main motor providing a puck drive to the takeup spool only. When the retraction is complete the ring stops and the DC motor is depowered. A

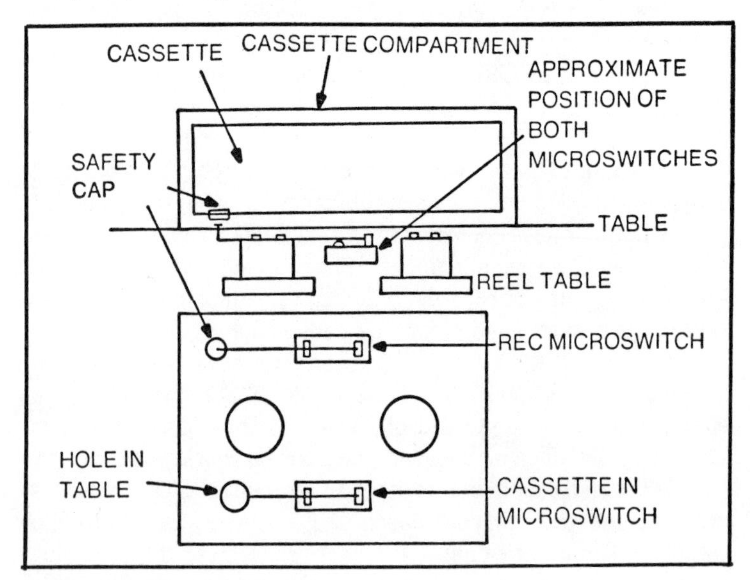

Fig. 8-5. Cassette in microswitch.

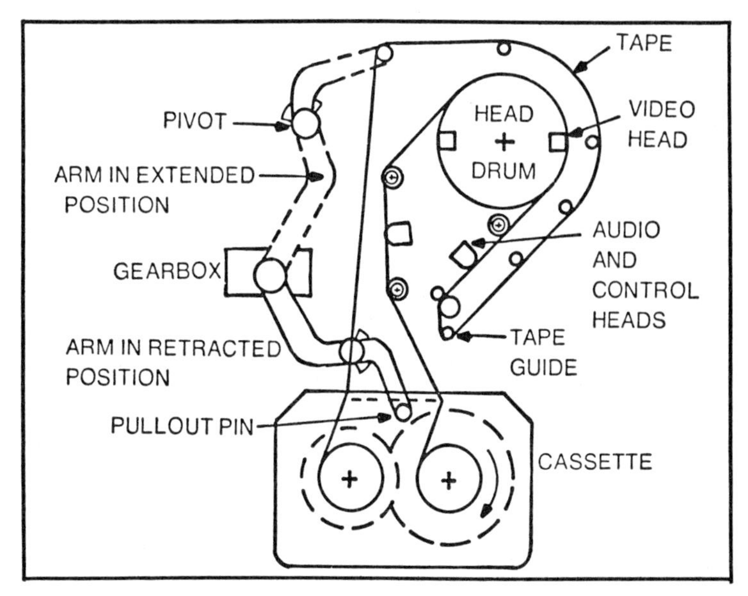

Fig. 8-6. Alternate threading system.

solenoid now cuts in and the main motor drive rewinds the tape for a few seconds back onto the supply spool. This is required to prevent a loss of material on the playback of the next section.

The length of this rewind is never an exact linear distance of tape, and it depends upon how far into the tape the previous play had progressed. During this retraction the warning light is again on, and nothing must be touched until it goes out. The main motor is then depowered and the **stop** button automatically released, either mechanically by a solenoid or electronically. The machine is now ready for another function to be selected or for cassette removal.

Most of the mechanical drive functions are provided by the main AC motor. A mixture of belts and pucks are used, and these are mounted inside the machine and are not always easily accessible. Figure 8-7 shows the underside of a machine where the belts are located for easy removal and replacement. Note that some of these belts have the dull side against the drive wheels and others use the shiny side. It is essential that the best be replaced with the correct side against the wheel.

The **rewind** and **fast forward** functions are conducted with the tape inside the cassette. It is not extracted and runs around the tape path. To enter these modes from either the **play** or **record** modes the **stop** button must be used as described previously. Once the tape is inside the cassette, the main motor provides a belt or puck drive to the reel tables. Each of these modes will then continue until the end of the tape, when the plastic leader will cause an automatic stop. This also will release the function buttons and depower the motor. Do not use these modes when the tape is already fully rewound or fully played because the automatic function will not work then. It is important to understand that the plastic leader and the light cells work on a *change* of light level and not on a definite level of light or darkness.

If the tape is played to its end, most machines will then cause an auto stop, tape retraction, and a full rewind to the start of the tape. Usually the cassette does not automatically eject and has to be removed manually. When the **stop** button is used to end a recording or a playback, the tape will retract, but it will then stop all motion. Any rewind or fast forward will have to be initiated with the buttons.

The location of the phototransistor and its lamp are shown in Fig. 8-8, along with the typical positions of the main solenoids and microswitches.

MAIN FUNCTION BUTTONS

In many of the machines these are a mechanical arrangement of buttons mounted in an interlocking assembly. Only one main button can be pressed down at a time, and this locks out all the others except **stop**. Also mounted on the assembly are several microswitches and printed circuit boards, and at the lower end is a solenoid which is for the automatic release of all the function keys at the end of the tape.

This is a complex mechanical assembly, and if it malfunctions it is best left to a qualified service engineer to fix. It is not easy to remove from the machine, and several linkages extend from it to meet with other internal levers.

THE TYPE 2 MECHANISM

This is an improved threading and transport mechanism which handles the tape much more gently and accurately than the older Type 1 mechanism. Fig. 8-9 shows the basics of the tape path, which has been slightly changed from the Type 1 machines. However, the tapes are still compatible with the earlier machines.

When the cassette is inserted the tape is pulled out to the position shown in Fig. 8-10. The threading motor drives a 'Geneva gear' which operates the two pull out arms, and the tape remains in this position until a function is selected. **FF** and **rew** are conducted

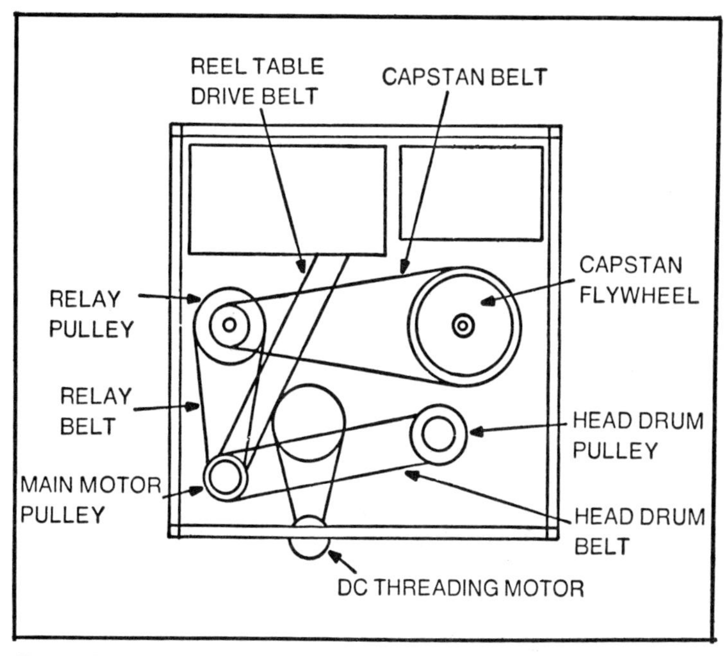

Fig. 8-7. Belts.

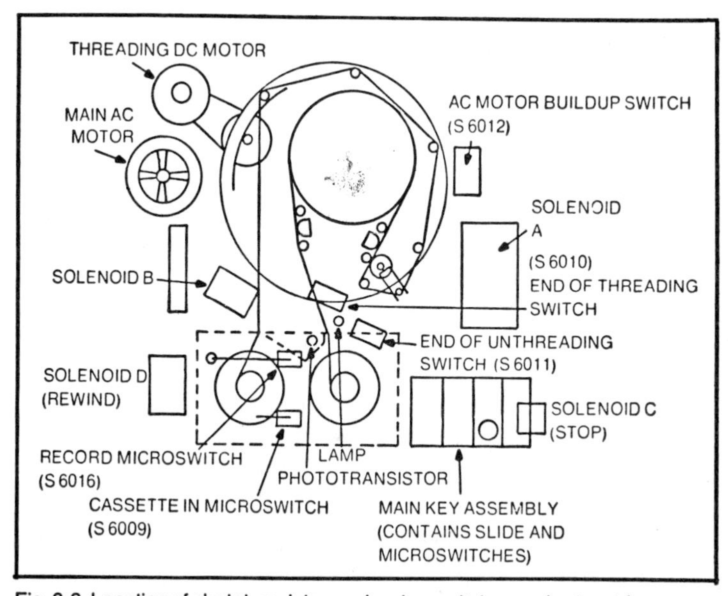

Fig. 8-8. Location of phototransistor, main microswitches and solenoids.

with the tape in this position, but for **rec** and **play** the tape is run out around the full threading path. This is accomplished by the dc threading motor now driving the ring. The ring has teeth on the underside which mesh with a gearwheel driven by the dc motor. This results in a very positive and accurate tape threading.

The first machines with this mechanism had a dc threading motor, and 2 AC motors to drive the heads, capstan and the tape reels. The later machines intended for broadcasting have 4 motors. These allow complete and independent servo control for the heads and capstan, and correct tension control for the tape. With these later machines the tape can be run in fast-forward or rewind to any spot within a few frames. This is essential for editing accuracy in broadcasting.

The head drum contains flying erase heads as well as the normal video record/playback heads. Also mounted in the drum are extra pole pieces and PG coils, which provide the switching pulse for the editing circuits. Although the FM is fed to the heads by rotating transformers, the erase heads are usually fed by slip rings. See Fig. 8-11.

The placing of the other heads is shown in Fig. 8-9. The audio and the CTL heads are unchanged, but the position of the main erase head, the CTL-2 head and the auxiliary head differ slightly from model to model. However, their distance from the point of contact of the video heads to the tape is always the same. The CTL-2 head is to read the CTL pulses on the tape during the **ff** and **rew** modes when the tape is being searched or positioned for editing. The auxiliary head records and reads the auxiliary track, which is usually used for editing time codes.

All the mechanical functions are engaged by solenoids, and not directly by levers from the function buttons. A function solenoid releases only certain slide bars, which are then pulled by the main solenoid to release or operate the other mechanical parts. These solenoids are controlled by the system control timing circuits, which ensure that only the correct parts move and only at the proper time.

The supply reel is driven by the tape pulling it, and the tape tension is controlled by an arm and tension band. This reel table is always braked, so it needs more attention than most of the other mechanical parts. It affects the editing accuracy and so it must be regularly checked and serviced. The brake and the take up reel are operated by the mechanical function slide bars.

The take up reel is driven by belts and idlers from the reel

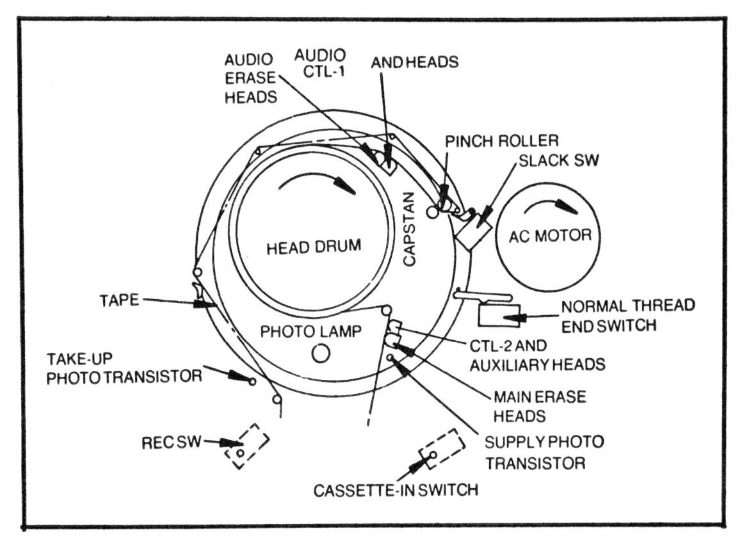

Fig. 8-9. Tape path in Umatic type 2 mechanism.

servo motor. In the **reverse play** mode; reverse play is accomplished by playing the tape backwards. It is NOT rewind. The tension must be accurately controlled to maintain edit accuracy. In this case the tape tension arm moves a small metal vane between and LED and a photo-transistor. This provides better tension control and maintains full contact with the tape heads to count the pulses and position the tape accurately.

THE BROADCAST UMATICS

Several manufacturers produce Umatics which are being used in broadcasting. These are either made for the broadcast industry, such as the Sony BVU200 series, or as high level industrial and educational machines such as the Sony VO 2850 and VO 2860. Several portable Umatics are also made for the broadcasting industry. Broadcast studio umatics are intended for studio use in broadcasting and in high level teleproduction studios. Although they work from the normal AC supply and are not battery operated, they can be used anywhere in the world on any voltage and frequency. This is because all the electric motors are dc operated, or are ac motors which are not dependent on the frequency.

Four main motors are used:

- A direct drive dc motor in the head drum.
- A dc motor which belt drive the capstan.
- A small dc motor for tape pull out and threading.

● An ac hysteresis motor which belt drives the tape reels. A voltage selector on the machine allows this to work anywhere.

The capstan can drive the tape in reverse, as well as forward, and at several speeds. This is very useful in editing, where the tape handling is accurate enough to make edits within a few frames of any chosen point. During the reverse play (not **rew**) the reel tension motor is used to control the tension in the tape by driving the take up table. The servo is controlled by a small metal vane on the tension arm which moves between a light source and a photo transistor. The circuit is shown in the servo chapter.

The full complement of heads are used, and these are shown in position in Fig. 8-9. To ensure edit accuracy the auxiliary track can be used for time codes, but most editors use the CTL-2 head to count the control track pulses.

In the **FF** and **rew** modes the tape is in the 'half out' position, where it is in contact with the auxiliary and CTL-2 heads. Thus the time code and the pulses can always be read by an editor unit. The tape can be controlled to about 2 frames for edits.

Later models of these machines have the **rf dub** feature added. This is for high quality transfers and copies, and is explained in the chapter on cassette electronics.

INDUSTRIAL AND EDUCATIONAL MACHINES

Although these use the type 2 mechanism, they are not really broadcasting machines. Only 3 motors are used, 2 AC motors and a

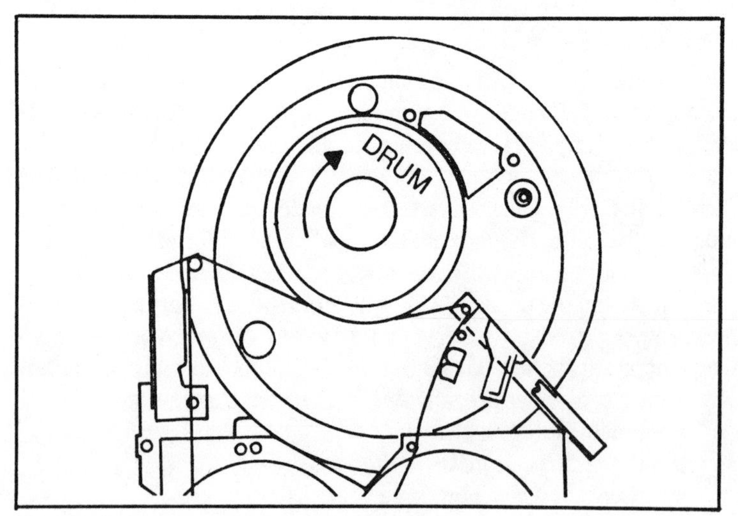

Fig. 8-10. Still position of tape in type 2 Umatic.

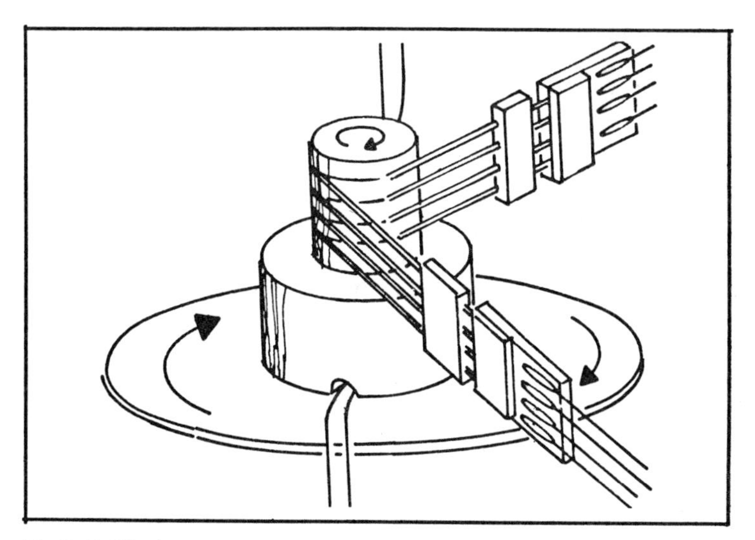

Fig. 8-11. Slip rings.

DC threading motor. Their performance and tape handling are superior to the earlier machines, and the solenoid operation allows remote control by edit units. However, the auxiliary head is not included and so all editing is performed using the CTL pulses and the CTL - 2 head. The tape can be shuttled and edits made to within about 10 frames of the desired point, which is good enough for most purposes but often falls short of precise production standards.

Typical examples of these machines are the Sony VO 2850 and VO 2860.

THE PORTABLE UMATICS

These are small battery operated units intended for mobile use, such as ENG (Electronic News Gathering) or hard to reach places. They are lightweight and are intended to be carried by one person with a camera. They contain a minimum of electronics, and in fact need additional items to be able to playback a tape in color. They all use the Type 2 mechanism, but the actual method of driving the ring and tape varies slightly. They use a smaller than normal 20-minute cassette which will fit other U-matic machines with an adaptor, and it is powered from internal rechargeable batteries which will give one hour of use. It also can be powered from an external DC source such as a car battery or from power lines via the AC unit which also recharges the batteries.

Simple cables provide the necessary hookups for all functions, and they are plugged into the side panel. The front—or top—has a series of pushbuttons for the main fuctions, which include **still frame** and **audio dub**. The machine can mount in any position and can be slung over the shoulder or backpacked. It will record from a portable color or monochrome camera—the DXC1600 is recommended—and a trigger control on the camera will control the tape motion. It also will record and play back color into a domestic TV set with its internal RF unit.

In the studio it can be used as a record or playback machine and can be used for copying and editing. Four separate motors are used, one each for the rotating heads, the capstan, the reels, and the threading.

The head servo is controlled either by sync from the incoming video or from an internal crystal controlled sync generator. This also provides the horizontal and vertical drives to the portable cameras, and these are counted down by integrated circuits from a 31.4658 kHz crystal. The whole machine is stable enough so that when it is used with a time base corrector, it can provide pictures which can be used satisfactorily for network newscasts.

Other facilities are an internal dropout compensator, metering circuits, and meters mounted on the front panel. The tape is still automatically threaded, and the usual control circuits are included.

The head drum has 3 pick-up coils and multiple vanes to produce tighter servo control. This helps to offset the 'gyro' error, which is uneven head drum rotation caused by excessive movement of the machine. How this works is covered briefly in the chapter on servos. One of the head coils is used to ensure the capstan is correctly phased to the head rotation. The capstan motor has its own internal FG coil for servo feedback.

These machines are about 13 × 17 × 14 inches in size, and weigh about 30 lbs with the batteries and cassette installed. The AC power unit is about 7 × 7 × 13 inches and weighs about 13 lbs. Fig. 8-12 is a photograph of a typical system.

ELEMENTS COMMON TO ALL CASSETTE MACHINES

In order for these machines to automatically and perfectly perform all required functions—such as to thread and transport the tape—a system of microswitches, relays, solenoids, mechanical clutches, pulleys, belts, and linkages are used. All these have been very precisely built, mounted, and adjusted to perfection.

Many of the mechanical and electronic functions must occur in the correct sequence, and to ensure this the mechanical components are interleaved and interlocked with each other in a complicated array of inhibits and lockouts which enable only the proper parts to move and the right electronic circuits to receive power. In this way the correct sequence and timing of events can be ensured to effect only the selected function. Examples of this are preventing the tape rewind when the tape if pulled around the threading path, keeping the bias oscillator depowered during the play mode, and preventing the pressure roller arm from closing if the tape threading is incomplete.

All these provisions have been designed to make the machines as foolproof and as automatic as possible, and in this they are quite different from the reel-to-reel machines, which have no similar counterparts within them. Although most of the linkages, rods, and levers are actuated by solenoids, the threading ring, heads, and capstan are driven by electric motors. Two motors are used in most machines. A small reversible DC motor is used for tape threading and unthreading. The main motor is usually an AC motor which drives the heads and capstan via separate belts; this is also used for the **rewind** and **fast forward** functions and to provide tape retraction. This is accomplished by a mixture of belt and puck drives to the reel tables, with the correct one being selected by the solenoids and other linkages.

Occasionally three motors are used. This is found in editing machines where the capstan is driven by its own servo-controlled DC motor.

These machines always should be treated carefully and with the utmost respect because damage cannot easily be repaired.

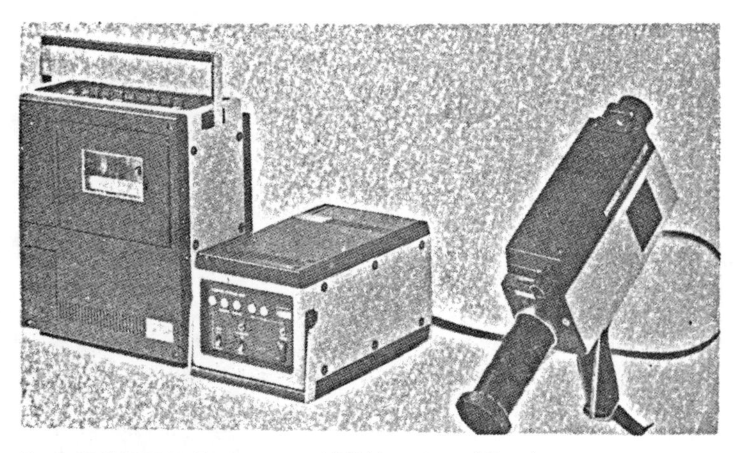

Fig. 8-12. V03800 with camera and CCU (courtesy of Sony).

Parts replacement is often not an easy or quick job and can be very expensive. Normally they will function perfectly unless mistreated, and despite their complications they can be relied upon to perform continuously. It is inadvisable to tamper with or to attempt to defeat any switch or relay in these machines. Disaster of both an electronic and mechanical nature can occur, tape can be ruined, and injury can be sustained. The tape guides and the whole mechanism are much more complicated than the open reel machines, and even the slightest maladjustment of the threading mechanism or guides is enough to completely ruin the playback of good tapes and to make a recording on the machine useless for playback on others.

The following review of the construction of these machines will not cover all the mechanical items necessary for the more exotic automatic functions, such as automatic rewind of the tape when it reaches the end of its play or how to continuously repeat the same section of tape. These are best learned from the service manuals of the specific machines because they are of more interest to a service person than to a user.

Common to all the machines are the various heads for recording and playback, and a few words about these are in order at this time.

Video Heads

The head drum can be in one or two parts and is very similar to those found in other videotape recorders. The heads are either mounted on a bar or plate which protrudes between two stationary sections of the head drum, or the lower section of the drum is still and the rotating upper section contains the heads. The tape wrap is such that the heads are in contact with the tape in excess of 180 deg, thus giving a small overlap in which the head switching can occur.

The signals to and from the heads are usually coupled through rotary transformers. The primary rotates, the secondary is stationary, and the coupling is through a small gap in a ferrite enclosure. Slip rings are used in some machines, but these are becoming less common. One, two, or three coils can be mounted in the head drum, with one or two rotating metal bars; these produce pulses to control the servos and the editing circuits.

The head drum is driven by a belt from the main motor, and a magnetic brake fed by the servo output is used to control the head speed and position.

The upper part of the drum has horizontal serrations to provide an air bearing for the tape. This prevents the tape from sticking and allows good tape tracking and smooth transport. A tape guide band is attached to the outside of the lower drum or is machined into it. This provides a track on which the tape rests as it traverses the drum and ensures that the video tracks are laid down in the correct place on the tape. Figure 8-13 illustrates these points.

To make the automatic threading easier, all the videocassette machines have adopted the two-head half-wrap system.

Other Heads

Several other heads are required in addition to the rotating video heads. Exactly what is provided differs from machine to machine, but the following four areas must be covered: audio heads, control tracks, erase heads, and the auxiliary track.

Audio Heads. The audio head is the most complex of the head blocks found in the machines. Two audio tracks are provided in the Umatics and they have slightly different operational characteristics. Track 1 is for normal recording and also for audio dubbing and so must work independently of the video system. Track 2 is usually used in conjunction with the video system and does not have a dubbing capability, but it can be used in the normal edit modes. The same head gap is used for both recording and playback, so simultaneous playback while recording is not provided. Basically these are no different from a normal audio recorder in their operation and function, and the main thing an operator needs to know is which is active in which operational mode.

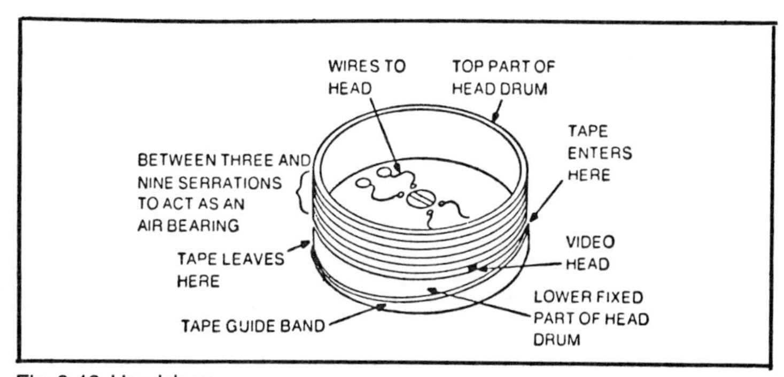

Fig. 8-13. Head drum.

In order to be able to record on Track 1 independently from the video and other tracks, a separate erase head must be used for this audio track. It is usually placed in the same housing as the record/play head, Track 2 is erased by the main erase head.

Control Tracks. A control track is necessary on a videotape in order to synchronize the tape position on playback with the rotating video heads. The control track consists of a series of pulses recorded at a 30 hz rate, and the same head gap is used for both record and playback. It is identical with an audio head and is often mounted in the same housing. The control track is erased by the main erase head in the normal record mode and also in the assemble edit mode. For reasons which will not be discussed here, the control track is not erased in the insert edit mode.

Erase Heads. A main erase head for all the tape tracks—except audio Track 1—is placed as shown in Fig. 8-9. This spans the whole tape except this audio track. It is the same as any other erase head found in an audio or open-reel machine. It is fed from the main bias oscillator, which is used for the audio recording and the other erase heads.

Erasure of the tape can cause some problems in editing. For assembly edits these are not usually a problem because the whole tape is cleaned by the main erase head and the new video and audio are put onto this cleaned tape. With insert edits a different situation exists because each video track must be treated individually. This means it must be erased individually, and the only successful way of doing this is with a "flying erase head." This is like another video record head, but it is mounted in the rotating head drum a few degrees ahead of the record play head. It scans the video tracks one at a time and will erase them one at a time. The erase signal for this head is a high-frequency signal which is usually provided by a separate oscillator and not by the main bias oscillator.

The Auxiliary Track. In some formats provision was made at the time of inception for the inclusion of other tracks. These were to be used for editing codes and other purposes.

The auxiliary track is suitable for the SMPTE time code.

In the U-matic machines this head is placed immediately after the full erase-head.

OTHER CASSETTE SYSTEMS

Several other cassette systems have been introduced with varying degrees of acceptance and success. Two early versions, the Cartirvision and the EVR have completely disappeared. Two others, the EIAJ single reel format and the Philips system, have limited acceptance in the US and are covered here briefly. Several small consumer formats have been introduced, but the two which have become the most popular are the Betamax and the VHS, which are described in later chapters.

The Single Reel Cartridge Format

In this system the tape is wound on one reel only inside the cassette— or cartridge as it is called. The free end of the tape is attached to a stiff plastic leader, the end of which rests inside the opening at the front of the cartridge.

The cartridge is loaded into the machine by pushing it firmly into the loading slot—or garage—where the front flap or door easily swings up to allow the cartridge to enter. It does not go in completely but remains with about ½ in. projecting from the machine. As it is pushed in, it pushes an arm which is mechanically coupled to the threading arm on the internal takeup spool, and it closes two microswitches which apply power to the motors. The cartridge is kept in place by three spring loaded guide rollers and two projecting guide pins. This is shown in Fig. 8-14, which also shows the record safety lever which is pushed back by the protective cap in the cartridge front.

The cartridge does not drop to a lower position. Instead, when the power comes on, the reel table rises up to allow three spring loaded pins to engage in the holes in the cartridge hub. The brakes are now removed from both reels, and the cartridge hub is driven in a counterclockwise direction to cause the plastic leader to come out of the cartridge opening. Both reels are belt driven to thread the tape.

Figure 8-15 illustrates the tape threading path. The stiff leader is guided by a metal channel around the path and then onto the takeup spool by the threading arm. The spool revolves clockwise to take up the tape. The stiffness of the plastic keeps it against the guide channel, and this allows it to follow the correct path, but when the tape comes out it falls away from this channel and rests against the tape guides and the head drum. Unlike the U-matic types, there is no threading arm which pulls the tape out, and the roller and guides do not move on a ring or arm—they are permanently mounted in position.

When the leader has engaged and wrapped firmly around the takeup spool the threading process pulls a few seconds of tape onto the spool, and then it jumps into the **play** mode as the roller automatically closes against the capstan. To stop this automatic

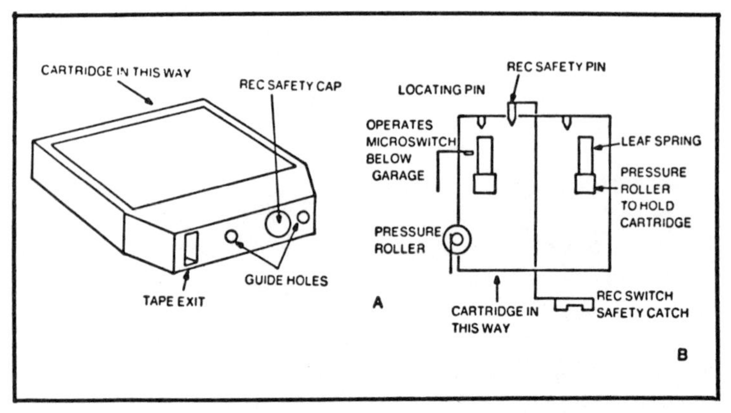

Fig. 8-14. Cartridge and cartridge insertion.

playing of the tape, the **stop** button must be used after the **automatic play** mode has begun. The **stop** button must be used between all modes from now on.

The takeup reel is lower than the cartridge of the head drum, and the tape path is a complete loop around the internal mechanics, although it is only in contact with the drum for about 180 deg. Most of the drives to the head, reels, etc. are by belt from the main motor. A gear mechanism is used in one particular instance. Figure 8-16 shows the belts used, but the array of solenoids and relays has been omitted.

The tape counter is driven from the feed spool by a belt and although it is not linear, it is very accurate and the minute markings can be relied upon.

The skew or tension control is a spring tension applied by turning a knob. This tightens and loosens a brake band around the feed spool.

Although the machine goes into an automatic play mode when the cartridge is inserted, all the other modes must be individually selected by the main function buttons. These are of the "piano-key" type, and they operate microswitches. They must be held down long enough for the function to be initiated; a quick, light touch is insufficient.

The **record, record safety,** and the **edit** buttons are separate from the others and are mechanical in their action, and they operate slide switches instead of microswitches. They are mechanically inhibited when the cartridge is out of the machine and also if the safety cap has been removed. Both of these conditions will prevent recording and E-E from occurring.

In the **fast forward** mode the brake is taken off the supply reel and the tape is driven by an idler which is belt driven from the main motor. The **fast forward** will not work until the cartridge has been inserted and the tape has been threaded. The **auto play** must be allowed to start and the **stop** button used, and then the **fast forward** can be used. Pressing the **fast forward** button upon insertion of the cartridge will not work.

The **rewind** mode can be initiated any time the tape is out of the cartridge and threaded in the machine. A belt drive from the main motor rotates the supply spool counterclockwise and pulls the tape back in. The tape will rewind fully, and then the cartridge will automatically eject from the machine. If this is not required the rewind can be stopped at any time with the **stop** button. If the tape is played to its end, it will stop and then rewind completely back into the cartridge and then auto eject. After an auto eject, to replay the tape the cartridge must be removed almost all the way out of the garage and then pushed in again.

The **stop** or **pause** button will keep the tape against the heads with the roller disengaged, thus producing a still frame picture. When the **stop** button is used the pressure roller is released and the tape motion stops, brakes are applied to the reels, and the tape does not retract back into the cartridge.

All of the functions which occur at the end of the tape are initiated by the clear plastic leader uncovering a cadmium sulphide

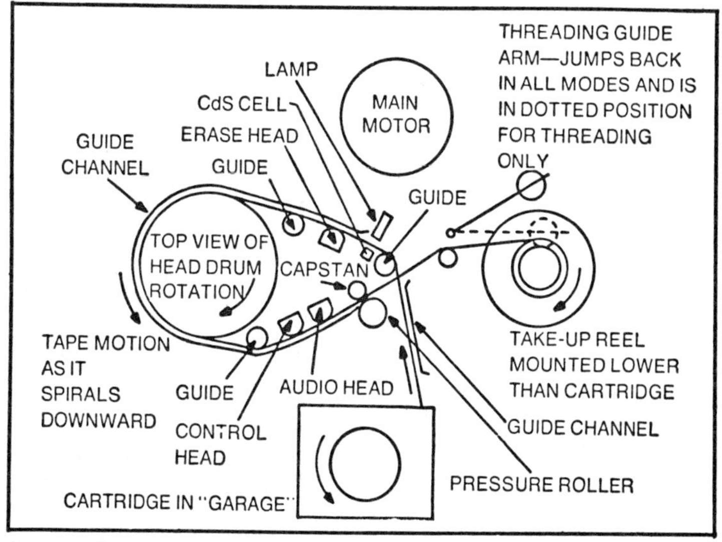

Fig. 8-15. Threading path.

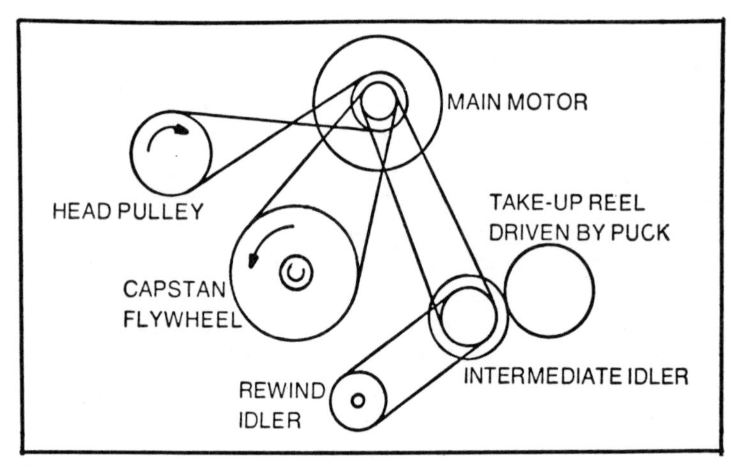

Fig. 8-16. Belts.

cell and alowing light to fall on it. Figure 8-15 shows its physical position in the machine.

Tape Tracks

The format used for this system is that of the common open reel ½ in. machines which use the EIAJ-1 tape format. In fact the tape can be rewound from an open reel onto a cartridge and vice versa. The tape is ½ in. wide and travels at 7 ½ in./second and the track layout is shown in Fig. 8-17.

The Cartridge

This contains one reel onto which the tape is wound, and it is different in construction from the open reel types.

The tape is attached by about 2 feet of clear plastic leader, called a "trailer tape," and it is the same width as the tape. The front end of the tape is attached to a shaped plastic leader as shown in Fig. 8-18. This has a curve which must be in the correct direction. It acts as a guide in the automatic threading process and aids in winding around the takeup spool.

The spool lays in the shaded part of the lower half of the cartridge and, when the top is put on, it is fixed with four screws. The cartridge is opaque, and no windows are provided.

Occasionally the tape will misthread. This is caused mainly by the leader not being securely attached to the tape, the leader being attached with the bend in the wrong direction, or the tape entangling in the works. It can be cured by physically removing the top of the machine, disentangling the tape or leader, and then opening the cartridge and repairing the tape.

One problem which may be encountered with this format is caused by the interchange of tapes from the open reel to the cartridge. If a high energy tape—such as the chromium dioxide types—are used, they may not play or record on a normal open reel machine without head current adjustment.

The Phillips System

This is basically a European system which has been adapted for the American and other markets. It has several attractive features and claims advantages not enjoyed by the other systems. Figure 8-19 shows the machine and cassette.

In this system the tape is attached permanently to two reels inside the cassette, but the reels are mounted one on top of the other, as shown in Fig. 8-20. Because of this mounting, when the tape is pulled from the cassette around its threading path a natural helical path results. This means that the drum can be mounted to

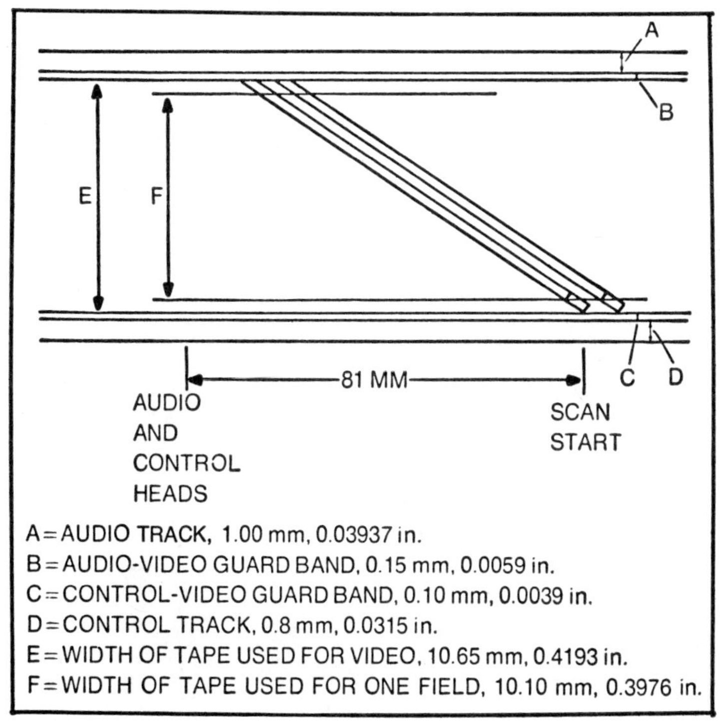

Fig. 8-17. Tape tracks for EIAJ system.

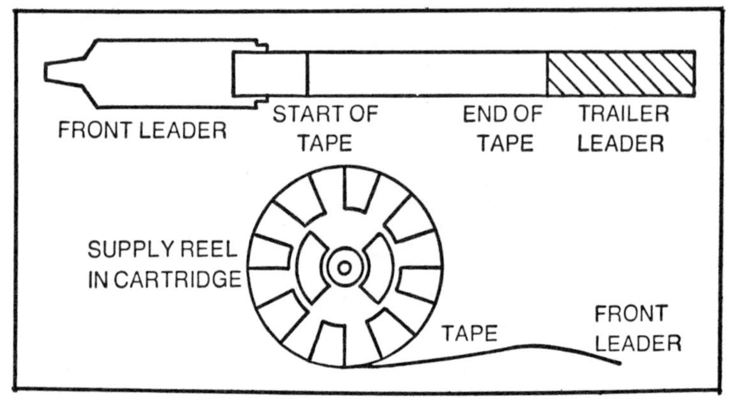

Fig. 8-18. Plastic leader.

rotate in a horizontal plane with its main spindle vertical. This is very simple to manufacture, and it puts all the pressure onto one singlebearing at the bottom of the main shaft, with no pressure on the sides. It is claimed that this produces a more stable and firm

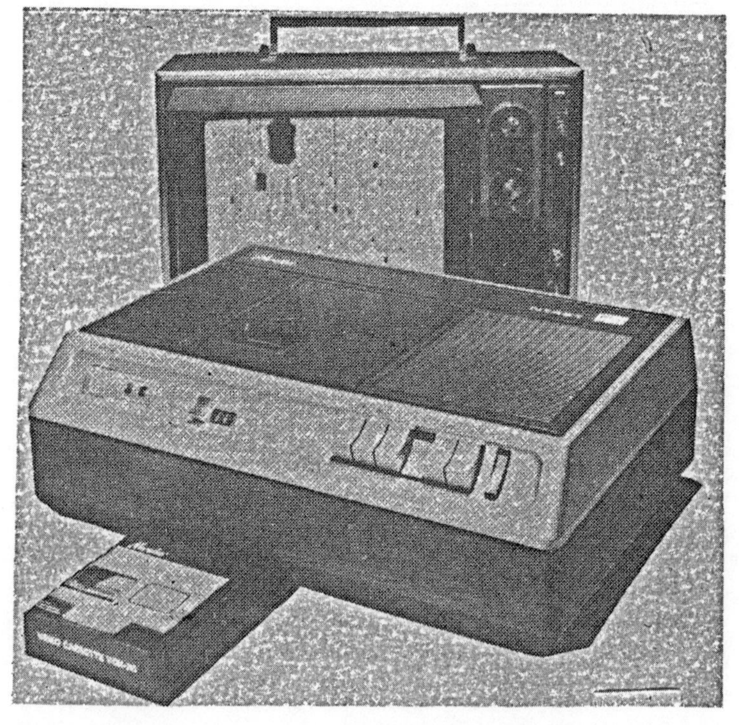
Fig. 8-19. Phillips machine and cassette (courtesy of Phillips).

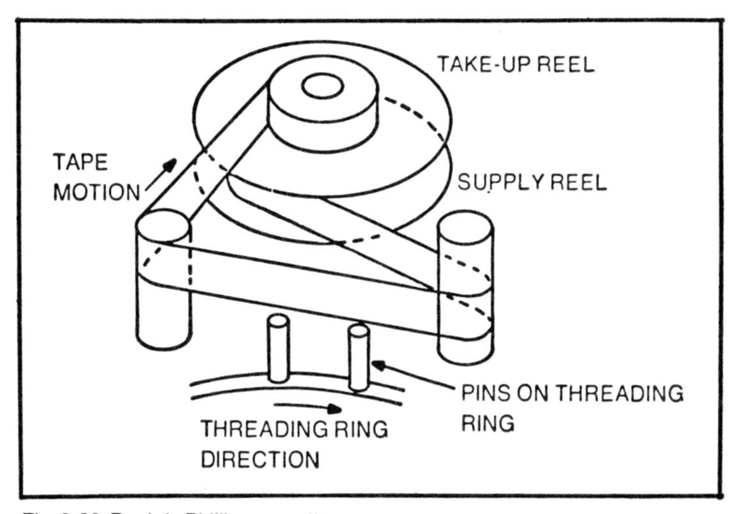

Fig. 8-20. Reels in Phillips cassette.

drive than the other systems, which must use a slightly slanted head.

The cassette is inserted into the "elevator," which is then pressed down. This lowers the cassette into the threading position and drops the tape so that two pins are placed to pull it out. This is explained in Fig. 8-21. When the **play** button is pressed, the threading procedure begins. The two pins are mounted on a threading ring which now rotates and pulls the tape out and halfway around the head drum, as in Fig. 8-22. Note the position of the pulled out tape, the capstan and drive roller, and the fixed heads.

When the capstan pressure roller closes, the tape is pulled from the lower reel around its path and onto the top reel. The heads

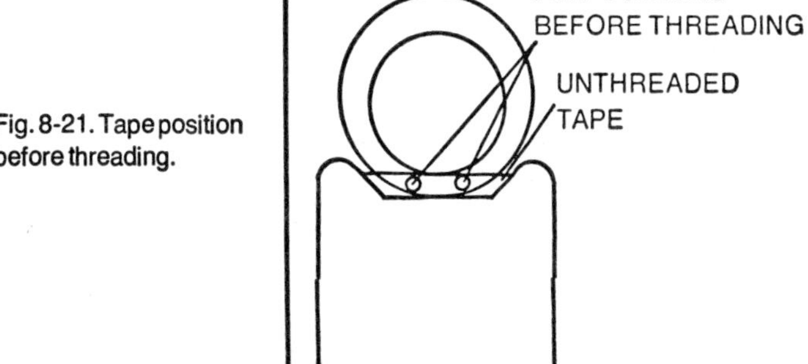

Fig. 8-21. Tape position before threading.

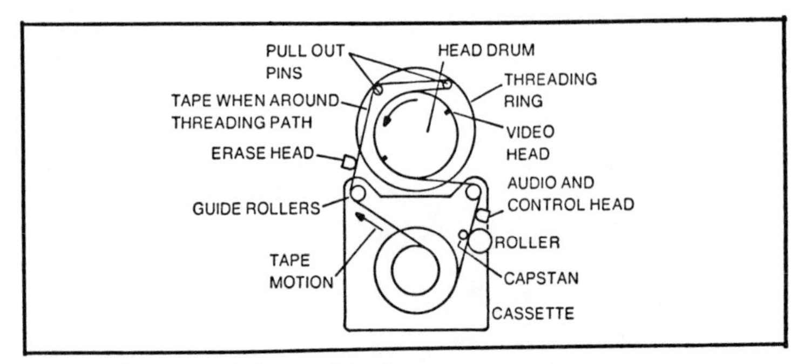

Fig. 8-22. Tape threaded around heads.

rotate in the same direction as the tape, which gives a smooth drive and tape transport.

All of the operational functions are provided, including **still frame, edit,** and **audio dub**. It is possible to record one program **off air** while watching another on the TV set without any extra external connections. Also the UHF antenna is internally connected and looped through to the output without the need for an external splitter.

The cassette consists of two flanged reels, one on top of the other, with the takeup reel on top. It is a very compact cassette and is much smaller than the other two reel types. Tape is provided in lengths of 20, 30, 50, and 60 minutes.

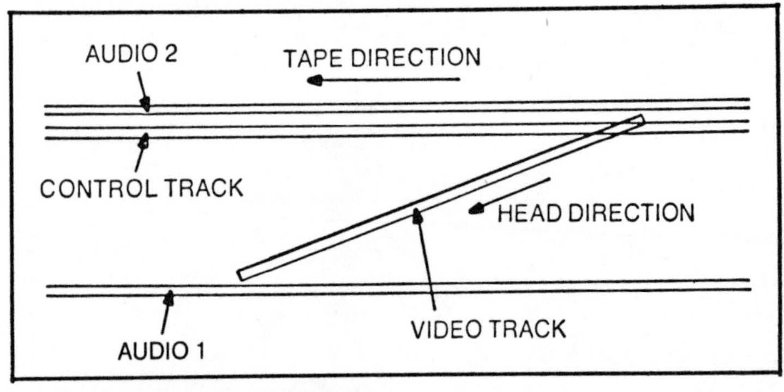

Fig. 8-23. Tape tracks in Phillips system.

The track layout is shown in Fig. 8-23. Electronically it is quite simple, but it uses different standards from the other formats.

One of the main advantages claimed for this machine is its very low weight and small size. Its 31 lb is about half the weight of the other machines.

9 Videocassette Electronics

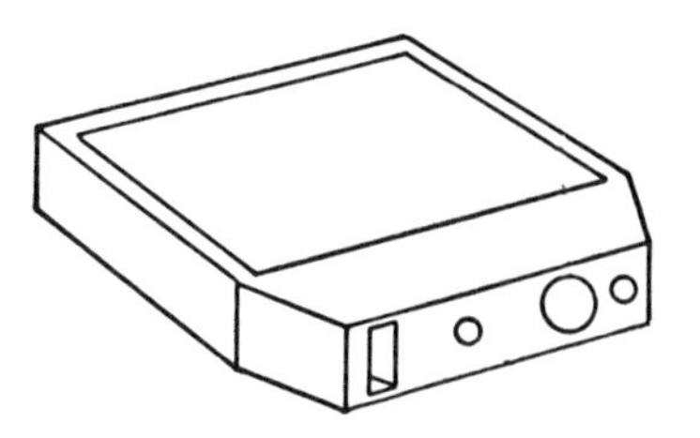

The electronic circuits in videocassette machines are very similar to those in other helical VTRs, and the same system of recording and playback is used. The video signal is made to frequency modulate a high frequency carrier which is recorded directly onto the tape. The color information is heterodyned down to a lower frequency and recorded onto the tape using the FM signal as a high frequency bias, just as in audio recording. The only difference from the other VTRs is the choice of the frequencies for the FM and the heterodyned color.

Most of the small cassette formats have only one audio track, but the Umatics have two. These are placed at one edge of the tape and are just the same as the audio tracks in an ordinary audio machine. The audio bias oscillator is also used for the main erase head, which is used to clean the video, control and audio tracks.

The earliest machines all used a very simple head servo and a belt drive to the capstan from the main ac motor. Later machines included a capstan servo and a dc capstan motor.

The main departures from the other VTRs are the incorporation of the TV tuner in the consumer models to allow off-air recording and the extensive use in all machines of system control circuits to control the automatic functions associated with the tape transport, and the selection of the required modes by remote editor panels.

One of the characteristics of the new videocassette machines is their small size. In part this is made possible by ICs which have been specially developed for VTR use.

Several advantages accrue from the use of these ICs.

- A reduction in the size of the PC boards, as less discrete components are needed.
- Overall performance of the machines is much more uniform from machine to machine, and is much improved over the older machines.
- The total power consumption is lower than with discrete components.

- Reliability is higher, and servicing is much easier.

Most of these ICs are from two manufacturers, Sony and Matsushita. See Taoles 9-1 and 9-2 for the part numbers and functions performed by the ICs. Fig. 9-1 is a block diagram of a Betamax, taken from a service manual, which shows how the ICs are used.

In several cases the contents of an IC, that is its actual circuit, has not been disclosed for proprietry reasons, and so some doubt exists as to its exact ciruit. This is reflected in the service manuals of certain machines where a slightly different circuit may be shown from that found in the manual of another machine. In some cases this doubt is compounded by what are obvious drawing errors, and slightly different descriptions of the circuit by persons whose normal tongue is not English. However, the functions performed by the IC usually are well described.

EMITTER FOLLOWERS, FETS AND MOSFETS

Extensive use is made of these components as switches which pass or stop the video signal.

The emitter follower is either powered or not powered by the record or playback 12v, thus routing a signal one way in one mode and another way in the other mode.

With the FETs and the MOSFETs, usually the signal is applied to the drain and it leaves at the source. A voltage at the gate turns the transistor on or off. Their main advantage is their very

Table 9-1. Special IC's by Sony.

IC	Functions
CX-131A	Y-AGC, Sync separation, Sync tip clamp, White clip, FM modulator.
CX-133A	Record chroma A.C.C., A.C.K. (Color Killer), Frequency conversion.
CX-135	Y-FM demodulator, Noise canceler, Y/C mix.
CX-136A	Playback Chroma A.C.C., A.C.K., frequency conversion.
CX-137A	Record/playback Chroma A.P.C., A.F.C. (44fH V.C.O.) 3.58 MHz crystal controlled oscillator, 3.57 MHz crystal controlled oscillator
CX-145	Sync separation, f/44 counter
CX-150	Burst ID, Carrier phase shifter switch
CX-134A	RF playback amplifier, RF switcher, Dropout compensator
CX-138	Servo: 30 PG pulse amplifier, CTL record amplifier
CX-139A	Servo: Gate circuit, DC amplifier, CTL playback amplifier, Sync separation
CX-141	System control: Auto stop circuit, Tape threading control, Video/audio muting, other protection circuits

Table 9-2. Special IC's by Matsushita.

Special IC's by Matsushita.
The Matsushita line of specialized ICs usually can be identified by the AN prefix, and the following is a list of the most common.
AN 301. Video amplifier, H and V sync separator, multivibrator, output amplifiers.
AN 302. AGC amp and detector, pre-emphasis, clamp, clippers, voltage stabilizer.
AN 303. Emitter follower, amplifiers, mixer, clamp, mute.
AN 305. Burst gate, AGC detector, AGC amp, balanced modulator (or frequency converter), killer amp.
AN 316. Drop-out detector, limiter, pre-amp, mixer, electronic switch.
AN 318. Multivibrators and flip-flops, gates.
AN 236. Phase detector, ACC detector, amplifiers, clamp, oscillator, pulse shaper.
6A753. A cascode pre-amp.

high off resistance and their quite low on resistance. The MOSFET often has two gates. These can be tied together and the switch pulse applied to both, or the signal can be applied to one gate and the switch pulse to the other. In this case the MOSFET can be used as an amplifier or a source follower.

Many ICs show in their functional diagram a mechanical switch symbol, as in Fig. 9-2. In fact this is usually 2 MOSFETs, operated by the same switching signal, as in Fig. 9-3. They are alternately ON and OFF, and provide excellent switching and isolation of the video signal.

The electronics can be divided into 8 main functions or sections. These are:

Luminance recording and playback	System Control Circuits
Color recording and playback	TV tuner and VHF modulator
Color correction	Audio
Servos	Power supply circuits

Here we will briefly review the standard method of recording and playing back the NTSC color signal, and introduce a few of the more important sections of the overall circuit. Then the basics of the system control, TV tuner and RF modulator, the audio and the power supply will be mentioned. The servos are covered in some detail in Chapter 10.

A further refinement in the cassette machines is the various methods used to correct the color timing errors which are so visible in playback. The earlier Umatics used the same simple system found in the early open reel machines, but later cassette

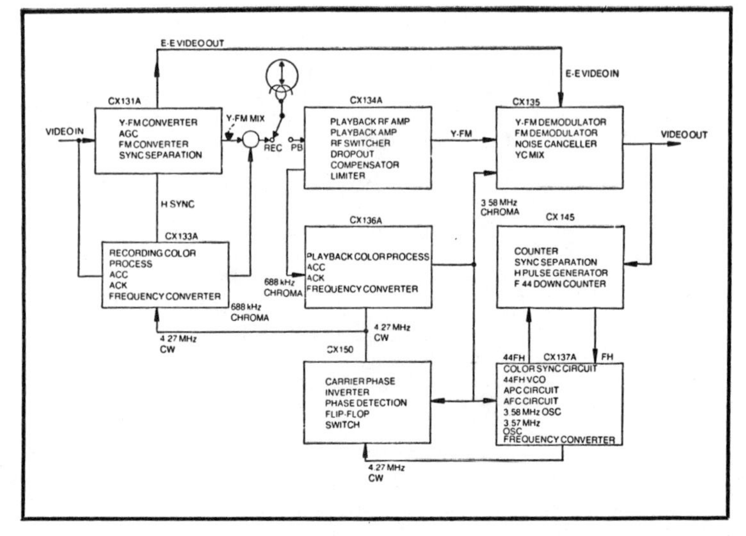

Fig. 9-1. The IC's in a Sony Betamax.

machines adopted several new ideas. Because of the range of ideas used, color correction is covered in Chapter 14.

LUMINANCE RECORDING

All VTRs use the FM method of recording the video signal onto the tape. The small helical machines, such as the cassette models, record the luminance separately from the chroma, so the input signal to the machine is separated into two main paths at the input. A basic luminance chain is shown in Fig. 9-4. This very simple block diagram is the basis of all cassette recording circuits, but refinements have been added to most models.

A low pass filter (LPF) removes the 3.58MHz subcarrier and leaves only the luminance components of the signal. A band pass filter (BPF) with a bandwidth of about 1 Mhz passes only the 3.58 MHz chroma. A third path for the input signal is to the sync stripper, but often the sync stripper is placed later in the luminance chain.

The input luminance signal is passed through an amplifier, which is automatically gain controlled to have a correct level of signal at its output. Sometimes the signal is split into two paths at this point; it is passed through a filter to restrict the bandwidth if a color signal is being used, or a direct path is used if the input is a monochrome signal.

Next the sync tips are clamped to a dc level. This is necessary

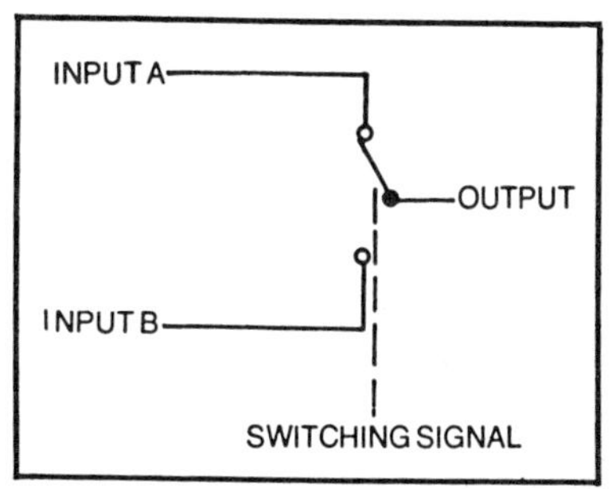

Fig. 9-2. Diagramatic representation of an IC switch.

as it sets a definite FM frequency for the sync tips. After this stage the signal is pre-emphasised, just as the audio in FM radio, to improve the signal to noise ratio of the high frequencies. The pre-emphasis causes overshoots which are now clipped by the white and black clip circuit; usually the clip points are adjustable.

Then the signal is applied to the FM modulator. This is an astable multivibrator whose frequency is varied by the applied video signal. A control sets the symmetry of the waveform because if the output is asymmetric with no input signal it will cause beats to be seen on the screen, even with a signal. Often an AFC circuit is used to ensure it stays on frequency.

The FM signal has a bandwidth as shown in Fig. 9-5. A high pass filter (HPF) is used to clean the spectrum below about 1.4MHz and then the signal is applied to one or two record amplifiers. Here the down converted chroma is added, as covered later, and the combined output is fed to the heads via the rotating transformers and the **rec/pbk** switch.

The rotating transformers are about 3 turns of wire in the secondary and the primary. The primary is stationary in the head drum and the secondary rotates with the heads. This provides a noiseless contactless transfer of the signal on both recording and playback.

In this record chain several circuits are of interest. Most of these are included inside ICs, and so are not available for discussion at the component level. However, the manner in which they perform their functions is worth knowing. These circuits are:

The AGC
The pre-emphasis
The clippers
The FM modulator
Filters
The rotating transformers

Recording amplifiers and head switching.
The automatic gain control can be one of three types.

Average video level.
Peak video level.
Sync type.

The main disadvantage of the first type is that changes in picture content, such as from light to dark, cause changes in the ratio of the sync to peak video level. These perfectly desirable changes now tend to be lost, resulting in a picture lacking in contrast and a signal with stretched sync pulses.

The peak type also has the above problems, but also when it detects the peaks of the video signal it reduces the overall signal to 1v. The disadvantage here is that it will take an overmodulated signal and reduce the sync level along with the video level, and the sync to video level is again lost.

The sync type is more complex than the others, but is becoming more popular as the circuits can be contained inside an IC. The sync is separated from the luminance signal and a pulse equal to the desired peak white level is inserted on the back porch. The peak of this is clamped and variations in the sync level due to input signal variations are now sensed and used to control the input amplifier.

Most modern cassette systems use an AGC system which is a combination of the second and third types.

Pre-Emphasis

A simple R-C network is used to pre-emphasize the video signal. This is shown in Fig. 9-6.

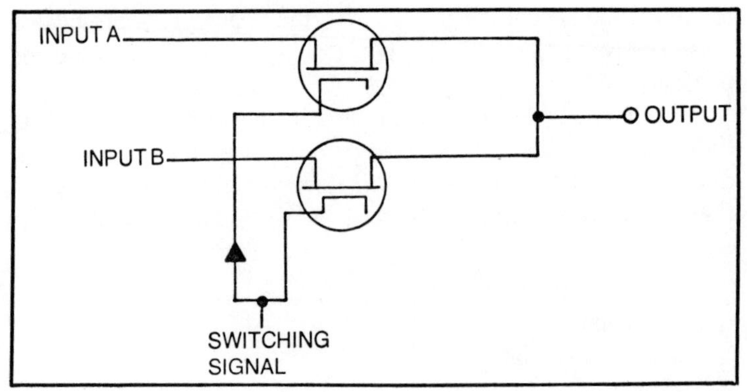

Fig. 9-3. Typical IC switching circuit.

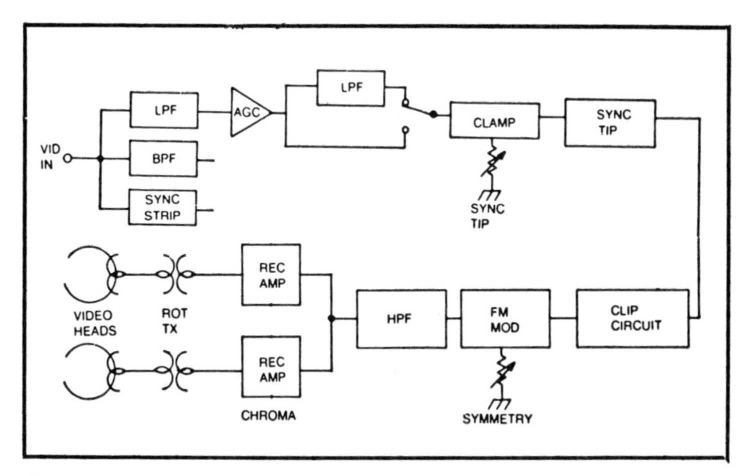

Fig. 9-4. Basic luminance chain.

In the long playing consumer videocassette machines a more complicated non linear pre-emphasis is used, which has a much more complex circuit.

Clipping

Figure 9-7 shows the simple method of clipping the overshoots caused by the pre-emphasis. The clip points can be set by the potentiometers.

FM Modulator

The basic FM modulator is a simple multivibrator circuit which is set to free run at the sync tip frequency. The varying dc level of the incoming video signal varies the output frequency. This same basic circuit is found both inside ICs and in discrete form in many VTRs. See Fig. 9-8.

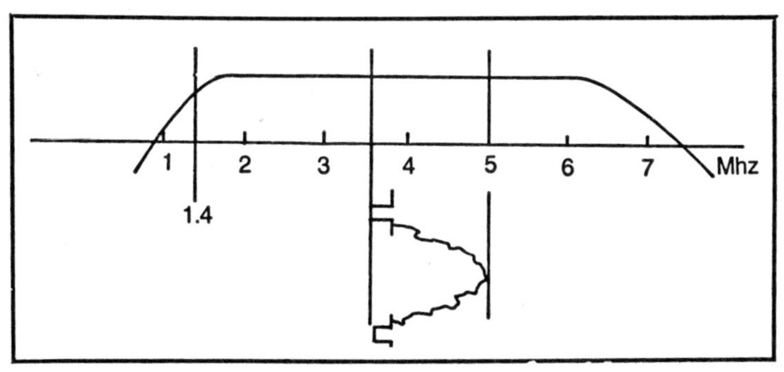

Fig. 9-5. FM spectrum.

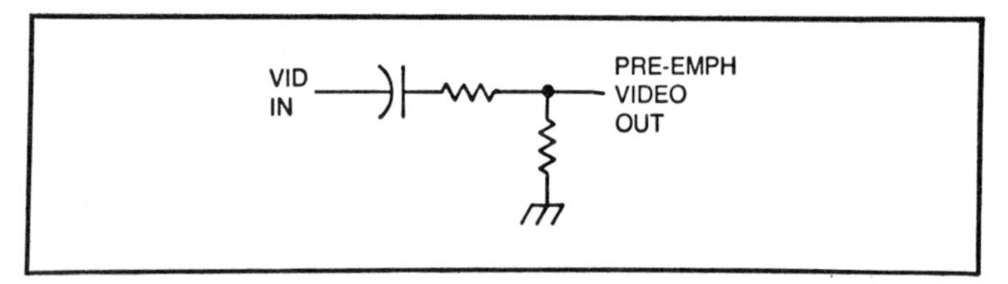

Fig. 9-6. Simple pre-emphasis.

Filters

Several types of filter are used to remove luminance, chroma, or high frequency noise. Mostly these are simple LCR circuits.

Rotating Transformers

These are usually about 3 turns of wire on a static part of the head drum assembly, and 3 rotating turns very close to the static turns. They provide a noiseless contactless transfer of the signal.

Recording Amplifier and Head Switching

Either one or two recording amplifiers may be used. The luminance FM input and the chroma signal are resistively mixed at the input and controls set the levels of the two signals. The output of the amplifier provides a current drive to the low impedance heads via the rotating transformers. A switching arrangement is used so that the heads can be used for both recording and playback. In most machines this is a relay operated by the power to the playback or record amplifiers, but in some machines electronic signal switching is used. Figure 9-9 shows a simple record amplifier.

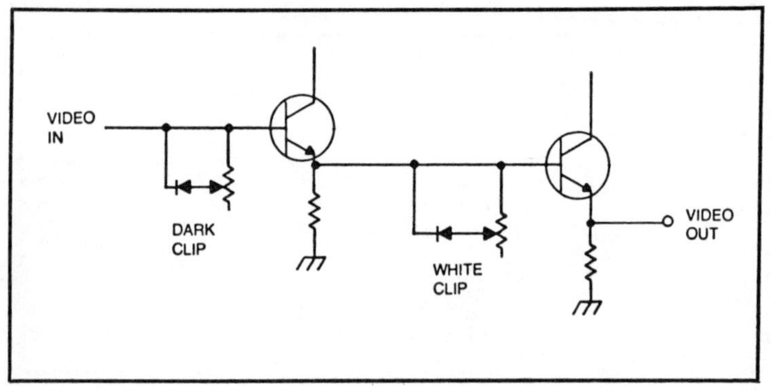

Fig. 9-7. Simple clipper.

COLOR RECORDING

The writing speed and time base stability of a small helical VTR or cassette machine is not good enough for the color to be recorded directly with the luminance as a normal NTSC signal. The universally adopted method of recording the color is to heterodyne the 3.58 MHz carrier down to a lower frequency and record it onto the tape using the FM signal as a bias, just as in audio.

To do this the video signal is passed through a filter which removes the luminance component and leaves only the 3.58 MHz chroma. The filter also restricts the bandwidth of this chroma to about 500 kHz above and below the 3.58 MHz. This is less than required for a full NTSC signal and is one of the reasons why these machines do not really meet FCC specifications for broadcasting. However, in most cases this does not seriously restrict the quality of the picture.

In Fig. 9-10, the signal is now fed into a fast acting AGC or ACC (automatic color control) amplifier. The feedback signal into this is the separated 3.58 MHz burst. Burst is used as it should not change in level, whereas the chroma level can change with the scene content. The AGCd signal is now mixed in a frequency converter or balanced modulator with the heterodyning frequency. In many machines this is 4.27 MHz, and for this discussion we will

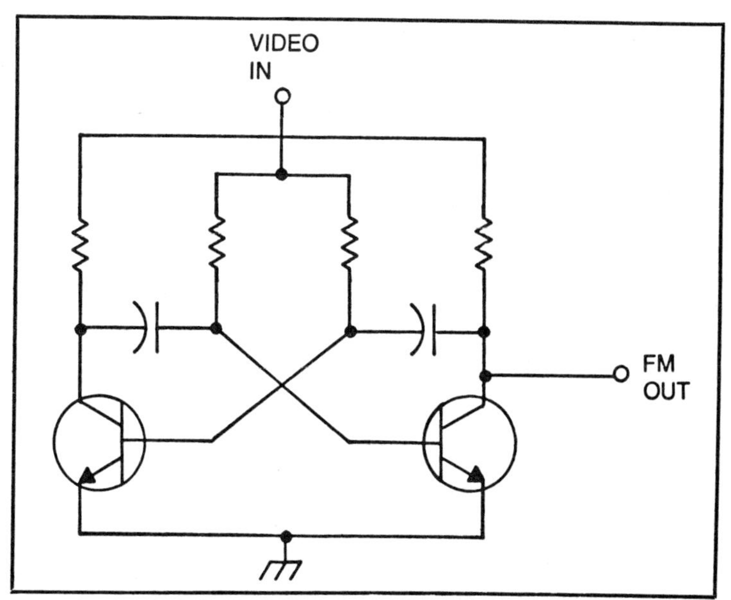

Fig. 9-8. FM modulator.

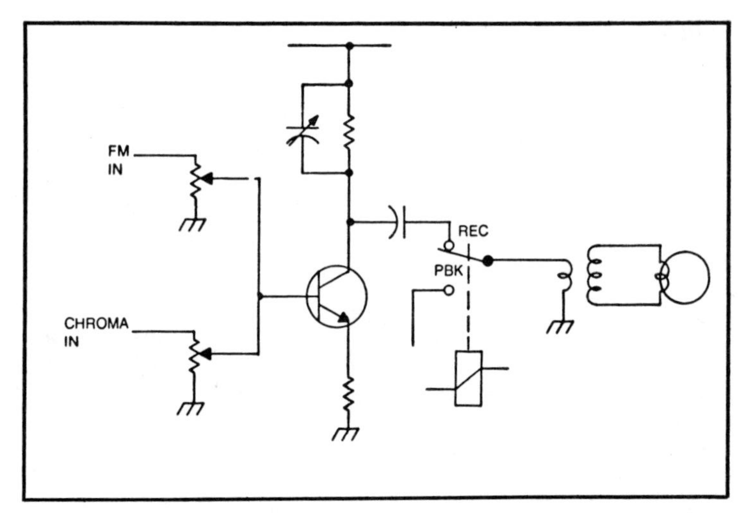

Fig. 9-9. Simple record amplifier.

use this frequency. The 4.27 MHz is usually produced in a crystal oscillator. The output of the frequency converter is passed through a LPF which allows only the difference frequency of 688 kHz to emerge. This contains all the color information in the original 3.58 MHz signal.

A band pass filter now limits the top of the range to below 1.4 MHz so that it does not intermodulate with the FM. The signal is now added to the FM by resistive mixing.

Although the 4.27 MHz oscillator is stable, it is not phase locked to the incoming signal. Consequently the 688 kHz recorded on the tape has no phase relationship to the horizontal sync pulse recorded onto the tape. This is why the term 'non phased color' is

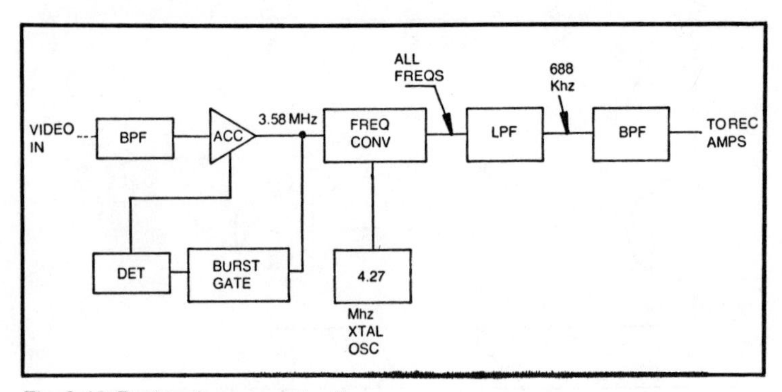

Fig. 9-10. Basic color record circuit.

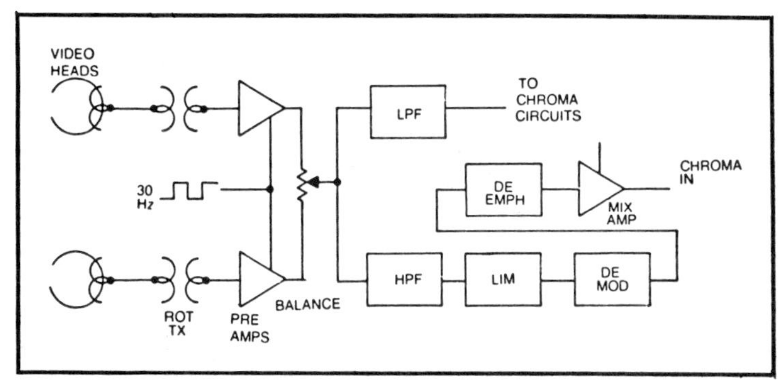

Fig. 9-11. Block diagram of luminance playback.

used to describe this type of recording. In the smaller cassettes, the 4.27 MHz is formed in a phase locked loop (PLL) type of circuit instead of a crystal oscillator. This is explained later in the Betamax and the VHS chapters.

LUMINANCE PLAYBACK

The basic luminance playback circuit is shown in Fig. 9-11, and most videocassette machines are very close to this. However, many of the separate functions are contained in ICs. The minute signal from the heads is coupled by the rotating transformers to the input amplifiers. These contain equalizers to match the head impedance and shape the frequency response of the pre-amplifiers. The outputs are switched on and off by the head PG or head tach pulses at a 30 Hz rate, and after a buffer or emitter follower the output are combined across a resistive mixer. The balance control feeds a continuous output FM signal to the luminance demodulation and the chroma playback circuits.

The feed to the luminance demodulation circuits passes through a HPF which removes most of the amplitude variations caused by the down converted chroma. Then a limiter chain amplifies and clips the signal and removes all amplitude variations due to noise. The output is now a clean FM square wave, similar to that put onto the tape. This is now demodulated to a video signal, de-emphasized, and applied to an amplifier where the upconverted chroma is added. The resulting NTSC type signal is now fed to the outputs of the machine.

This basic luminance chain is usually enlarged by the addition of other circuits, all of which are for the purpose of improving the final video signal. The most common of these extra circuits are:

A drop-out compensator
Noise canceller
Muting and still frame vertical drive.

Figure 9-12 shows where they are usually placed in the playback chain. A brief description of some of the more important items is now given.

The Drop-Out Compensator

A drop out appears as an annoying horizontal flash on the screen. Drop outs are caused by loss of oxide on the tape, or by minute particles of dust which cause the tape and head to lose contact with each other. The drop out compensator senses the drop outs and replaces them with a repeat of the last good horizontal line, so that the flashes are no longer visible on the screen. Figure 9-13 is a simplified block diagram.

The detector rectifies the FM signal to a dc, and the loss of the FM from the tape produces a change in the dc output. This dc is fed to a Schmitt trigger or similar circuit which operates an electronic switch. The switch normally selects the direct FM path from the tape and feeds this to the next stage in the playback chain, but when a drop out occurs the Schmitt trigger causes it to select the delayed path. Here the previous horizontal line is delayed by the 1H delay line, and this is again sent to the output. About 5 repetitions can occur before the signal degradation becomes serious. When normal FM level is restored, the Schmitt trigger reverts the switch to the direct path.

Usually the trigger circuit will have some hysteresis built in so that the FM must be restored to almost normal level before the direct path is again used.

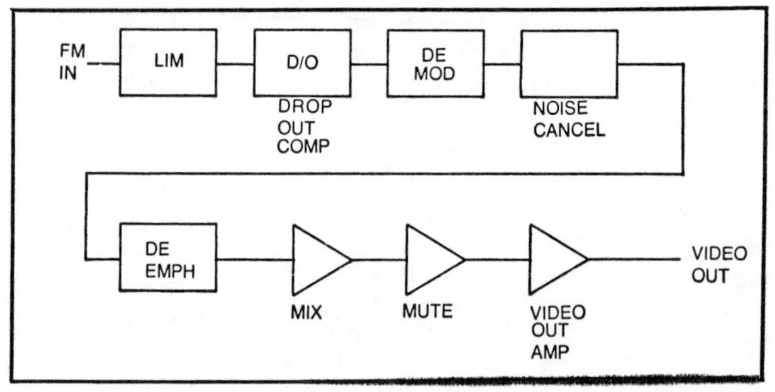

Fig. 9-12. Luminance playback chain.

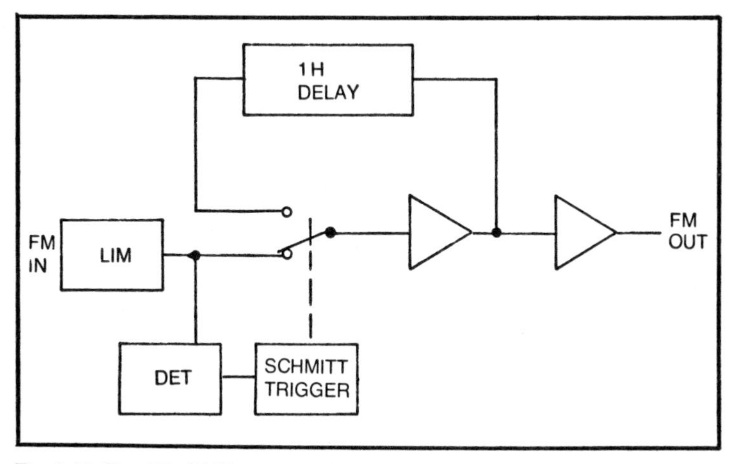

Fig. 9-13. Simplified D/O compensator.

The Noise Canceller

The high frequency noise can be largely removed from the output signal with a simple noise cancelling circuit, as in Fig. 9-14. The video signal is split into two paths. A HPF allows the noise to pass but it blocks the lower frequency components of the signal. The noise is now inverted and limited and then added to the direct signal. This cancels the noise components and leaves a fairly clean video signal. The limiter rejects the large signal noise components, which tend to be rare. Small signal noise is always present.

Muting

At certain times it is expedient to mute the signal. Examples are during threading, FF and REW in those machines where the tape remains around the head, and during speed changes in the dual speed machines. Muting is easily accomplished, and it is usually applied to the final video and audio output stages. Figure 9-15 is a simple but practical muting circuit. A positive dc is applied to the base of Q3 to turn it on. This grounds the base of Q2 and shuts it off. The dc level to Q3 is either from the system control or the capstan servo, depending on the model.

Still Frame Vertical Drive

If the machine is capable of producing a still frame picture, then an artificial vertical pulse is added to the final output video signal. This is necessary to ensure the still picture on the screen does not jitter up and down, which it will do if the heads do not

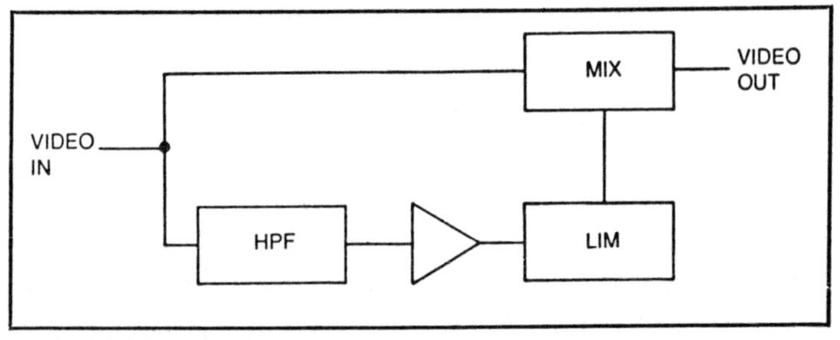

Fig. 9-14. Noise canceller.

exactly track over the recorded vertical interval. The circuit is very similar to the mute circuit. A negative pulse is applied to the video signal by Q3 in Fig. 9-15, but this is applied only during the vertical interval and acts like an artificial vertical sync pulse.

FM Demodulator

Several methods of demodulation are available, but with the introduction of ICs into the small VTRs a digital method has become popular. The FM signal is treated as a digital pulse train and a simple digital circuit is used.

The playback FM is fed through a single limiter to remove the largest amplitude variations and then is split into two paths. The first path feeds the FM directly into a comparator, as in Fig. 9-16. The second path is through a delay circuit, and then into the other

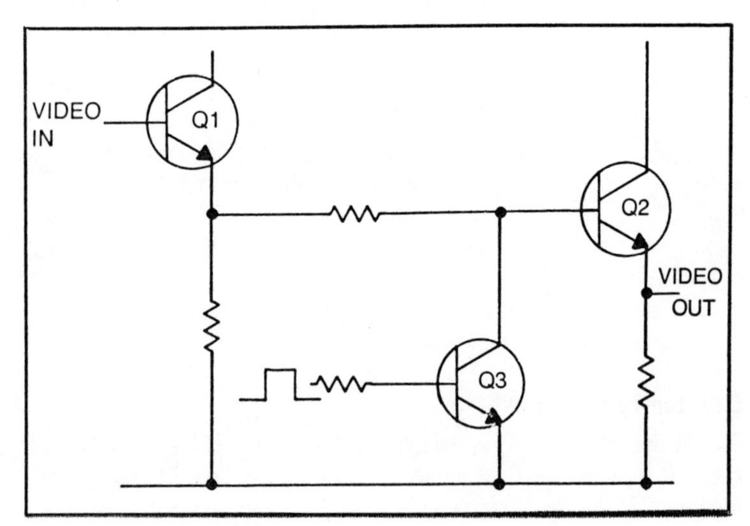

Fig. 9-15. Muting circuit.

input of the comparator. Several delay designs are used; the delay line, a phase shifter or an injection locked multivibrator. All have the same function of producing a constant delay. Any of several types of digital phase comparator can be used. An exclusive OR gate is a common choice.

How this demodulates the signal is best explained with the waveforms in Fig. 9-17. Three different input frequencies are shown. The comparator output can only go positive when the inputs are different. So the output is a series of constant width pulses with a spacing apart that depends on the input frequency.

These pulse trains are now rectified to a dc level, which is in fact the video signal.

The advantage of this type of demodulation is it is very insensitive to amplitude variations in the incoming signal, and thus is not affected by noise and thus allows a simpler limiter to be used. Also, because of its simplicity, it can easily be built into an IC.

COLOR PLAYBACK

During playback the signal from the heads is amplified by the pre-amplifiers and the two outputs combined to make a continuous signal. This signal is now fed to the luminance and the chroma playback chains. Fig. 9-18 shows this latter path.

The signal is passed through a LPF which passes the down converted chroma (688kHz) but not the higher FM frequency and its sidebands. This chroma signal is now automatically gain controlled in an ACC amplifier and the output is heterodyned in a frequency converter with 4.27 MHz. The difference frequency of 3.58 MHz is extracted with a BPF and then sent to the output amplifiers where it is combined with the demodulated luminance signal. The burst is gated out from the 3.58 MHz output signal and is used in the ACC feedback loop to control the level of the input chroma signal. Only burst is used, as it is desirable that the output burst level should remain constant while the scene chroma level should be allowed to change.

Although this process is very simple and is now often completely contained in ICs, it does not correct the color timing errors which show up as bands of color and continual changes of hue on the screen.

No video tape recorder will play back a color signal directly, all need extensive correction of the color signal. This is true even of the broadcast machines, which do record the chroma directly onto the tape with the luminance and do not convert to a lower

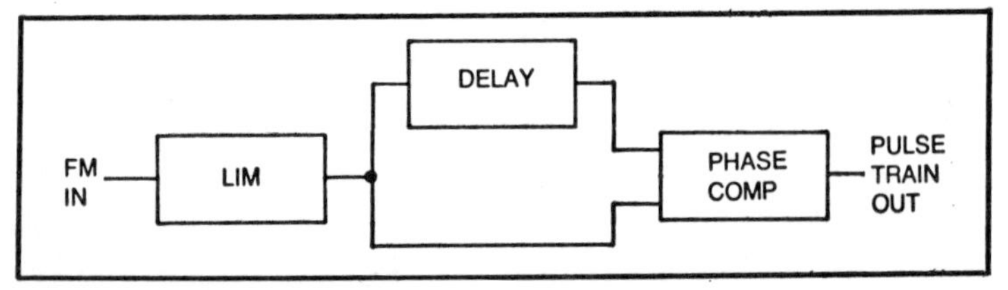

Fig. 9-16. FM demodulator.

frequency. The reason why the color playback signal must be corrected is that it contains 'time base errors'. These are slight variations in the relative timing of the signal, and they are caused by factors such as tape stretch, mechanical instabilities like wow and flutter, slight variations in the speed of the rotating video heads, and the minor mechanical imperfections of the machine. All these are impossible to control mechanically or physically, and electronic methods must be used.

Because of its importance to the playback process, and because it contains much of the circuitry found in the modern VTR, and because the cassette machines have some very different and interesting methods for correcting color; color correction is covered separately in a later chapter.

THE TV TUNERS AND THE RF MODULATOR

The smaller cassette machines that are intended for consumer use are provided with a VHF and UHF tuner and antenna inputs. A switching arrangement permits the viewing of one program on a TV set while a different one is simultaneously recorded on the machine. This is explained in Chapter 5.

The TV tuner section in the machine contains exactly the same tuners, IFs and demodulator sections found in TV sets, usually with AFT and AFC included. They feed the selected video and audio to the record circuits in the cassette machine.

To playback a tape into the TV set antenna terminals the video and audio outputs are fed into an RF modulator. This is a miniature TV transmitter which is switchable between channels 3 and 4. It is usually a sealed unit and is classed as 'unservicable' by the manufacturers. This is to prevent unauthorized persons contravening FFC regulations.

AUDIO

The audio sections of the machines are exactly the same as in a normal audio recorder. Usually three inputs are used; TV, mic and

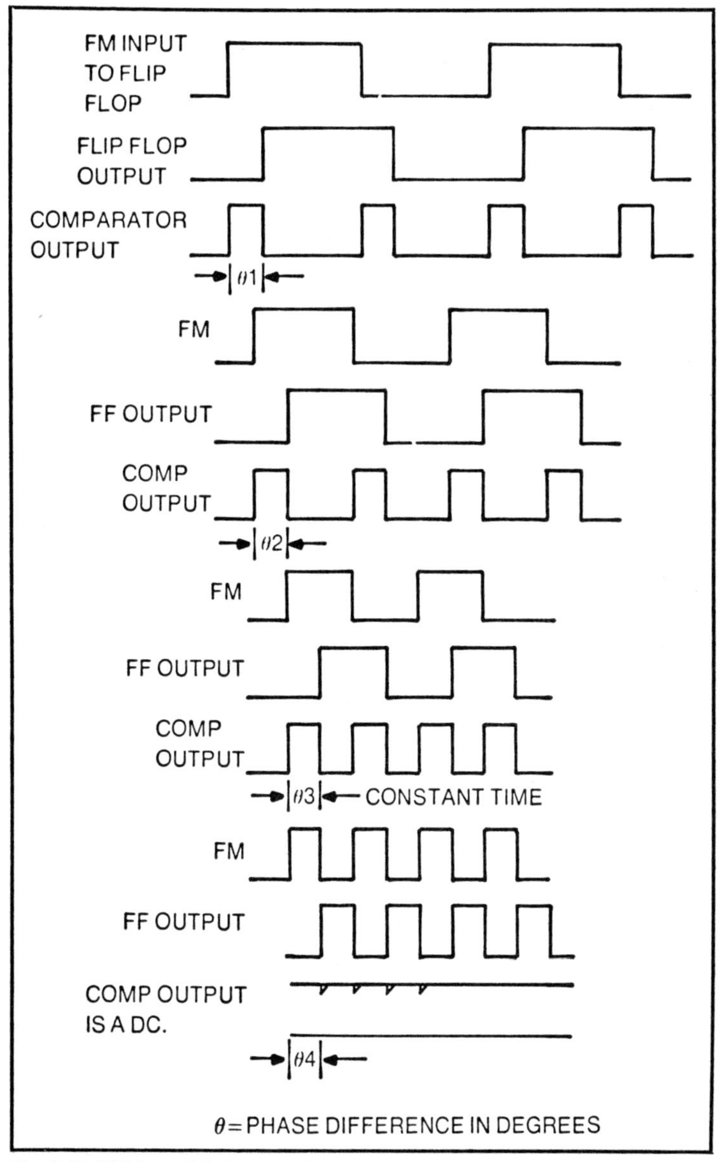

Fig. 9-17. FM demodulation waveforms.

line, and they have an order of precedence. The outputs provided vary with the model, but usually there is a line out, a feed to the RF modulator, headphones, and sometimes a meter. In many circuits the AGC has been discarded in favor of a limiter. This merely clips

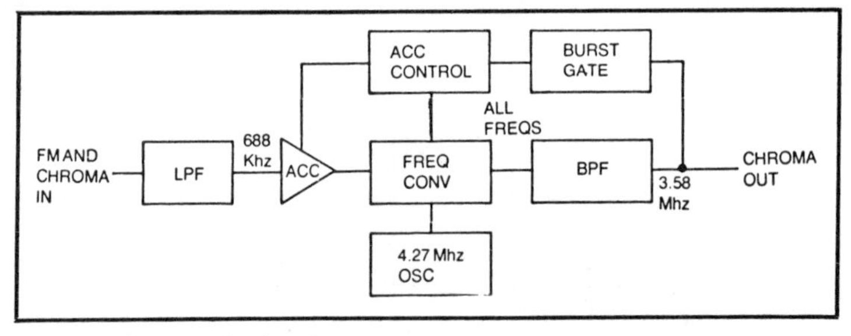

Fig. 9-18. Color playback path.

excessive peaks and leaves the rest of the signal alone. Occasionally an AGC will be used to lift a low level signal out of the noise region.

RF DUB

The purpose of **RF dub** is to improve the quality of the tape to tape copies.

Older machines, especially the broadcast quads, had an **RF out** from the FM limiter which could be directly connected to the record amplifier inputs of another machine. This provided a direct FM or RF transfer of the signal without the degradation introduced by the complete playback demodulation, or the record remodulation electronics in both machines. Both quads could be locked to station sync, so the capstan and heads would be properly controlled and phased, and the time base errors eliminated. Improvements in tape, heads and electronics have made this an outmoded method of copying.

In the recent Umatics, a 'half way' situation exists. The FM is demodulated to the luminance signal and this is fed to the luminance circuit of the second machine. However, most of the processing, such as the noise canceller, is not used. The 688 kHz chroma is separated from the playback FM and is fed directly to the second machine as 688 kHz. It is then added to the FM in the record amplifiers. Note that the 688 kHz is *not* heterodyned back up to 3.58 MHz, and again most of the signal processing is bypassed.

The result is much cleaner video on the copies without some of the errors which are likely to be introduced by the continual up and down conversion of the chroma.

The main disadvantage is that the signal cannot be time base corrected, and hence this process is good for only the first generation copy if broadcasting is envisioned.

Figure 9-19 is a block diagram of the process.

SYSTEM CONTROL CIRCUITS

All the mechanical functions concerning the tape transport and the operating keys are controlled by a system of levers, links, solenoids, etc. which are driven by control circuits. Eight definite modes can be distinguished after the power has been turned on, and these are listed in the manuals as follows:

Play
Manual **stop** during the **play** mode
Auto rewind at the end of the tape during **play rewind**
Manual **stop** during **rewind**
Auto rewind at the conclusion of play or **fast forward**
Fast forward
Manual **stop** during **fast forward**
Auto stop at the end of the tape during **fast forward**

The service manuals contain separate simplified circuit diagrams to illustrate and explain the principles and actions for each of these modes. Although the circuits are usually a complex system of interlocks and the like, they are made up from simple mechanical switches and simple transistor switches. The circuit action is easy to understand; the main problem is in identifying the overall action of the system and locating the parts inside the machine.

The system control consists of four main circuits. The output of these control the operation of the threading motor, the auto stop, the brake system, and the muting circuits.

Several inputs are used, all of which engage normal operation of the machine or inhibit operation for a specific protective reason. Each of these inputs are described briefly in the following paragraphs.

Cassette-In switch and the function buttons. Depending upon which function is selected, the appropriate control circuit and output is actuated.

Drum revolution detector. This uses the head-tach pulses to sense drum rotation. Should the drum stop, the auto stop circuit pulls in the stop solenoid to arrest tape motion.

Condensation detector. This senses condensation on the head drum which can cause the tape to stick to the head and result in tape damage. Condensation is common if the machine is moved to a new environment or a tape is not allowed time to acclimatize. The detector is between the output of an oscillator circuit and a

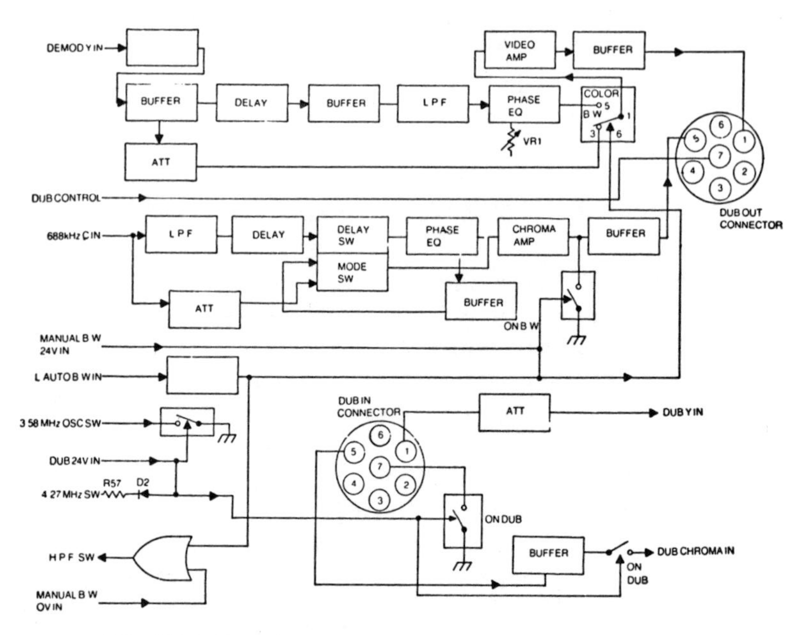

Fig. 9-19. RF dub block diagram.

switching transistor. It passes or inhibits the oscillator signal, causing the transistor to operate the auto stop.

Tape slack sensor. This is a microswitch held on by a correctly tensioned tape. If the tape slackens, the microswitch releases and actuates the stop function.

Muting. The audio and video circuits are muted by the function buttons and the lack of control pulses. As control pulses occur only when the tape is running, this circuit prevents noise on both outputs—when the tape is stopped or when it is not in the machine.

Pause. The pause control mechanically disengages the pressure roller and closes a microswitch to actuate the brake circuit. The head drum still revolves, but no output is seen as no control pulses are produced. Hence, there is no still frame picture. The circuit has been modified in the industrial version to produce a still frame output.

End of tape detector. The Betamax uses inductive sensing. There are two small coils, one for each end of the tape, which are in permanent contact with the tape while it is running. The tape is attached to its reels with aluminum tape. This is pulled from the cassette at the end of the tape to contact the coils. The coils are part of the same oscillator circuit, which is detuned by the inductive action of the aluminum tape. The output of the oscillator operates the auto stop. Most other cassette systems use an optical end of tape sensing. A light shines through a clear plastic leader at each end of the tape to produce a pulse output which initiates the stop mode.

Counter memory. This stops the machine in the rewind mode when the counter displays 000. It is a switch on the counter located in the same oscillator circuit as the end-of-tape sensor.

Power failure sensing. If the AC power fails or it is interrupted, the reel tables are immediately braked, preventing excessive tape slack and tape damage.

Threading motor protection. If the threading motor continues to run after the tape has been fully threaded, it will burn out. A timer circuit ensures power cutoff after a few seconds to prevent this.

For the purpose of illustration the **play** and **auto stop** are briefly covered, and then two minor but important control circuits, **pause** and **muting** are shown.

Play. In this mode several internal items are controlled. When the **play** or **forward** button is pressed, the tape threading is started, the head and capstan motor is powered, the pinch roller is closed, tape motion is commenced, back tension is applied to the tape, the cassette is firmly locked in place, all the function buttons are locked out except **stop**, and in the final threading phases the **stop** button is temporarily electronically disabled.

Three separate sequences of internal operation put the machine into the playback mode:

- When the **play** button is pressed, Q4, Q5, and Q6 in Fig. 9-20 pull in RY-1. This starts the DC motor, which drives the threading ring and threads the tape. Simultaneously Q7 and Q8 pull in solenoid A. This locks the **play** button, releases the supply spool, and locks out the other functions except **stop.**
- Near the end of threading, the ring closes microswitch S6012 shown in Fig. 9-21. This pulls in RY-2 and powers the main motor. The heads and capstan now turn, and the eccentric begins to move.
- S6010 is opened by the ring, and this stops the DC motor at the point where the tape is fully threaded. The contacts of this switch then apply the ground to Q1 and Q2 in Fig. 9-22. This allows the **stop** button to become active again.

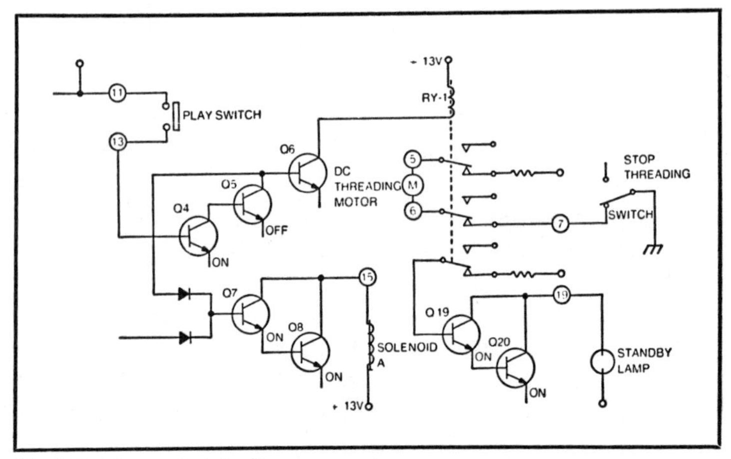

Fig. 9-20. Play control circuit, sequence 1.

During the unthreading cycle, it is desirable that the **play** button be inhibited; Figure 9-23 shows the circuit for this.

Auto Stop. This is used to end all modes of operation when the end of the tape is reached; in some cases it is used to initiate another mode after the automatic stop. For example, at the end of a playback a full tape rewind is initiated, but at the end of a rewind only a stop is required. Although the circuits are quite simple, the logic interconnections are somewhat complicated and differ from machine to machine.

Most of these modes are initiated by a small lamp which throws light onto a photosensitive element when the clear plastic

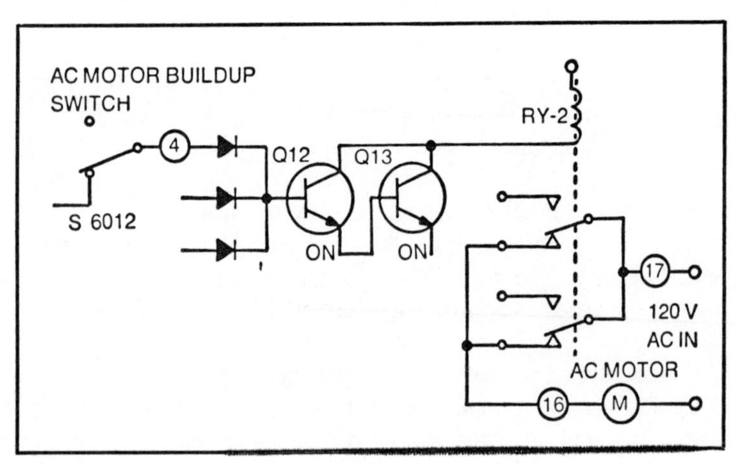

Fig. 9-21. Play control circuit, sequence 2.

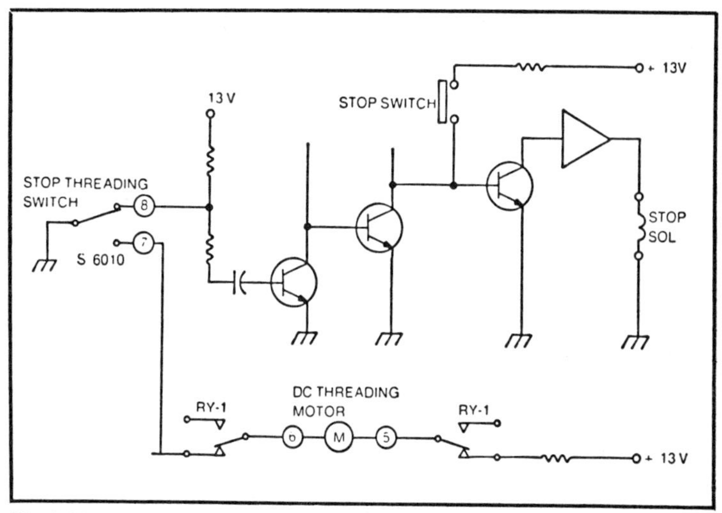

Fig. 9-22. Play control circuit, sequence 3.

leader at the end of the tape passes between them—normally the opaque tape prevents the light from falling on the photo device.

Figure 9-24 shows the circuits for the automatic stop. The light falls on the phototransistor, and its output pulse is amplified to

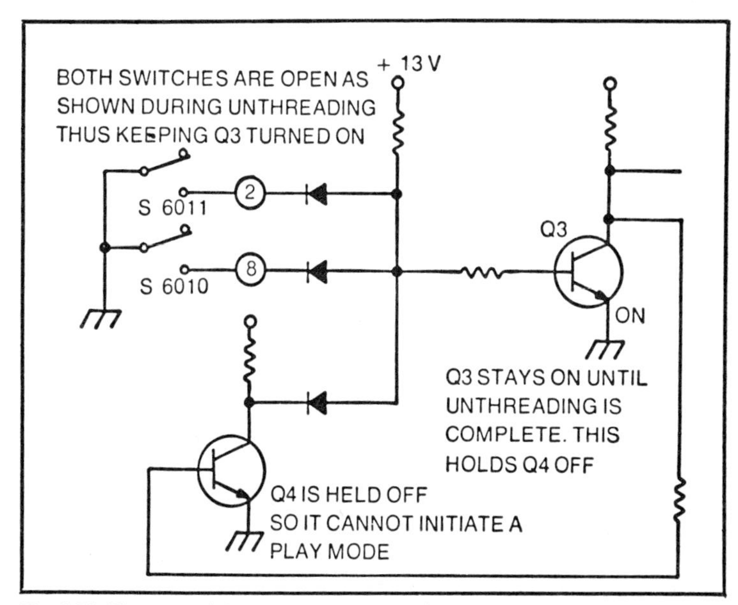

Fig. 9-23. Play control circuit, unthreading mode.

the point where it triggers the "one shot." The output of this is then used to start the various mechanical functions needed to stop the tape, release the pressure roller, rewind the tape, and also lock out those functions which would damage the tape if used at this time.

Note the capacitor C in the circuit. This shows that the light pulse operates the one shot on an AC basis and explains why a continuous light on the transistor does not work. This is why if the **rewind** button is pressed and the tape rewinds fully and automatically stops, the auto stop will *not* work again if the **rewind** is inadvertently pressed.

Transistor Q3 controls the **stop** solenoid through an amplifier, and it is switched on from two parallel sources Q31 and Q32. This is why the **stop** button need not be used in the **play** mode, because eventually the auto stop will provide that function.

Pause. The **pause** mode is very important in videocassette operations, and in those machines where this facility is provided it is very complex and does several essential things inside the machine:

- The main AC motor is stopped, so the capstan, heads, and take-up reel cease to rotate.
- A solenoid is depowered, thus engaging the supply reel brake, releasing the tension arm, and disengaging the pressure roller.
- Another solenoid is energized to engage the take-up reel brake.

This provides other useful features for machine and tape protection:

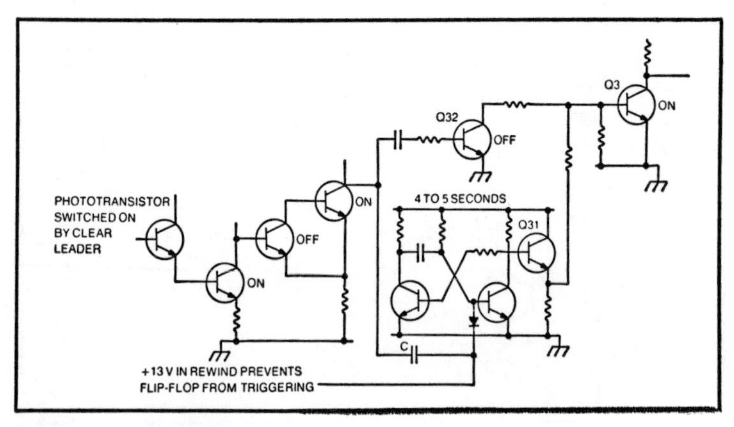

Fig. 9-24. Auto stop circuit.

● If the **pause** button is used while the tape is stopped, then when the **play** button is pressed the tape will thread and then stop in the **pause** mode.

● After the tape is run out and threaded, a timed inhibit circuit prevents the pause from occurring for at least two seconds.

● When the **pause** is used, 0.5 seconds later the DC threading motor is driven for a very short time in the threading direction. This ensures a small amount of tape slack to prevent the tape from sticking to the head drum. Note that this is one of the differences between **pause** and **still frame.**

● The circuit also ensures that the **pause** is turned off or released; it cannot be used again for another two seconds.

MUTING. During the threading and unthreading and the times when the tape is coming up to speed, the audio and video outputs are muted.

These examples were taken from the early Umatics, and they demonstrate the principles involved. The broadcast Umatics have more complex circuits mainly because they have more functions and mechanical parts to control, but the ideas are the same.

An interesting system control is that of the consumer Betamax. Most of it is contained in a single IC, with diode gates and driver circuits made up with discrete components. See Fig. 9-25.

When the cassette is inserted the cassette- in switch closes and provides 12v to the circuit. This causes the dc motor to turn and thread the tape. The thread stop switch is operated by the threading ring and it breaks power to the motor at the completion of threading. The thread control circuit is basically a delay circuit to protect the tape and the mechanism. Normally threading is completed in about 3 seconds. If the mechanism jams, the charge on the capacitor at pin 20 builds up and depowers the threading motor and operates the stop solenoid after 10 seconds. The supply reel brake is released during threading and unthreading when the 12v is applied.

During the two FWD modes - **play** and **rec** - the **pause** can be used. This mechanically pulls the pressure roller away from the tape to stop its motion. It also reapplies the reel brake by operating the solenoid. The pause function also starts the timer circuit, which pulls in the stop solenoid after 2 to 3 minutes to ensure the safety of the machine, the heads and the tape.

The drum rotation sensor works when the cassette is in and when the heads are rotating. The 30 Hz square wave from the head servo is turned into dc which keeps the stop solenoid circuit active.

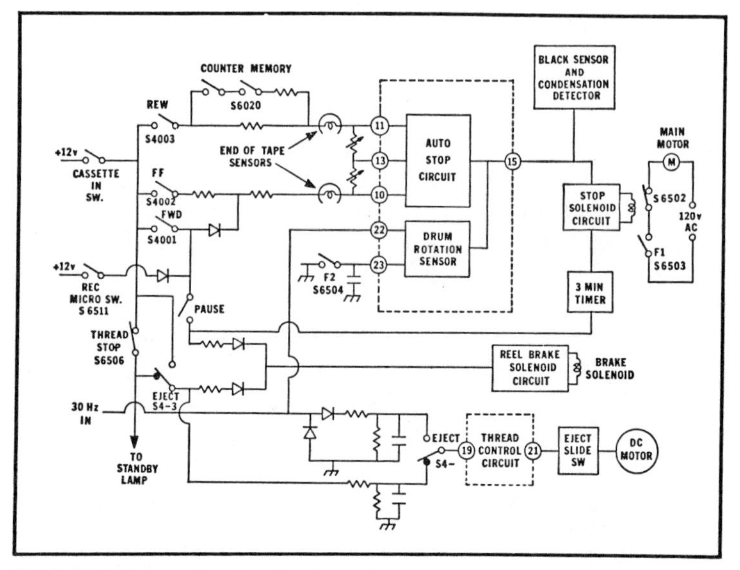

Fig. 9-25. Betamax system control circuit.

If the heads stop turning (broken belt, etc.) then the 30 Hz square wave disappears and the dc drops to 0v. This pulls in the stop solenoid.

FF and **rew** energize the auto stop oscillator through their respective sensing coils. When the metal foil at the end of the tape is pulled out of the cassette it detunes the oscillator as it passes close to the coil. The oscillator output drops and this loss of signal is used to pull in the stop solenoid. The **rew** button powers the oscillator through the coil placed near the other side of the cassette, this senses the beginning of the tape after a rewind. This also contains the counter memory circuit, which stops the oscillator and pulls in the solenoid when the counter reads 999.

The tape slack sensor and the condensation sensor are external circuits which operate the stop solenoid driver directly.

Physically the stop solenoid opens a microswitch to break power to the main ac motor, and it pulls the stop bar in the function button mechanism to release all the function keys.

10
Servos

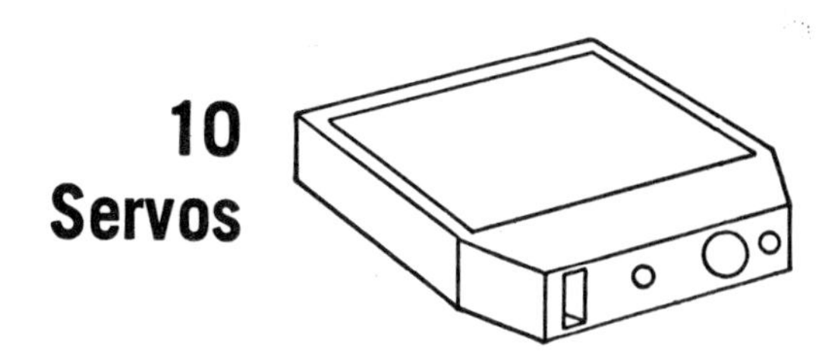

A servo is an electronic circuit that controls some piece of mechanical or electro-mechanical machinery. The rotation of an electric motor is a common example, and as this is the main use for a servo in a VTR; this chapter will be limited to electric motor control.

In a VTR the rotating video heads must scan the tape at a constant speed, and the scan must start at an exact point in the video signal. The only way to guarantee this happens is to servo control the heads.

The tape must be driven at a constant speed, and for this a constant speed motor is good enough. But for editing, the tracks on the tape must be aligned or phased with the incoming video, and so a capstan servo is needed.

Servos are essential to VTRs, and this chapter will briefly cover their principles, and introduce a few simple circuits from current models to illustrate these ideas.

GENERAL PRINCIPLES

For the purposes of a VTR a servo can be considered a circuit which controls an electric motor. The output can be either an AC or a DC, but this discussion will be limited to DC as this is the most common in cassette machines.

A servo needs two inputs. A feedback from the motor is needed to indicate how fast it is turning and its angular position at any time. And a reference input is needed to compare with the feedback to tell if the motor is fast or slow, and if its position is correct. The two inputs are not usually clean pulses, and so multivibrators are used to produce sharp pulses to feed to the comparator circuit. Figure 10-1 shows this.

One input is now made into a narrow pulse, and the other into a ramp. The pulse then 'samples' the ramp. The output of the sample gate or comparator is a pulse which has the same width as the sample pulse, and with an amplitude equal to the voltage level of the ramp at the sample point. See Fig. 10-2.

The output pulses are now integrated to dc by a capacitor, and this dc drives the motor. Variations in motor speed cause the ramp to be sampled higher or lower, resulting in different dc output levels. The DC then corrects the motor speed.

In some motor control circuit, 2 servo circuits or 'loops' are used. One is for speed control and the other is for position or phase control. The outputs both feed the motor drive amplifier; or, the phase servo output is used to refine the control of the speed servo, while the speed servo output drives the motor. See Fig. 10-3A.

HEAD SERVO

All VTRs must have a head servo. In record the heads must be controlled so that the video tracks are written evenly and contain one field only. Usually the vertical interval is placed near one end of the track and it must be in this position in every track. In playback the heads must be controlled so that they begin their scan when the tape is in the correct position around the head drum, and the video tracks already on the tape must be followed accurately by the heads. Slight variations are known as 'mistracking' and produce picture disturbances.

In record the reference input for the head servo is the vertical interval of the incoming video signal. The feedback information is taken from coils in the head drum. Metal vanes or magnets pass over these and produce an output pulse. Two of these PG or head tach coils are used in a two head machine. Although only one is required for servo control, both are used to tell the head pre-amplifiers when to switch outputs during playback.

The output of the servo is dc which controls a direct drive dc motor in the head drum, or a magnetic brake on the main head shaft.

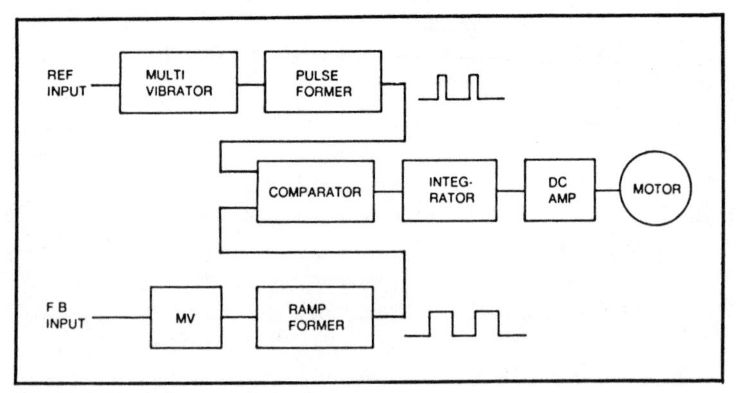

Fig. 10-1. Block diagram of a basic servo.

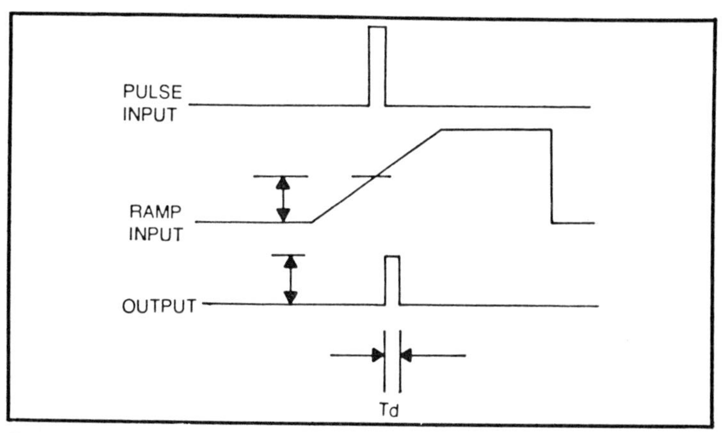

Fig. 10-2. Ramp and sample pulse.

In this last case, the heads are driven by a belt from the main ac motor in the unit, as in Fig. 10-3B.

The vertical interval is also shaped into a 30 Hz square wave and recorded onto the control track of the tape. These act like 'electronic sprocket holes' and are used in playback.

In playback the servo works the same, but the reference used can vary. In most small machines it is the CTL pulses on the tape. Occasionally the vertical interval from the tape playback may be used, but often a 30 Hz pulse is developed from an internal crystal oscillator or from the incoming ac power line.

In the simplest machines the head servo contains a tracking control in the playback reference circuit. This is a multivibrator which delays the reference input slightly. The amount of delay is varied by the tracking control on the front panel, which varies the width of the output pulse from the multivibrator. This enables minor mistrackings to be corrected by altering the sample point on the ramp slightly.

CAPSTAN SERVO

The simplest machines do not use a capstan servo, since a belt drive from the main motor will provide a constant tape speed in both record and playback.

In recording, when a servo is used, the reference input is again the vertical interval of the input video signal. The feedback is usually taken from a toothed wheel and a pick up in the capstan motor. This is usually called an FG coil. This produces a high frequency output which is formed into the ramp and sampled by the reference input.

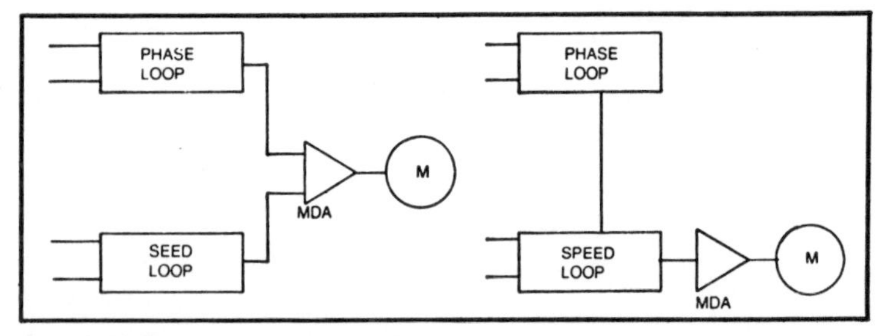

Fig. 10-3A. Two common servo systems.

In playback, the reference input is usually the CTL pulses, and the FG pulses are again used as the feedback input.

The capstan servo must do two jobs. First it must maintain a constant tape speed, and secondly it must place the tracks at the correct point on the drum during playback.

In editing a 3 way alignment is required. The incoming video, the heads, and the tape tracks must all be coincident in time and in position around the head drum, as in Fig. 10-4. This cannot be achieved without a capstan servo. Often a capstan servo is divided into two loops, one for speed control and the other for phase control.

SERVO FEATURES

Several interesting features are now being found in cassette machine servos. Both the Betamax and the portable Umatics have

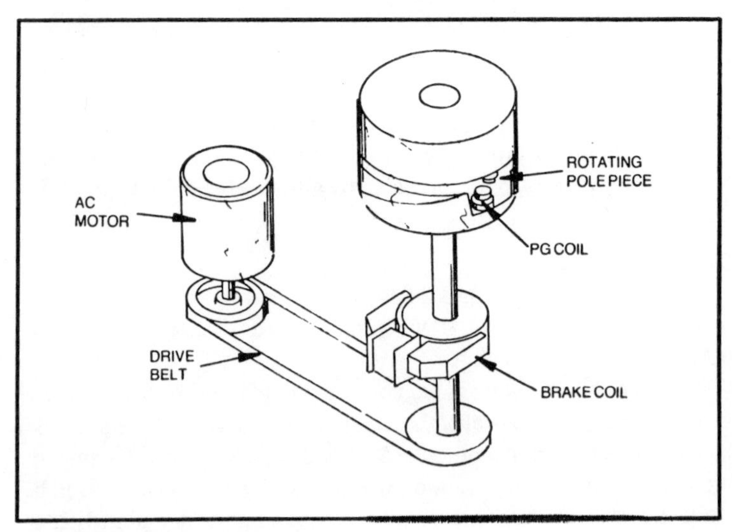

Fig. 10-3B. Simple belt drive for head.

multiple PG coils and vanes in the head drum to help avoid the 'gyro' errors. A tone wheel, or FG generator, is being used for feedback information and for speed control of capstans and heads. An injected frequency is used in direct drive dc motors for speed and phase control. And special servo ICs are being used in many machines.

The principles of each of these four items will be covered here. Circuit details will be kept to a minimum as these can be obtained from the service manuals of each machine.

Multiple PG Coils

Many of the small cassette machines are used in mobile situations, where they are subjected to excessive movement. This produces large 'gyro' errors in the head drum, which are erratic changes in the rotational speed. These can be severe enough to cause faulty tracks to be made during recording, which are too severe to be properly tracked during playback.

A method of overcoming this is to use several PG coils and vanes in the head drum, as in Fig. 10-5. These are all used as inputs to the head servo and they act several times per rotation to correct speed and phase variations. A single coil can only act once per revolution.

In Fig. 10-5 there are 6 magnets which pass over two closely spaced coils. Each coil now has an output of 180 Hz instead of 30 Hz which a single coil would produce. One coil output forms the sample pulse and the other forms the ramp. As they are spaced an exact distance apart, there is an exact timing difference between the coil outputs when the heads are revolving at the correct speed. Variations in speed produce an immediate difference in the timing which immediately causes the servo to correct the speed varia-

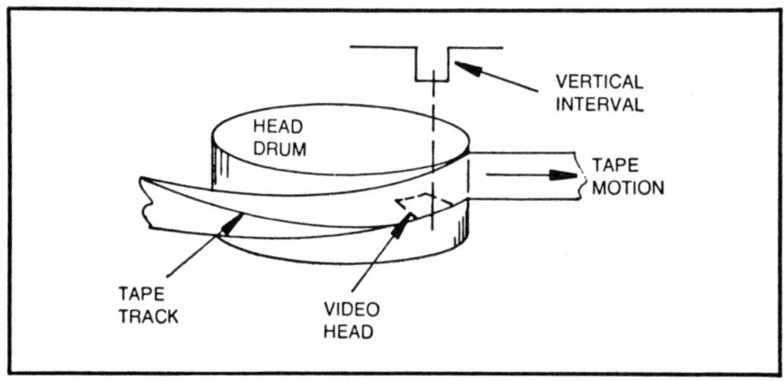

Fig. 10-4. Three way alignment of signal, head and tape.

tions. The result is a much more accurate speed control of the head, and a great lessening of the gyro errors.

Single Input FG Signal

The FG coil in a motor (usually in the capstan circuit) produces a high frequency output - high compared to rotational frequency of the motor. This is often used to form the ramp in the servo circuits, and it is sampled by the reference pulse.

However, in many cases it forms the *only* input to the servo and is fed into both the feedback and the reference inputs. It is then used to form both the ramp and the sample pulse. The circuit of Fig. 10-6 and the waveforms in Fig. 10-7 explains how this works.

The FG coil output is amplified to a square wave and applied to both inputs. This is waveform 1. The delay MMV is triggered by the falling edge, and it produces a constant width output pulse—waveform 2. The falling edge of this triggers the second MMV, which produces a constant width output pulse—waveform 3. The width of waveform 3 is set by the dc level control. The falling edge of this pulse starts the ramp, and the rising edge resets the ramp—wavefrom 4. The gate pulse—waveform 5, is also generated by the falling edge of waveform 1 and it samples the ramp.

If the motor speeds up, the input frequency rises, so the pulses in both waveforms 1 and 2 become closer together. But those in 2 remain the same width. Because of this, the time between the end of one ramp and the start of the next remains constant, but the length of the ramp shortens. Also, it is sampled lower.

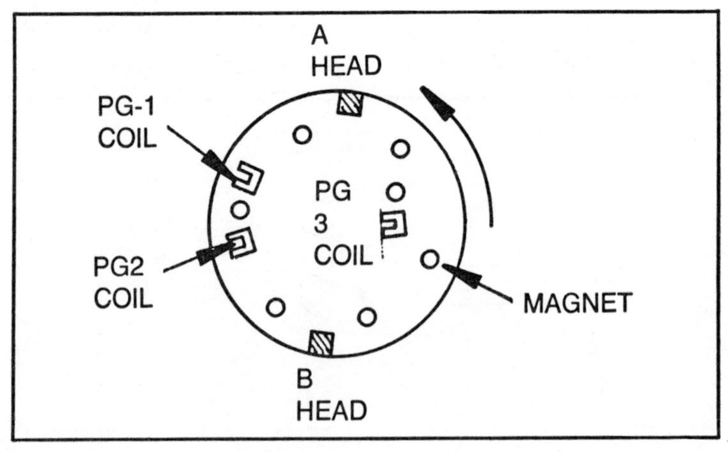

Fig. 10-5. Multiple coils and magnets.

Conversely, if the motor slows down, the ramp will be sampled higher.

The reason why this servo works with only feedback information as its input is because the reference is provided by the two internal constant width pulses instead of a separate reference input. Note that this type of servo control can only be used for speed control. For phase control an independent reference is essential.

Direct Drive DC Motor

These are finding increasing use in many VTRs for both head and capstan servo drive. They have the advantage of using electronic commutation instead of brushes, and so are easy to control and are noiseless. Figure 10-8 shows a sketch of the motor layout.

The coil former and the coils do not rotate. Three drive coils are used and each consists of two windings 180 degrees apart. Another 6 smaller coils provide positional and speed information. An 8 pole magnet is mounted on the outer ring, and this rotates with the position rotor in the center.

In order to apply the correct drive current to the main coils, the position of the magnetic ring must be known. This is indicated by the position rotor and the 6 position coils. A 65 kHz signal is applied to coils a.b and c. The 4 notches in the rotor produce amplitude variations or modulations of the signal which are picked up by the coils x, y, and z. The position of these variations is used to control the servo which drives the motor.

In a 2 head machine the head drum motor rotates at 1800 rpm, or 30 rps. With 4 indents on the rotor, each output signal will be amplitude modulated 4 times per revolution, ie, at 120 Hz.

Figure 10-9 shows the principle of the drive and control circuit. The position indicator coils provide a 65 kHz input with 120 Hz amplitude modulations to the position signal detector. This

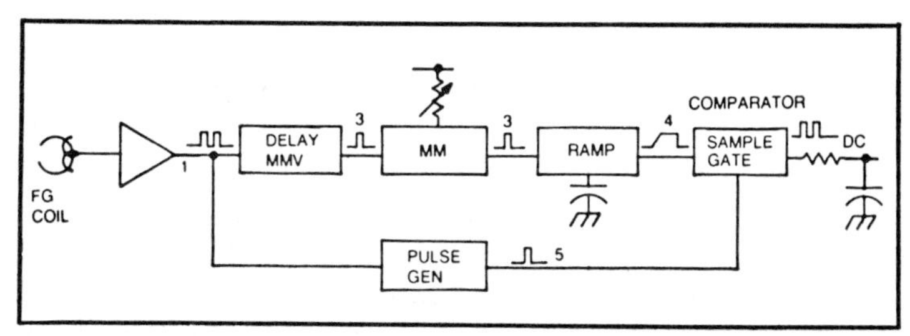

Fig. 10-6. Block diagram of a single input servo.

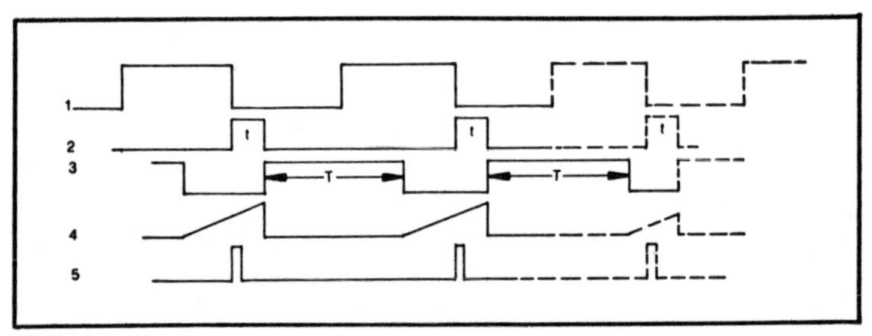

Fig. 10-7. Waveforms for the single input servo.

rectifies the 65 kHz to a varying dc or pulse waveform, which is applied to the switching circuit. The output of the switching circuit turns on 2 drive transistors simultaneously and keeps the other 4 off. This causes current to flow through 2 of the main coils. By selecting the correct two main coils at the correct time an even rotation is produced. The actual switching timing is controlled by the main servo circuit and is refined by the position coils. A feedback signal to the servo is also taken from the common emitter resistor of the drive transistors.

THE SERVO IC

An example of an IC designed specifically for servos is the Sony CX 143A, which is used in several of the cassette machines for head and capstan control. Figure 10-10 is a function block diagram.

It accepts 4 pulse inputs and provides a dc output which can be used to drive a magnetic brake or a dc motor, via a power amplifier. The dc output is provided by an op-amp with inverting and non-inverting inputs. The non-inverting input is driven by the speed loop and the inverting by the phase loop. Each loop has 2 inputs, but often the speed loop is used with only one input. The circuit layout of each loop is quite different.

The phase loop is the simplest. Usually the input to pin 17 is a square wave, which is amplified and integrated to a ramp by the capacitor between pins 18 and 20. The pulse input at pin 19 is buffered and it samples the ramp in the sample gate. The pulses out of the gate are coincident in time with the sample pulses, but their amplitude corresponds to the dc level of the ramp at the sample point. The capacitor at pin 21 integrates these pulses to a dc level. The dc output from the op-amp appears at pin 23 and provides feedback to pin 22. The function of the phase loop is to control small

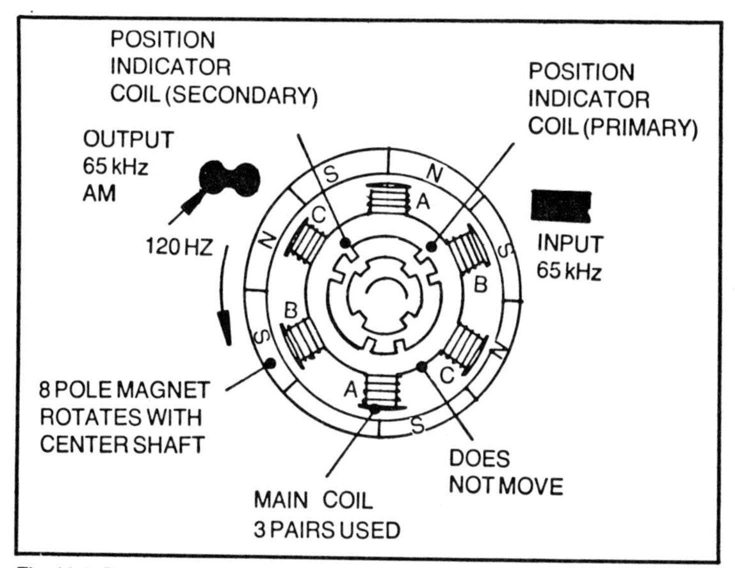

Fig. 10-8. Direct drive dc motor.

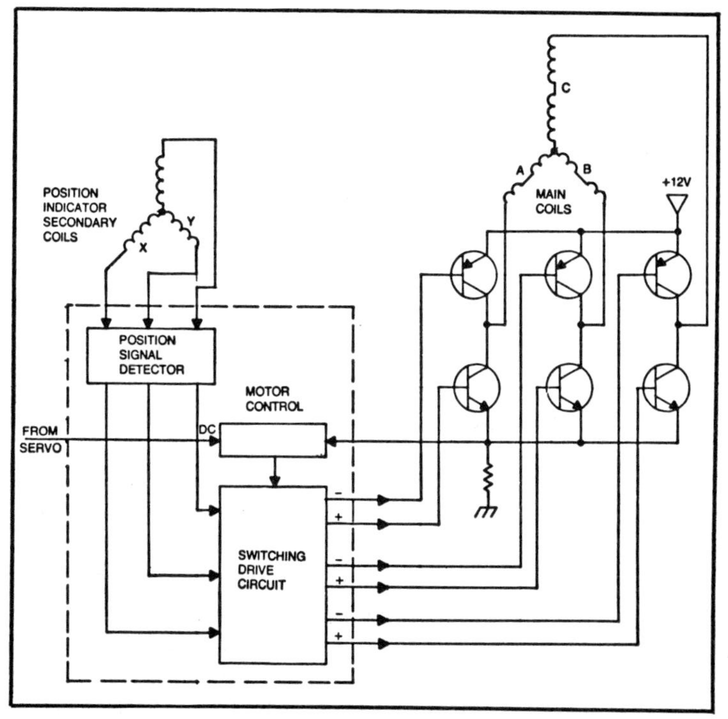

Fig. 10-9. Direct drive motor circuit.

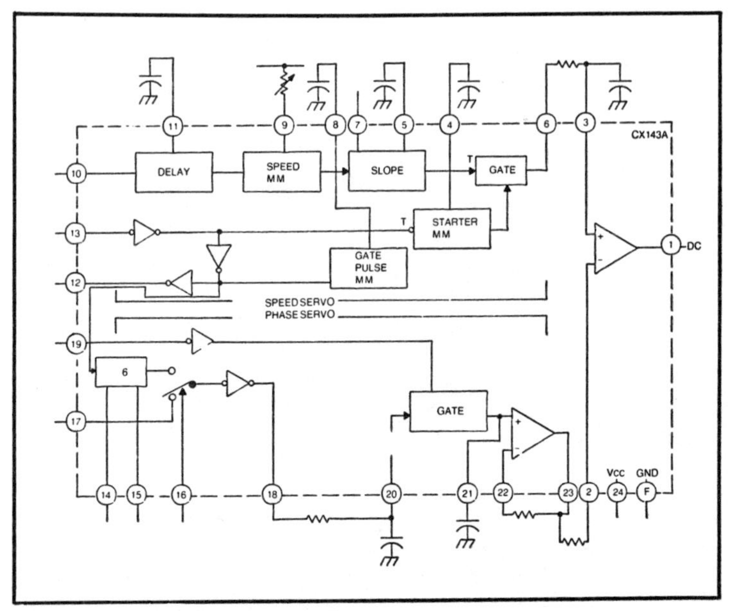

Fig. 10-10. Block diagram of servo IC.

variations in motor speed once it is up to its correct operating speed; and to ensure that rotational position is correct.

The speed loop is more complicated than the phase loop. A pulse input to pin 10 triggers the delay monostable, and the output width of this is set by the capacitor at pin 11. This feeds the speed monostable, and the dc level at pin 9 determines the width of the output pulse from this monostable. This output is integrated to a ramp waveform, and the slope of the ramp is determined by the capacitor on pin 7. The Pulse input to pin 13 is buffered and fed to the gate pulse monostable and also to an amplifier. The gate pulse monostable output is a narrow pulse set by the capacitor on pin 4. This pulse samples the ramp in the sample gate, and the output pulses are integrated to dc by the capacitor at pin 3. This dc provides the non-inverting input of the main op-amp.

At point A the input is buffered and split 3 ways. A buffer feeds an output to pin 12. A counter divides the signal by 6 and feeds this lower frequency to the phase loop. How these two paths are used depends on the actual servos in which they are used. The third path is to the starter monostable. The function of this is to keep pin 6 at or near at 11v until the motor comes up to speed. As the motor comes up to speed the capacitor at pin 6 charges and inhibits the starter and allows the servo to operate normally. These wave forms in Fig. 10-11 will explain its operation. The incoming signal is shaped to a series of negative pulses. The leading edge triggers the starter and causes its output to drop to 0 v. The capacitor on Pin 5 determines the time (T_d) that the output stays low. The output then jumps back to 11v until the next input triggers it low. This continues until the frequency increases, which is caused by the motor coming up to speed. Eventually the input pulses re-trigger the starter before it has 'timed out' and so its output stays low. When the output is low, the output impedance is very high. The charge on the capacitor at Pin 6 is no longer continually replenished and the charge and dc level is now determined by the output of the main servo circuit.

REEL SERVOS

In the broadcast Umatics the tape can be played in reverse at 1/20th, 1/5th, x1, and x2 normal speeds. This is used in setting up the tape during editing. During the reverse modes the supply reel is servo controlled to keep the tape tension constant, thus ensuring it maintains contact with the CTL-2 and time code heads.

An induction motor drives a puck which contacts the supply reel, in the reverse mode, and this motor is controlled by the reel servo. Figure 10-12 is a simplified block diagram. The tension arm, which is operated by the tape, has a shading plate which moves between an LED and a solar cell. The LED is driven by a 3.8 kHz oscillator, and the output from the cell is thus a 3.8 kHz sine wave whose amplitude depends on the position of the shading plate. Hence, tape tension changes appear as changes in level out of the cell. The cell output is amplified and rectified in the envelope detector to dc, where the gain and tension controls set the dc level. After buffer and phase compensation, the dc now controls the conduction of the motor driver.

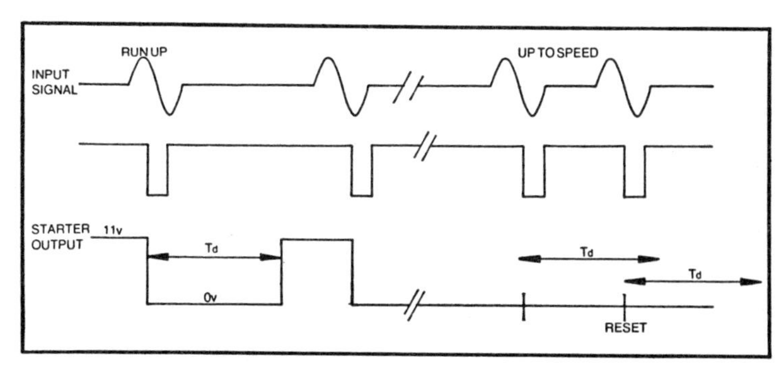

Fig. 10-11. Starter circuit waveforms.

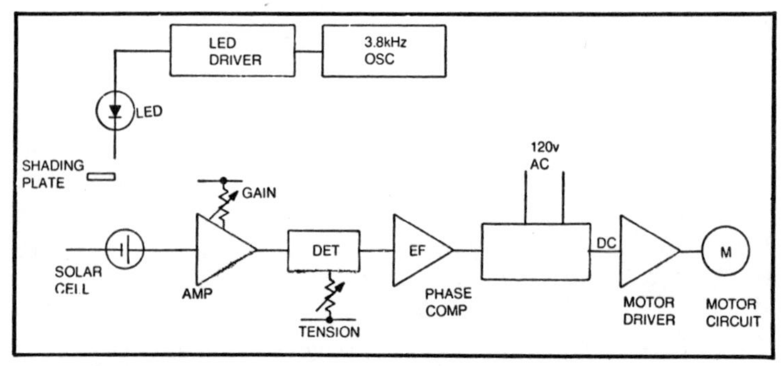

Fig. 10-12. Block diagram of reel servo.

The reel motor is driven by a 60 v AC signal, as shown in Fig. 10-13. When Point A of the transformer secondary goes positive then the current path is through D1, the main output transistor, D2, and then to the motor and to Point B of the secondary. When Point B goes positive the current path is through the motor, D3, the transistor, D4, and then into Point A. The ac current is reversing normally through the motor, but flows in one direction only through the transistor, hence the amount of current in both half cycles is controlled by the single output stage.

HALL EFFECT DEVICES

The *Hall Effect* is a property possessed by semiconductors. The voltage across a semiconductor and the current through it are

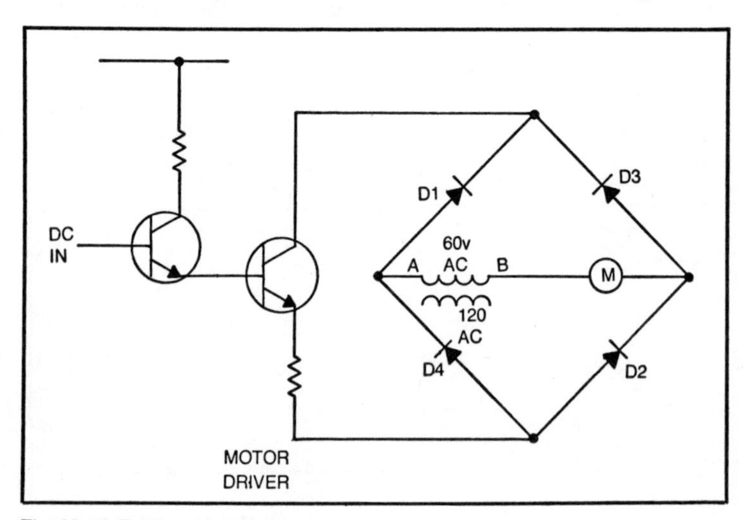

Fig. 10-13. Reel motor drive circuit.

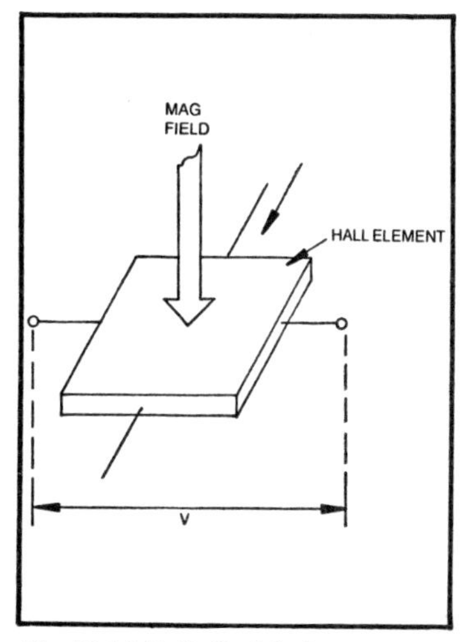

Fig. 10-14. Hall effect device.

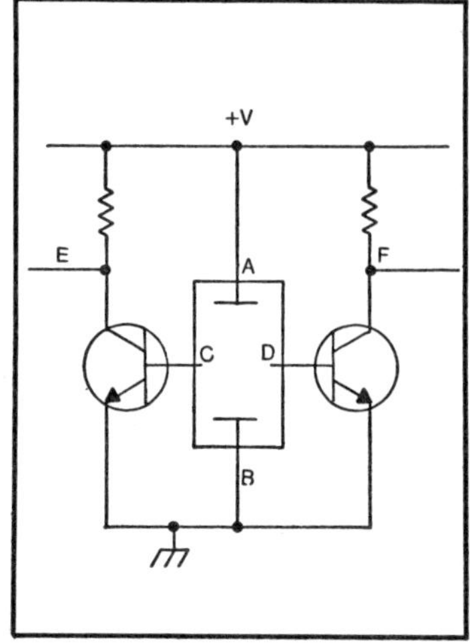

Fig. 10-15. Hall effect circuit.

both affected by an external magnetic field. This is known as the Hall Effect. If a semiconductor has a constant voltage held across two terminals, then the current flowing through the semiconductor between the current terminals can be varied by an external magnetic field. Note that the voltage, current and magnetic field are all mutually at right angles to each other, as in Fig. 10-4. Conversely, if the current is held constant, then the voltage across the device will vary. Figure 10-15 shows an IC circuit which uses the Hall Effect. By holding a constant voltage between A and B, a constant current will flow between these two points. Changes in magnetic field, which occur perpendicular to the page will cause a slight voltage difference between points C and D. The differential amplifier now amplifies these slight changes to produce an output voltage at E and F.

A small magnetic field changing in sine wave fashion will produce a sine wave output. This is found in dc motor controls. A single magnet passing close to the IC will produce a pulse output. This is used in simple motion sensing. The VHS uses a simple Hall Effect IC to sense the motion or rotation of the take up spool.

11 The One Hour Betamax

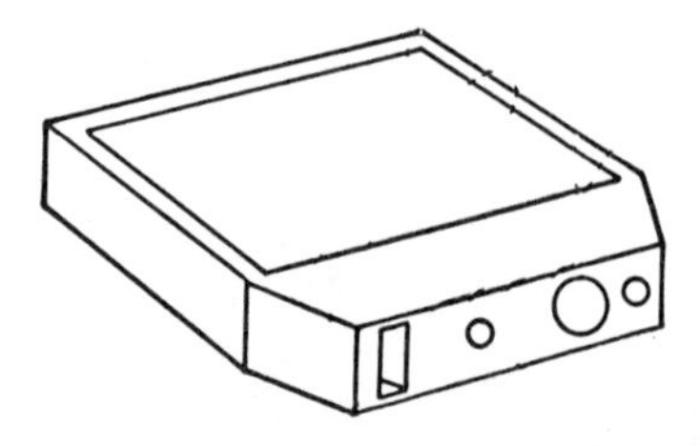

Although the earlier videocassette machines were intended to enter the consumer market, they did not succeed until the introduction of the Sony Betamax. The first model was the SL 6200, followed by the SL 7200 and 7200A.

The distinguishing features of these machines are their small size, the very small cassette, and the completely automatic nature of the tape loading and other functions. The small size was achieved by some radically new ideas in tape format and electronic processing of the signal, and the introduction of specially designed ICs which contained many of the standard circuits required in VTRs.

Although these earlier machines were later superseded by the long playing and dual speed machines, these later models are mainly variations, modifications and additions to the basic ideas found in the one hour Betamax. So, the one hour Betamax provides a good starting point for understanding the other machines. Also, the later industrial and educational models of the Betamax are one hour machines with an improved mechanical system. These are briefly described at the end of this chapter.

Some of the features which make the Betamax so attractive are:

- Their small size and low cost.
- The small and inexpensive cassette.
- Easy off-air recording. They can be used with an ordinary TV set with no modifications required.
- A timer permits recording when away from home.
- It is possible to record one program while viewing another.
- Microphone and camera inputs are provided, so home movies are possible.
- All the important controls are automatic.
- Automatic selection of monochrome or color.

Very important features are the automatic tape threading when the cassette is inserted, the simple function buttons, and the complete lack of controls and dials. These make the machines

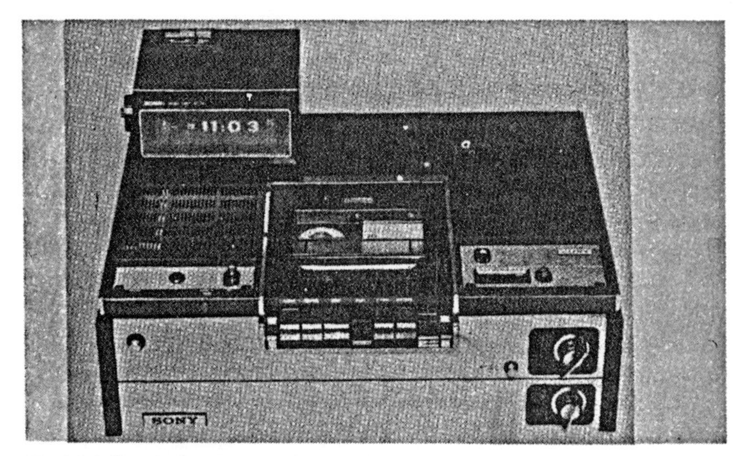

Fig. 11-1. Sony's Betamax model SL7200 with timer.

extremely easy to use and the user does not feel that he has to be a trained engineer. The cassette is about the same size as a paperback book, and in fact looks similar, thus helping to make the whole cassette system fit into the home. All the models are produced in simulated wood cabinets and look like a piece of hi-fi furniture suitable for the home, rather than a piece of electronic equipment removed from a TV studio. See Fig. 11-1.

THE BETAMAX CASSETTE

The cassette looks like a Umatic but is much smaller and has a few other differences. The photo in Fig. 11-2 compares the two. The tape is mounted on two double flanged reels which do not overlap. These are pushed down against the bottom of the cassette by leaf springs, which internally brake the reels whenever the protective door or flap is closed.

Upon insertion the protective door flips up to expose the tape and release the reels for threading.

A plastic tab is used to protect the cassette against re-recording or erasure. A small plastic window exposes to view part of the supply or feed spool, thus affording a check of tape motion while in the machine; or indicating the need for rewinding the tape.

THE BETAMAX FORMAT

The main characteristics of all the consumer machines are their compact size, small cassette and their great economy of tape. This section will decribe how this was achieved.

The most obvious way to save tape is to run the tape at a slower speed, and to use thinner heads to produce narrower tracks.

In all previous VTRs the video tracks on the tape have been separated by a guard band, as in Fig. 11-3. This has always been required to prevent crosstalk between the tracks, and interference in the picture due to mistracking. Mistracking is the head crossing or scanning two tracks instead of one.

In the Betamax the guard bands have been eliminated and the adjacent tracks are touching, as in Fig. 11-4. New techniques have been introduced to eliminate the crosstalk and tracking interference and to produce a clean color picture. A very small sized head is used to produce the narrow tracks and a further refinement is the use of a very small head gap - in the range of 6 microns or 28 microinches. This has allowed a very low writing speed while maintaining a fairly good frequency response. In the Betamax the head gap of 6 microns allows an FM frequency range of 3.5 to 4.8 MHz to be used with a writing speed of approximately 272 inches per second. This has allowed a very small head drum with a diameter of about 3 inches to be used, thus further reducing the size of the machine.

The effect of the zero guard bands, narrow track width and the low writing speed has been to permit a low linear tape speed, a small head drum size, and a short tape path to be used. These in turn have allowed much easier tape handling and thus a thinner backing to be used for the tape. The overall result of these factors is a very small cassette which uses a minimum of tape to provide 60 minutes of program time.

Fig. 11-2. Comparison of Umatic and Betamax cassettes.

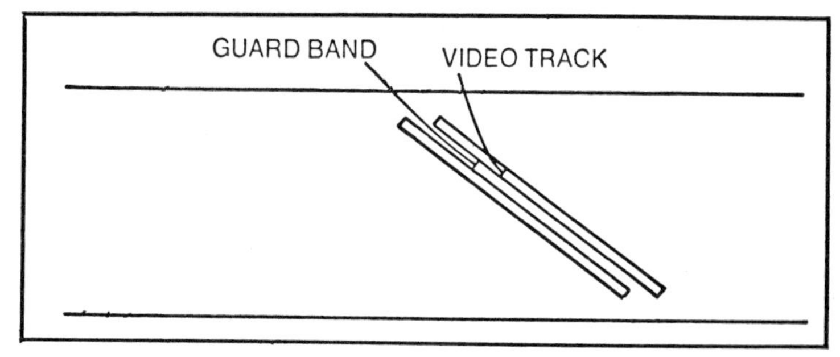

Fig. 11-3. Simplified diagram showing video tracks separated by a guard band.

HEAD AZIMUTH

Since the tracks are touching, the heads will occasionally stray slightly over the adjacent track during playback, this is known as mistracking and is quite common on all machines. Mistracking introduces crosstalk from the adjacent tracks, which shows up as an interference in the picture. To obviate this mistracking noise in the Betamax, a technique known as 'azimuth recording' is used. This is not a new idea, but the technology to use it only became available about the time the videocassettes were first introduced.

Anyone who has serviced an audio recorder knows the importance of having the head gap at 90 degrees to the direction of tape travel. The same is true in video, and in all previous VTRs the video head gap was at 90 degrees to the video track, as in Fig. 11-5. With a small azimuth error, the high frequency losses become very noticable. But if a track is recorded with a slight azimuth offset and

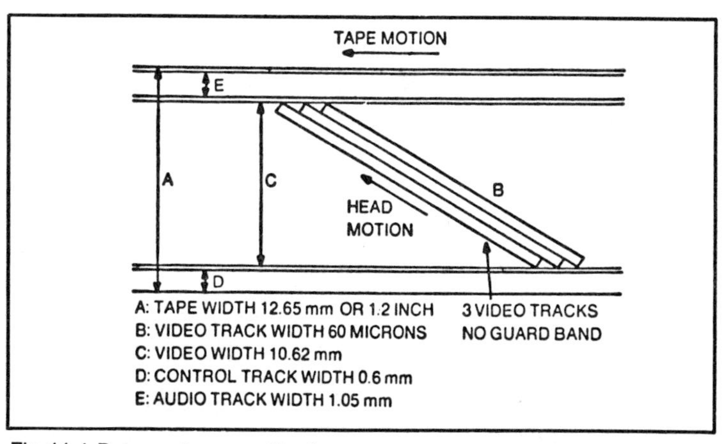

Fig. 11-4. Betamax tape specifications.

then played back with the same offset, the results are just as good as with a perfect 90 degree azimuth. This is true only for small angles.

The signal loss with an azimuth error is more serious at high frequencies than low ones.

Figure 11-6 plots the signal loss in dbs against the azimuth angle, and the separate lines are for different frequencies. Those shown here are typical for a small helical VTR, and are for a head gap of about 0.6 microns. With a very small azimuth difference between record and playback a large loss of signal occurs. For example, a 2 degree difference produces about 12 db of loss at 3 MHz. A 12 or 14 degree difference produces a loss that is so large it almost cannot be measured. At a frequency of about 650 KHz, less than a db is lost with a 2 degree offset. A difference of 12 to 14 degrees produces about 12 db of loss.

This shows that the azimuth offset in the video heads will easily render the FM signal on a track invisible to the head which did not record it. However, although the lower frequency chroma signal is attenuated, it is by no means lost, and other methods must be adopted to get rid of the chroma crosstalk.

In the Betamax the video heads are set so that the A head has an azimuth angle of +7 degrees and the B head is −7 degrees. Hence, the total azimuth difference beween the tracks is 14 degrees. If one head strays onto the other track the azimuth

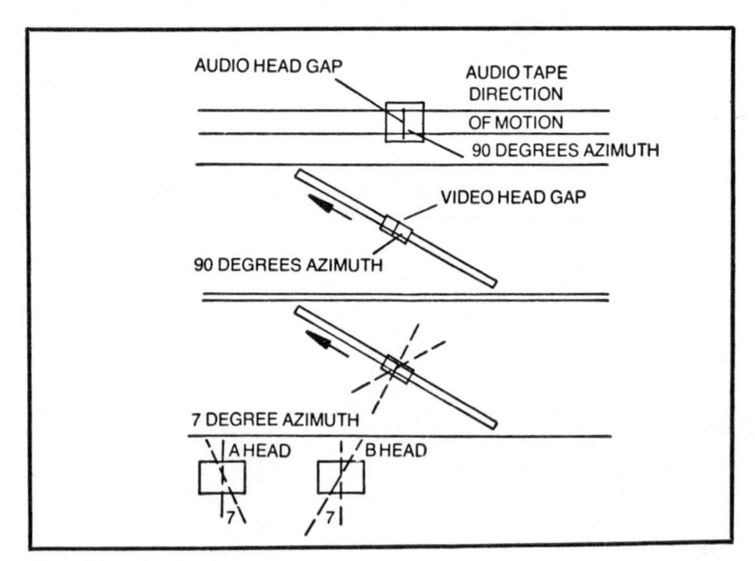

Fig. 11-5. Azimuth alignment.

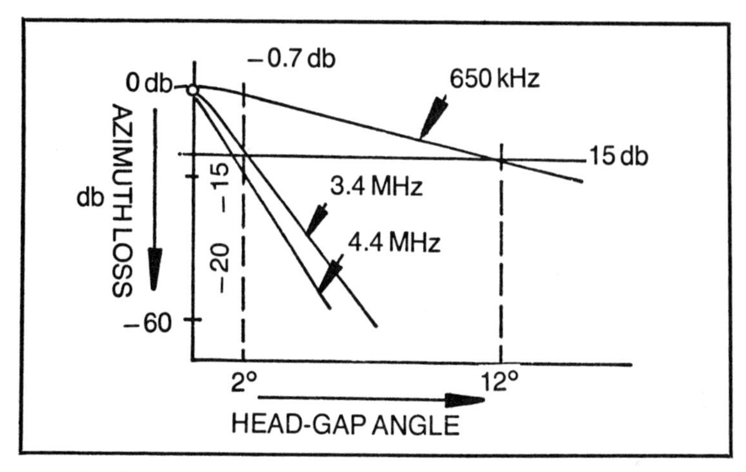

Fig. 11-6. Characteristics of the azimuth loss.

misalignment is enough to totally eliminate the crosstalk of the FM signal. In effect, the adjacent tracks appear as guard bands as far as the luminance FM is concerned.

Each head must play back the track it recorded because of this azimuth offset. This is achieved by the physical placing of the head tach or PG coils and the electronic circuits, which identify each head and cause it to scan only its own track.

The chroma, which is heterodyned down to a carrier frequency of less than 1 MHz, is too low to be ignored by the head which did not record it, and so chroma crosstalk between adjacent tracks occurs on playback. To eliminate this a complicated and unusual system of altering the chroma phase is used. This phase alteration is introduced into the recording process when the 3.58 MHz is heterodyned down, and the phasing is corrected back to NTSC when the color under subcarrier is heterodyned back up to 3.58 MHz.

The object of the phase changing is to get the crosstalk all of the same phase during the playback correction. This way the signal can be inverted and passed through a comb filter to remove the crosstalk while strengthening the main components. How this is achieved is covered in the chapter on color correction.

BETAMAX MECHANICS

The Betamax consists of two separate parts enclosed in the same case. The smaller part is the TV-tuner section. This is a V and U tuner, IF, demod and audio sections, as used in the triniron

TV sets. It provides a video and audio feed to the Betamax electronics for off-air recording.

The larger section is the actual mechanics and electronics of the cassette machine, which will record and playback on its own format of tape.

Mechanically the machine is quite simple: this simplicity is largely due to the half inch tape, the relatively short tape path, and the low transport speed. All of these permit easy and gentle tape handling.

The top cover lifts off by removing seven screws to expose the inside, as in Fig. 11-7. This does not interfere with any functions, and allows the mechanism to be viewed while working. It should normally be kept on at all times.

Figure 11-8 shows the tape threading path, direction of tape travel, and the direction of rotation of the heads, which are the opposite to that used in the U-matics.

Two motors are used. The main motor is a 60 Hz synchronous motor, which provides belt drive to the capstan, head drum, and the tape reels in the cassette. A smaller dc motor is used to drive the threading ring, which belt drives a small toothed wheel that engages the rim of the threading ring.

The threading mechanism is much simpler than that of the Umatics, with a much shorter tape path. The metal ring has teeth

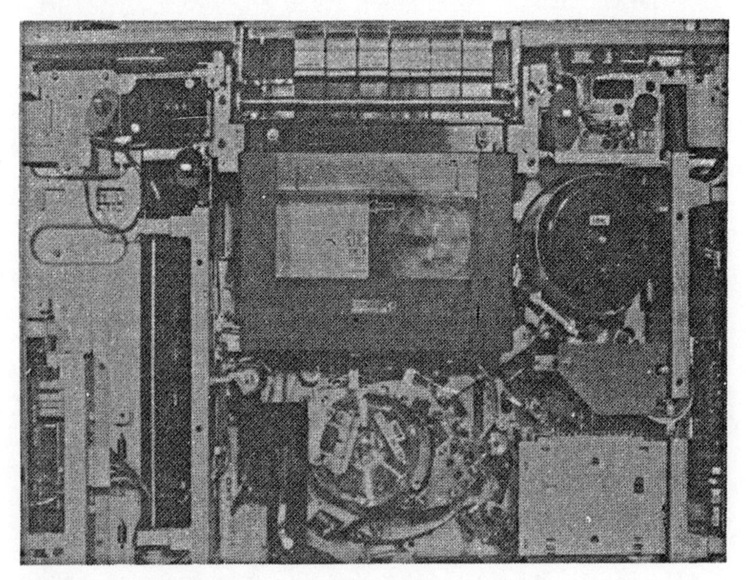

Fig. 11-7. Top view of Betamax, cover removed.

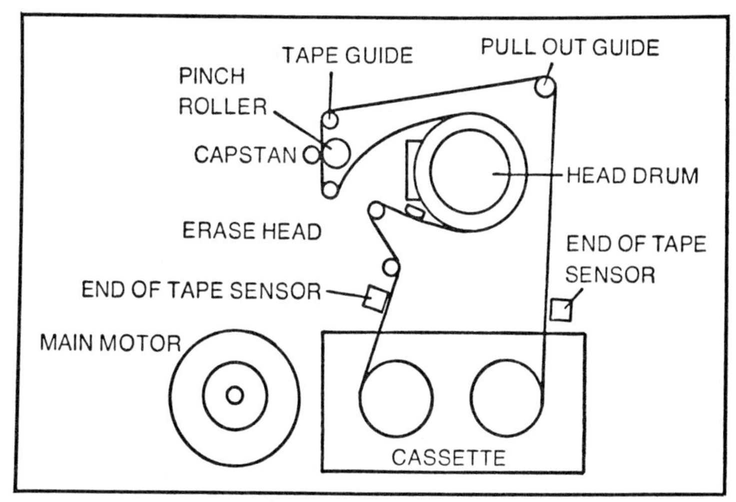

Fig. 11-8. Betamax tape threading path.

which mesh with a small gear wheel. This is driven by a dc motor to provide both threading and unthreading. During threading a pin on a double jointed arm pulls the tape out of the cassette. This pin is positioned by a plastic guide to provide a tape path which has tape guides only at the entrance and exit of the head drum, around the drum, and past the audio/CTL head stack.

The threading ring has notches and indents on its outer rim which engage with moving bars and sliders under the cassette reel tables. This enables it to be locked in position before a cassette is inserted, and to operate the end of the threading switch and other mechanical parts as it rotates. The rotation of the ring is guided by three small white plastic rollers, which provide a minimum of surface contact.

During threading the supply reel is unbraked to allow tape to be pulled out. To retract the tape back into the cassette during unthreading, the take up reel is driven by an idler.

The tape is not retracted back into the cassette for **ff** and **rew**, these are conducted with the tape fully threaded around its operating path.

The main function buttons are mechanical interlock types with a solenoid mounted underneath for automatic stop. The stop operates when the end of the tape is reached, the head drum stops, the power fails, or excessive condensation occurs on the heads.

Figure 11-9 gives the layout for the main sections of the machine. The head drum is in two stationary parts with the heads mounted on a rotating disc inside. The heads protrude through the gap between the two halves to make contact with the tape. See Fig. 11-10.

Also mounted on the head drum are the lower tape guide and the condensation detector; two springs on the upper part help guide the tape.

The heads are mounted on a disc which is attached to the central spinning shaft by two bolts; it can only fit one way, so it is impossible to mount it incorrectly.

The head drum is much smaller than that in most other cassette systems. Because of this, the linear speed of the head tach poles over the tach coils is too low to produce a useful output. This has been obviated by using coils wound around permanent posts, which vastly increases the output from the coils.

Two separate coils and pole pieces are used, and each rotating vane produces a pulse from one coil only. In this way the two heads are uniquely identified by the servo and head switching circuits.

Tape-end sensing is magnetic, not optical. Two sensors are used, one for each end of the tape. These are small coils placed as shown in Fig. 11-8. They are part of an oscillator circuit which is detuned by the aluminum foil at the end of the tape to initiate auto stop. Part of this same circuit is also the tape memory counter switch.

The very simple system control uses two solenoids only. One is the stop solenoid, and the other is a braking solenoid.

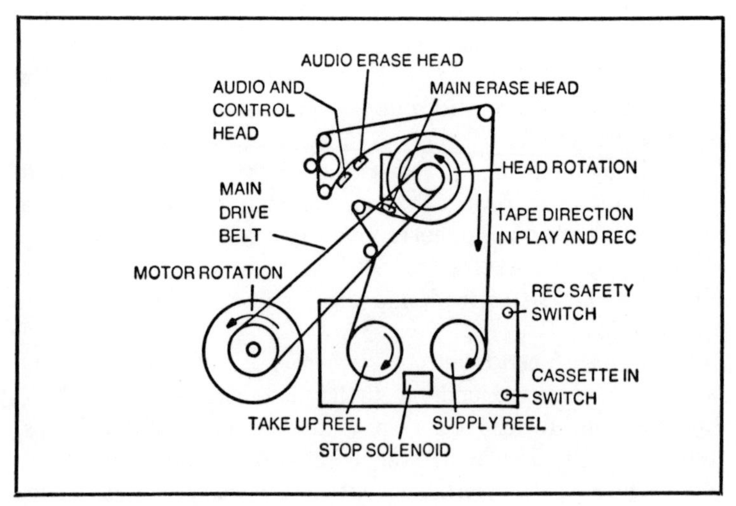

Fig. 11-9. Simplified pictorial diagram of the device.

BETAMAX ELECTRONICS

The basic record and playback electronics of the Betamax are very simple, and very close to the circuit block diagrams shown in Chapter 9, Figs. 9-4 and 9-11. The Betamax uses many of the special ICs and this makes the circuits very easy to follow and to service.

The main difference in the Betamax when compared to previous machines are found in the chroma circuits. Here the 4.27 MHz heterodyne frequency is not formed a crystal oscillator, but a phase locked loop (PLL) type of circuit is used. The reasons for this are to enable satisfactory copies to be made from other machines, and to provide an easy method of eliminating the color timing errors and the color crosstalk.

Here the basics of the circuits will be covered, and the details of the chroma correction will be covered in Chapter 14.

Luminance Recording

The luminance component of the video signal is treated exactly as in any other video tape machine. It is made to frequency modulate a high frequency carrier. The main difference is that the frequencies used are 3.5 MHz for the sync tips and 4.8 MHz for peak white. These frequencies are different from the U-matic's and the deviation is less.

The incoming video is split into two main paths. One goes directly to the luminance circuits and the other through a 3.58 MHz

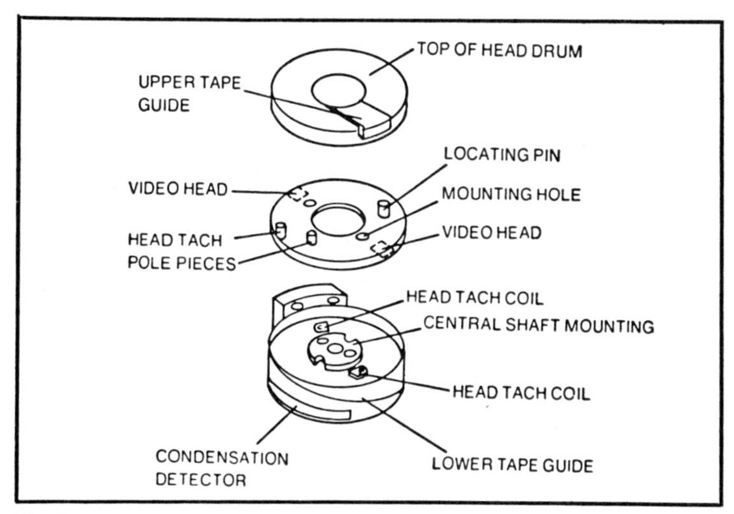

Fig. 11-10. Head drum, exploded view.

bandpass filter to the chroma circuits. A full NTSC signal is applied to the luminance circuits where it is automatically gain controlled (AGC). The AGC circuit has two outputs. One is fed to the output of the machine and serves as an E-E signal in the record mode; the other path is through a 3.58 MHz filter and into the modulator and sync separator. The FM output from the modulator is then applied to the heads via a record amplifier, while the sync is used to control both the AGC and the chroma circuits.

Luminance Playback

The output from the rotating heads is fed through rotary transformers to the playback preamplifiers. Then follows a path similar to that of other machines.

It is separated first into the luminance and chroma components. The FM luminance signal is limited, demodulated, amplified, delayed, and finally has the recovered 3.58 MHz chroma added to become a color video output from the machine. During this process the sync is stripped from the final luminance signal, and is used in the chroma playback circuits.

Figure 11-11 is a simple block diagram of the luminance path.

The E-E Circuit

The E-E signal at the output of the machine is not a true E-E signal, as it has not passed through the full modulation and demodulation process found in other machines. One output of the AGC is fed to the output video amplifier through a 3.58 MHz filter. This allows the input signal to be monitored on either a TV set or a TVmonitor.

The purpose of the filter is to reduce the chroma level to the TV set. If a distant station is picked up, the color on a TV set may be weak and noisy. To prevent recording an inferior color signal the Betamax will record a weak signal in monochrome only. While recording, the TV set would display a noisy color picture, then later playback a monochrome picture. This will cause the viewer to think the Betamax machine was faulty. To prevent this erroneous conception, the filter in the E-E path prevents the weak color signal from reaching the TV set during recording. So in both record and playback modes, the TV set displays a monochrome picture only.

The Chroma System

The method of recording and playing back the color information makes the Betamax unique. Several very new and different ideas have been used—necessitated by the zero guard band technique. The head azimuth offset will not satisfactorily remove cross talk at the lower frequencies of the own converted chroma, so electronic methods have to be adopted for this.

In this section the basic signal path of the chroma component is covered in both the record and playback modes (the principle of the chroma treatment is given in a later section).

The 3.58 MHz is converted down to 688 kHz, as in the U-matic, but two new ideas are incorporated into this conversion: (1) The 688 kHz is phase inverted every other line in one field. (2) The 688 kHz is phase locked to the incoming horizontal sync.

The first step to eliminate the cross talk is to invert the phase of the 688 kHz color signal every other line during recording, but to do this with only one field. This phase inverted signal is applied to the A head only; the B head receives a normal signal. On playback the phase alternation is restored before the recovered 3.58 MHz is added to the luminance component. The significance and method of achieving this are explained later.

The 688 kHz is produced by heterodyning the incoming 3.58 MHz with a stable 4.27 MHz signal. But in the Betamax, unlike previous machines, the 4.27 MHz is phase locked to the incoming horizontal sync pulses. Thus, the resulting 688 kHz is also phase locked to the incoming horizontal sync.

This has the advantage of improving the time-base correction on playback and allows a "coherent" or "phased" color signal to be recovered from the tape. Both these ideas are effected simultaneously in the same circuits.

Chroma Recording

The incoming 3.58 MHz chroma component is separated from the input video signal, converted down to 688 kHz by heterodyning

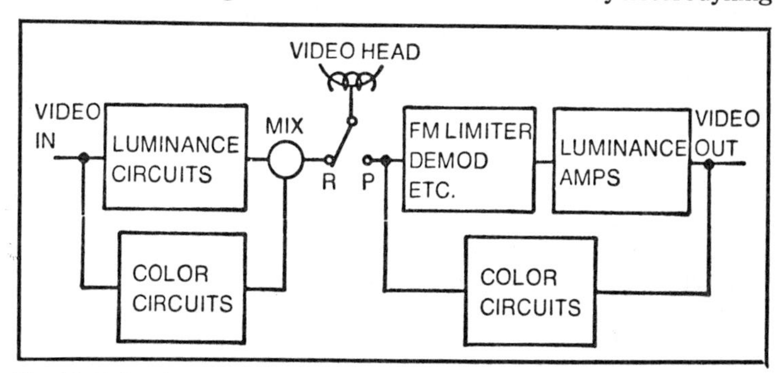

Fig. 11-11. Luminance path block diagram.

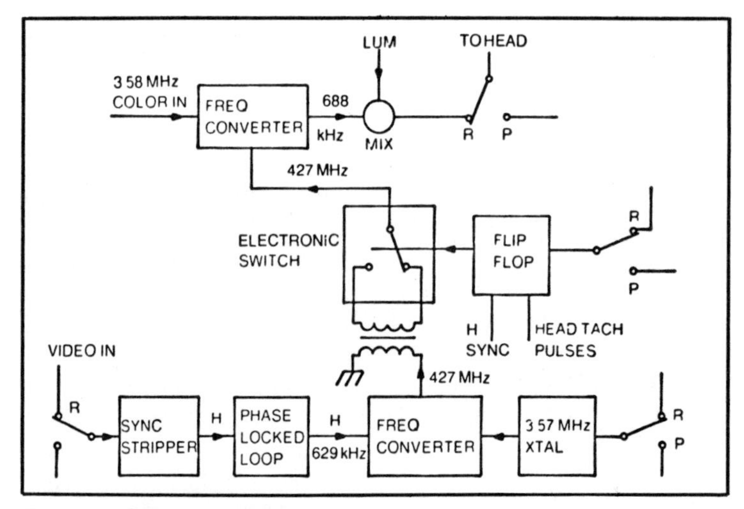

Fig. 11-12. Chroma recording circuit.

it with a stable 4.27 MHz signal, and added to the FM luminance signal and recorded onto the tape.

This differs from all other systems in how the 4.27 MHz is formed. This provides the phase inversion on alternate lines and keeps the chroma signal phase locked to the horizontal sync.

In Fig.11-12 a crystal 3.57 MHz oscillator is heterodyned with 692 kHz from a voltage controlled oscillator (VCO) to produce the 4.27 MHz. The 692 kHz is produced in VCO as part of a phase-locked loop. The input to the phase-locked loop (PLL) is the horizontal sync pulses which have been separated from the incoming video. The 692 kHz output of the PLL VCO is divided by 44 to give a signal at the horizontal frequency. This is compared to the incoming horizontal sync pulses in the phase comparator part of the PLL. The dc output from this controls the VCO. This phase locks the 692 kHz to the horizontal sync, and the 4.27 MHz produced is also locked to the horizontal sync. This in turn produces a 688 kHz chroma signal which is locked to the horizontal sync pulses.

The 4.27 MHz is either applied directly to the chroma signal converter, or it is inverted in an electronic switch to produce a 180 deg phase shift. The output of the 4.27 MHz circuit is fed to a transformer, which has a tapped secondary. Both ends of this pass through an electronic switch controlled by a flip-flop. The flip-flop is triggered by horizontal sync pulses and thus alternates the position of this switch, changing the phase of the 4.27 MHz on each line. The flip-flop can be inhibited or enabled by a head-tach pulse taken from the servo, and in this way the phase alternations occur only when the A head is recording.

Chroma Playback

The chroma playback signal follows almost exactly the same path as the record signal. The input in this case is the chroma signal off the tape instead of the input-signal chroma. The horizontal sync pulses are also separated from the playback signal, and these produce a 692 kHz which is phase locked to the playback horizontal sync. This in turn produces a 4.27 MHz which is also phase locked to the horizontal sync, and contains the same time-base errors as the playback signal. The same 3.57 MHz crystal oscillator is used in this heterodyne process, but it contains a minor modification explained later.

The 4.27 MHz passes through the same switch and is phase inverted every line during the A field. When this is heterodyned with the playback 688 kHz chroma, having the same time-base errors and phase reversals, it produces a 3.58 MHz which has most of the serious time-base errors removed. It is also corrected back to NTSC phasing.

At this point the signal is passed through a comb filter where the chroma cross talk components are removed, as described shortly. The output of the comb filter now has its burst separated and this is compared in phase to a stable 3.58 MHz osc signal. The resulting signal is used in the automatic phase control (APC) loop and the burst ID circuit. The output of the APC loop is a dc which is used to "fine tune" the 3.57 MHz oscillator; this removes the finer jitter components from the output signal.

THE INDUSTRIAL BETAMAX MODELS

Several models of the Betamax are produced for the educational and industrial market, as distinct from the consumer market. These all use the one hour mode and an improved Type 2 threading and transport mechanism. Figures 11-13A, B and C/are photographs of the three most common types.

Their mechanical construction is much superior and in some ways simpler than the consumer models, and is very similar to the Type 2 Umatics. All are solenoid operated and can be controlled remotely by several editing consoles. Because of this the system control circuits are much more complex, relying on digital circuits and principles to perform their functions much more than the other machines.

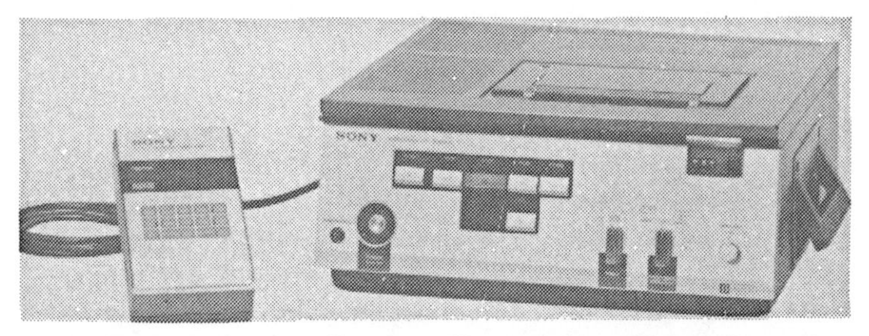

Fig. 11-13A. Industrial model Betamax with automatic search control.

The electronic circuits are basically the same, and use the same ICs. Editing is possible, but the edits are not true vertical interval edits that would satisfy a broadcaster.

The main differences in these machines from the consumer models are:

- A plastic threading ring with static guide posts is used.
- The head drum has an integral dc motor instead of belt drive from the ac motor.
- A dc capstand motor with a belt drive is used.
- Solenoid operation for the functions
- Extra PG magnets and coils are used in the head drum.

The Threading Ring

This is basically the same as the earlier types, with the difference that the tape rests on posts placed around the ring and the double jointed arm is not used. The ring is notched differently, as most of the switches and linkages are in different places from the earlier machines. The dc motor with a gear drive is placed next to the ring, and a small belt from this is used to drive the take-up spool during eject.

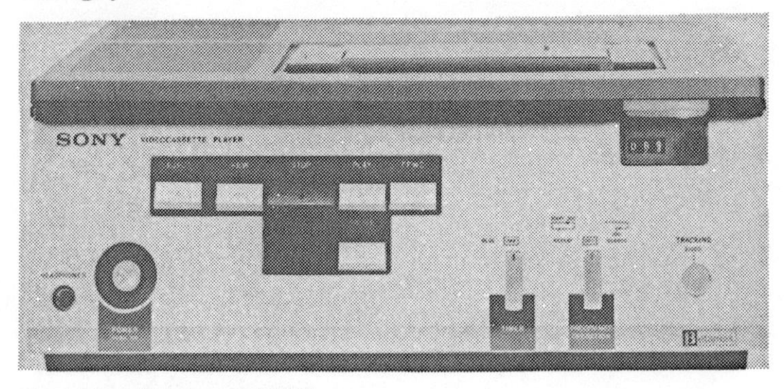

Fig. 11-13B. Betamax SLP 300.

Fig. 11-13C. Betamax SLO 340.

Head Drum

The video heads are driven by an direct drive dc motor which is part of the head drum. A belt is used from this to provide a very light drive to the reel tables in the **play, ff** and **rew** modes.

Capstand Motor

This is a dc motor, which provides a belt drive to the capstan flywheel. The motor contains an 'FG' coil inside its housing which is used to provide feedback to the servos.

Solenoid Operated Functions

All the mechanical interlocks, brakes, etc are operated by a system of sliding bars. Four bars, one for each function, are

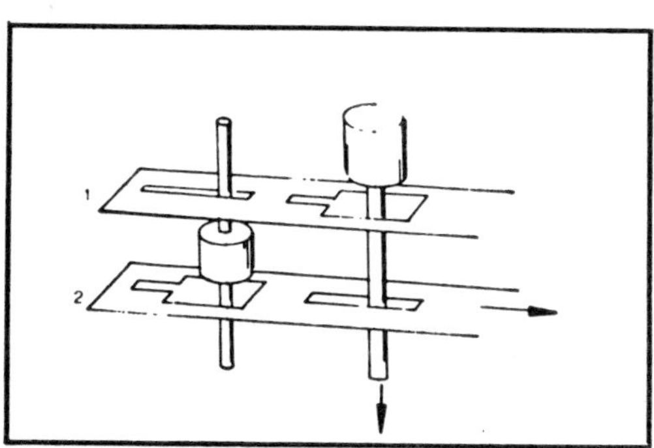

Fig. 11-14. Function button slider bars.

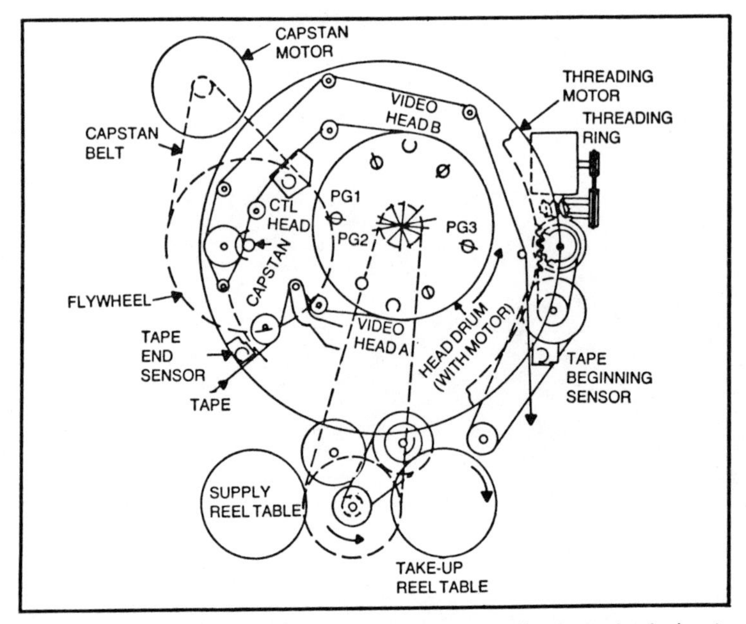

Fig. 11-15. Betamax multiple head coils and magnets and main mechanical parts.

mounted close together and are pulled by the main solenoid. When a function is selected, the function solenoid pulls a connecting pin through a series of slots and holes in the bars. See Fig. 11-14. This way only the correct bar can move while the others are locked in position. For example, if Pin A is pulled between the bars, then Bar 1 can slide but Bar 2 cannot. If Pin B is pulled, then Bar 1 is held and Bar 2 moves. A separate function solenoid is used for each of the pins.

PG Coils and Magnets

Figure 11-15 shows 6 PG magnets mounted near the periphery of the rotating head mechanism. Five of these are embedded in the mounting plate and the other is in the head disc. These pass over the PG 1 and PG2 coils. As the head rotates at 30 rps, the output from each coil is a series of pulses at 180 Hz. The PG 3 coil is mounted closer to the center and only the inner magnet on the head disc passes over it; hence its output is a pulse at the 30 Hz rate.

The 6 coils and magnets are used in the head speed servo loop, and provide feedback information for the rotational control of the heads. Using several magnets and coils enables the 'gyro' errors in the head to be much reduced.

12
The Two Hour Betamax

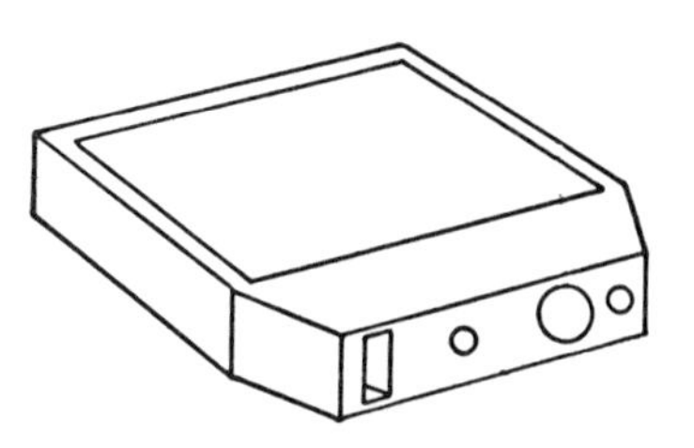

The 2 hour Betamax and the other long playing models were introduced so the home user could have at least 2 hours of program time on one cassette.

Basically the machines are the same as the 1 hour Betamax but with several additions and modifications to accomodate the longer program time. In order to pack 2 hours of program into the same sized cassette, three main ideas are used:

- Halving the tape speed from the standard speed.
- Using a narrower head to get a narrower video track.
- Using a thinner tape backing so that the greater length of tape can be packed into the same sized cassette.

These ideas are also used by the other formats to obtain an extended play time.

Operationally the machines are just the same, but all have a switch to select the desired speed in the record mode. In playback the correct speed is sensed from the CTL track timing and is automatically selected. See Fig. 12-1.

The format of the 2 hour machines is slightly different from the 1 hour, and this has caused several other changes to be introduced into the electronics. The format will be covered next, and then the changes that were needed. Many other ideas and problems described here also apply to the VHS and the other formats. They will be covered in some detail here and then referred to when the other formats are described.

THE BETAMAX 2 HOUR FORMAT

The Betamax indicates the 1 hour mode with the legend X1 and the 2 hour mode with X2. These designations appear on the speed select switch on the top panel.

To obtain greater density of tracks on the tape the 2 hour machines have a tape speed of 2 cms/sec. Naturally a narrower head must be used, but here a problem arises. At the time these machines were introduced technology limited the miniaturizing of the heads, and the smallest practical head that could be made was

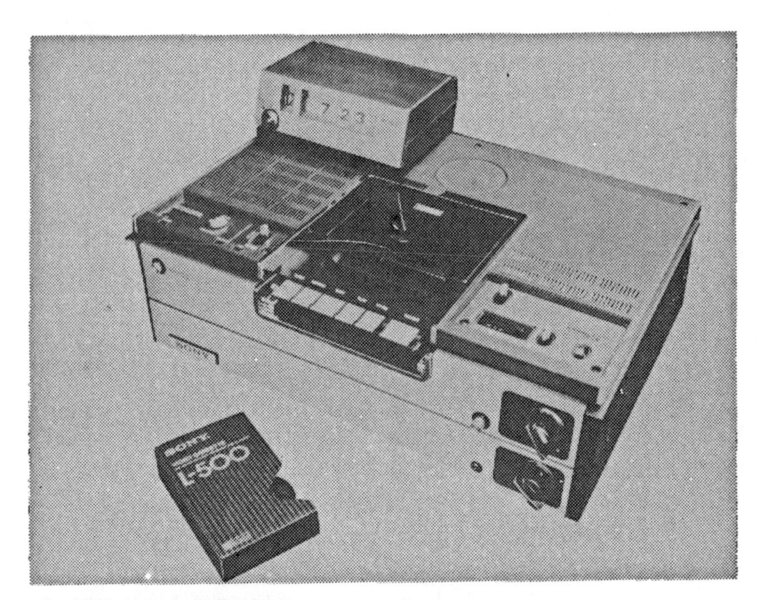

Fig. 12-1. A two hour Betamax.

about 38 microns thick. By using the X-type construction, instead of the H-type used in the 1 hour machines, the actual gap can be reduced to nearly 38 microns. The head gap is still about 0.6 microns. These are shown in Fig. 12-2.

With this size head and a speed of 2 cms/sec, the tracks overlap as they are recorded. See Fig. 12-3. The overlap is about 8-9 microns. On playback this causes an interference pattern in the picture. This appears as a wavy vertical line about one quarter in from the right hand side of the screen. It is eliminated electronically as described later.

On playback the tracking is offset a small amount so that the heads follow a slightly different path, as in Fig. 12-4. This improves the signal on playback and helps to overcome the noise introduced by the overlap.

When the 2 hour machine records at the 1 hour speed, the track layout becomes as shown in Fig. 12-2B, and guard bands are produced. X1 tapes can be interchanged between the older 1 hour machines and the newer dual speed machines. Tapes recorded on the older machines will play on the dual speed machines as the narrow head travels within the wider tracks. Tapes recorded on dual speed machines will play on the single speed machines, but the video signal to noise ratio is slightly degraded as the wider head extends into the guard bands.

Tape recorded at the slower X2 speed will not play on a machine with only the faster X1 speed.

THE MECHANICS OF THE 2 HOUR MACHINES

The dual speed and X2 only machines have exactly the same tape path and almost the same tape transport as the earlier single speed machines. The main differences are:

- A DC capstan motor and a capstan servo are used to drive a smaller sized capstan.
- The X type heads are used.
- A slightly narrower tuner unit on the right hand end is used. This allows the capstan servo board to be mounted without increasing the size of the machine.
- A different function button block is used, using one button only to enter the record mode.

The head drum and head disc are the same, but the disc should not be interchanged with those in the earlier models because they are different. The reel tables under the cassette, the threading mechanism, and the other major parts are the same.

THE ELECTRONIC CIRCUITS

Several different ideas have been introduced into the electronics of the 2 hour Betamax. All of these are necessary to offset some picture fault caused by the extremely narrow video tracks and the track overlap. Most of these changes have been introduced into the luminance record and playback. The chroma circuits are unchanged.

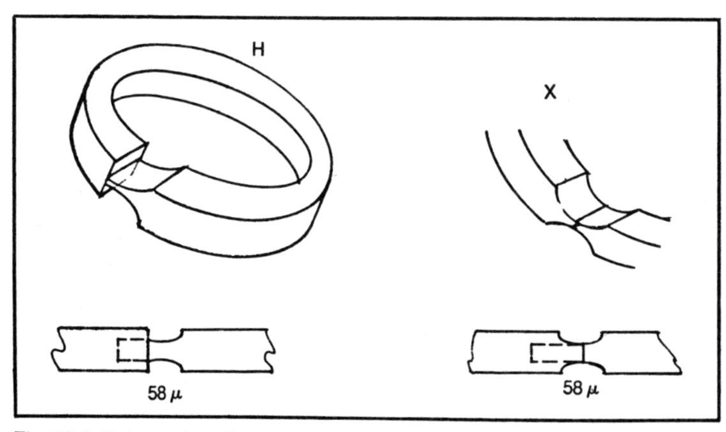

Fig. 12-2. Betamax heads.

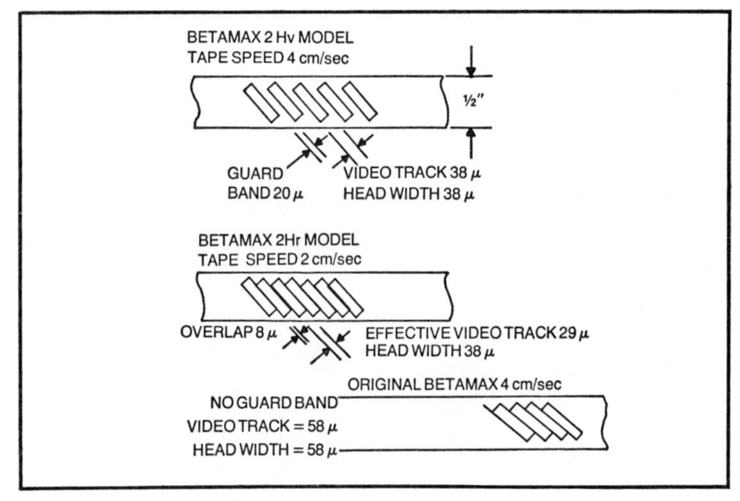

Fig. 12-3. Relative track positions for the Betamax.

Luminance Recording

The basic luminance recording method is the same as for the 1 hour machines, but several new circuits and processes have been included. These are:

- The comb filter.
- The non linear pre-emphasis.
- Differential recording.
- The ½H FM shift.

Figure 12-5 is a block diagram of the record circuit, and Fig. 12-6 is the playback circuit after the pre-amps and switching.

Comb Filter. A comb filter is used to separate the luminance and the chroma components of the input video signal. This has the advantage of not restricting the bandwidth of either, nor introducing unwanted phase errors. The circuit used is the same one used in the chroma playback, and it is switched electronically from one circuit to the other as the mode changes from record to playback. In record it removes the chroma and leaves the luminance component of the signal. In playback, it removes the chroma crosstalk. A brief description of how it works is given in the chapter on color correction.

Non Linear Pre-Emphasis. Because of the worsened signal to noise ratio, due to the narrow tracks, the 2 hour Betamax uses a non linear pre-emphasis instead of the normal pre-emphasis used in the one hour mode. This adds increasing pre-emphasis to the

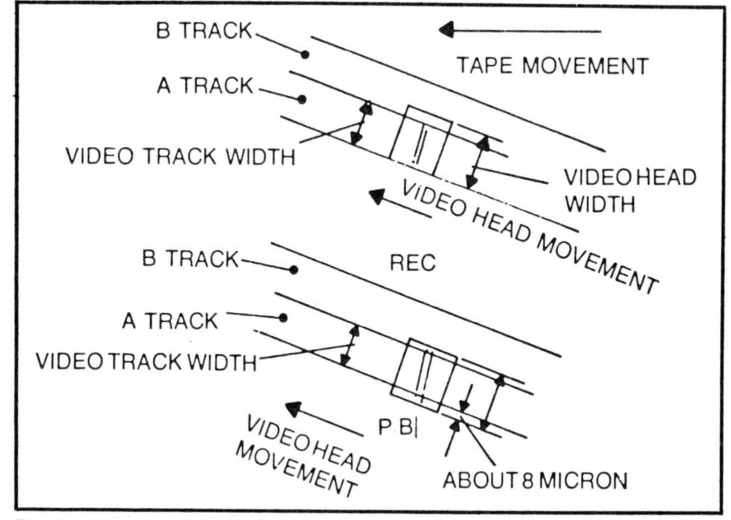

Fig. 12-4. Head tracking offset during playback.

video signal as the frequency of the signal increases, and also the amount of pre-emphasis depends on the level of the signal. It is similar to several audio systems which have become popular in the recording industry and in FM radio. Basically the signal is pre-emphasized and limited, as in Fig. 12-7. The actual circuit is quite complex and will not be covered here.

Differential Recording. The non linear pre-emphasis can cause excessive video level at times, and this causes overmodula-

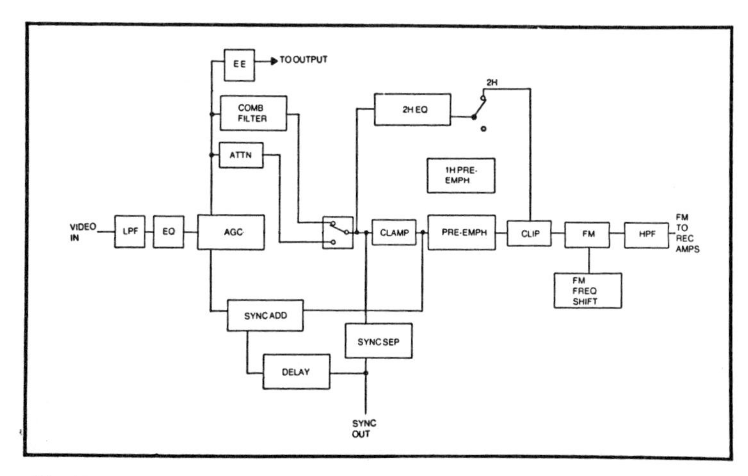

Fig. 12-5. Block diagram of the record circuit.

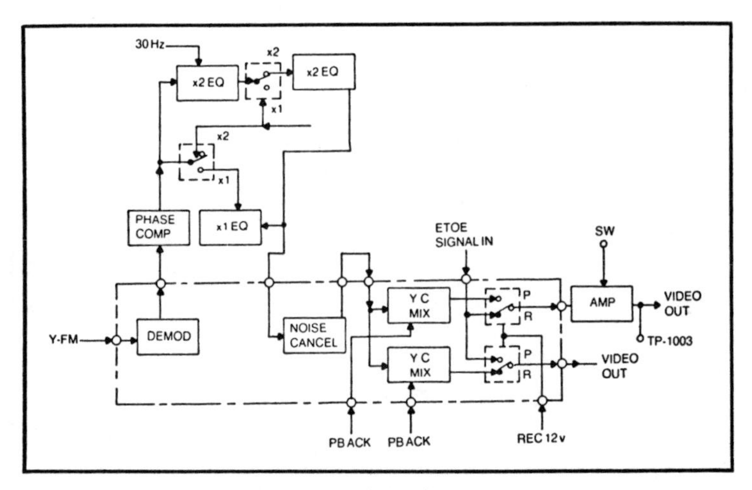

Fig. 12-6. Block diagram of the playback circuit.

tion and overload. To prevent this the amount of FM energy put onto the tape is reduced by using what is called a differential FM signal. The FM signal is passed through a HPF and changed, as shown in Fig. 12-8. As the amount of energy in a waveform is represented by the area under the curve, it can be seen that there is less energy in the differentiated waveform. But as the zero crossings are still preserved, the essential FM information is also preserved. The peaks of the waveform are high enough to still act as a carrier for the down converted chroma.

The ½H FM Shift. In the 2 hour mode the tracks have a slight overlap when they are recorded, as in Fig. 12-3. Because of the azimuth offset and the non alignment of the H sync pulses, the FM frequency which corresponds to the sync tips causes an interference pattern to be seen on the screen about one quarter of the way from the right hand side of the picture. The manner in which this

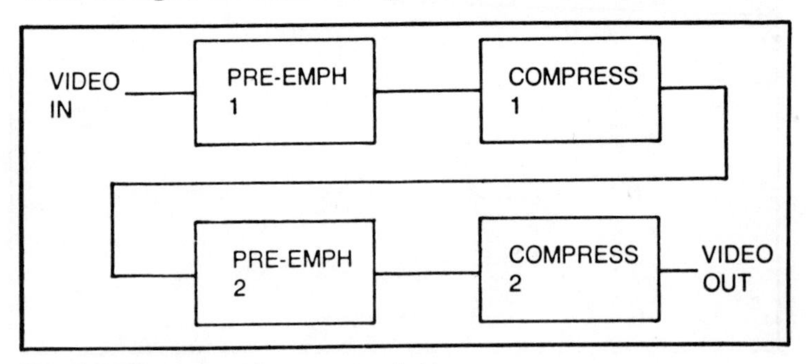

Fig. 12-7. Betamax non-linear pre-emphasis.

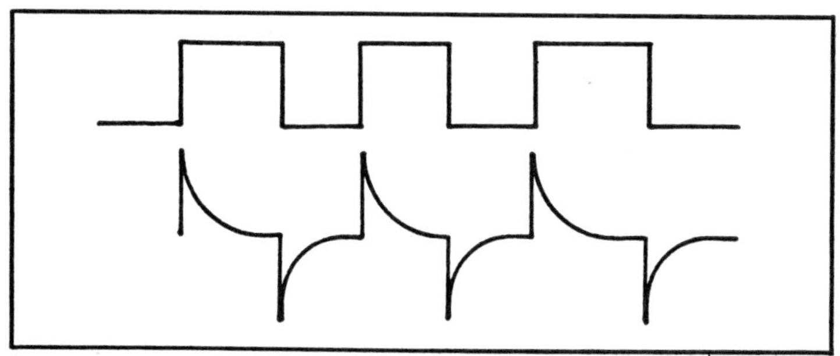

Fig. 12-8. Differential recording waveform.

interference is lost is to alter the FM frequency when the B head is recording.

Figure 12-9 shows how the FM frequency which corresponds to the H sync pulses crosses into the picture area of the adjacent track. The H sync pulses do not lay on top of each other. The entire overlapped video portion of the track causes beats, but the beat caused by the H sync is the most noticable. This is because the FM frequency corresponding to the picture is constantly changing but the FM frequency of the H sync does not change.

The beat produced is a new frequency that was not present in the original picture. It is a frequency that was not recorded by either head in the first place. This beat frequency has no true azimuth and so it will be picked up equally by both heads.

The beat is cancelled visually by a method called 'interleave recording'.

An inherent characteristic of the NTSC system is that any frequency that is an odd multiple of half the horizontal frequency

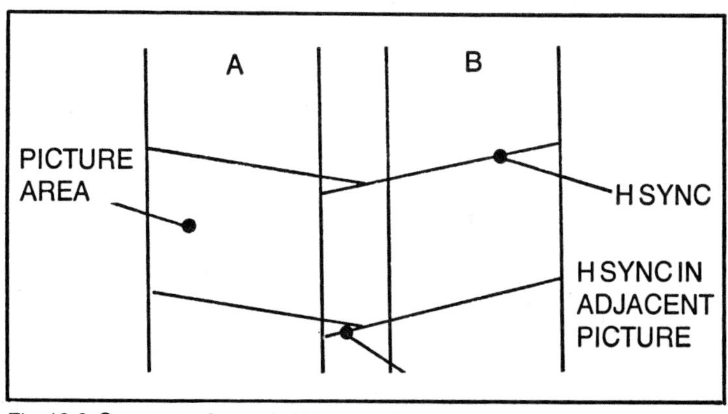

Fig. 12-9. Crossover of recorded H-sync pulses.

will 'interleave' on the screen. This is why 3.58 Mhz was chosen for the chroma subcarrier - it interleaves. An interleave frequency is 180 degrees out of phase on adjacent lines within each field. Also when the same field is repeated (dotted lines in Fig. 12-10) the signal is 180 degrees out from the previous field. This causes visual cancellation on the screen.

The beat caused by the H sync interference is handled in this manner. It is made into an interleave frequency by altering the FM frequency when the B head is recording. When the B head is recording, The FM frequency is increased by an amount corresponding to ½H. Thus the sync tip frequency changes from 3.5 to 3.5 + ½H MHz and peak white changes from 4.8 to 4.8 + ½H MHz.

The frequencies are now out of phase on every other field, and are always ½H different from the A head frequencies, hence, interleaving has been achieved, and along with it, visual cancellation of the interference pattern.

Luminance Playback

The playback block diagram is shown in Fig. 12-6, and is very similar to the 1 hour mode. The pre-amplifiers raise the signal from the heads to a level which can be limited by the limiter chain, and the result is the usual square wave signal which is then demodulated. No special attention need be paid to the differentiated FM waveform, as the limiting process amplifies and clips this to a square wave.

The main difference is the non linear de-emphasis used to offset the non linear pre-emphasis used on recording. This is shown in the block diagram of Fig. 12-11, where the process is the exact reverse of the recording process. The actual circuit is quite complex and will not be covered here.

The ½H FM frequency shift used in recording visually covers up the interference pattern. But changes in FM represent changes in brightness of the video signal, and so the FM shift will result in alternate frames having an overall difference in brightness; one

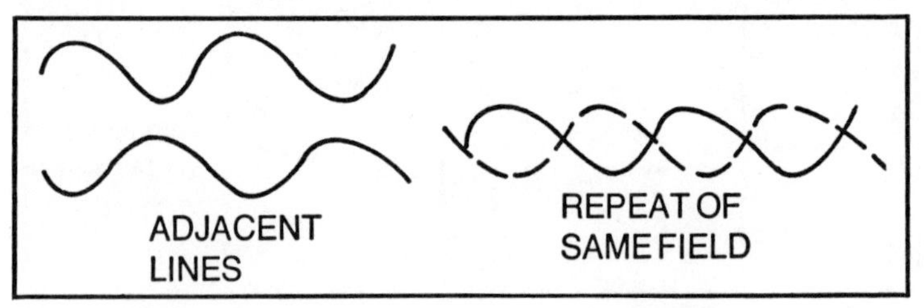

Fig. 12-10. Out of phase signals causing visual cancellation.

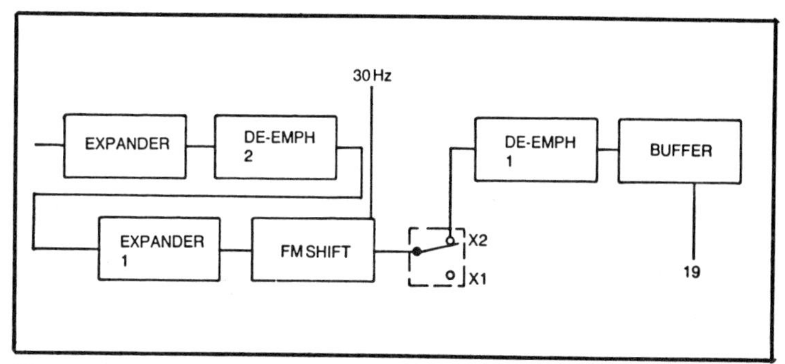

Fig. 12-11. Block diagram of non-linear de-emphasis.

will be slightly brighter than the other. The eye will perceive this as a 30Hz flicker. To prevent this flicker, the video level is adjusted on playback on alternate frames. See Fig. 12-12. The video level is changed by the 30 Hz head tach pulse alternating the overall dc level at the emitter.

Color Recording and Playback

The color circuits are exactly the same in the 2 hour machines as in the 1 hour machines. The only real circuit difference being that the comb filter is shared between the luminance record chain and the color playback chain. As these are never in operation at the same time this expensive circuit can be shared quite easily.

In Fig. 12-13 the luminance input is via transistor Q1 and the color input is via diode D1. During recording the diode is back biased, and during playback the transistor is off. At the output the luminance is taken from one side of the resistor bridge and the chroma from the other, so switching is not required. How the comb filter works is covered in Chapter 14.

THE CAPSTAN SERVO

Because of the need to change speed in these machines a capstan servo was introduced. The tape is driven by a smaller capstan than in the 1 hour machines, and the dc motor provides a belt drive to the capstan flywheel.

The capstan uses both a speed and a phase loop, and most of the circuitry is contained in one IC. This is described fully in the chapter on servos.

However, the capstan servo system contains other unusual ideas which will be covered briefly. The block diagram is shown in Fig. 12-14.

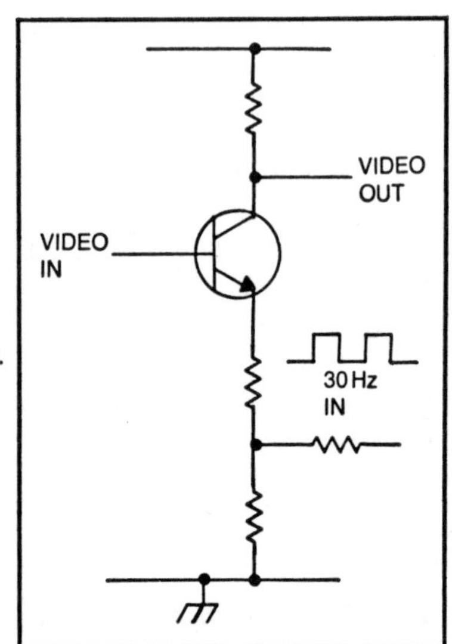

Fig. 12-12. ½H level shift on playback.

The block containing the interval V drive and tracking control is interesting and is shown in Fig. 12-15. The input to this is can be either 50Hz or 60 Hz. This was considered desirable as both frequencies are found in Japan. The circuit is basically a 300 Hz oscillator which is locked to the incoming frequency. A feedback loop is used, in which a ramp and sample pulse are formed, very similar to a normal servo, and the resulting dc keeps the 300 Hz multivibrator locked on frequency. The output is divided by 10 in

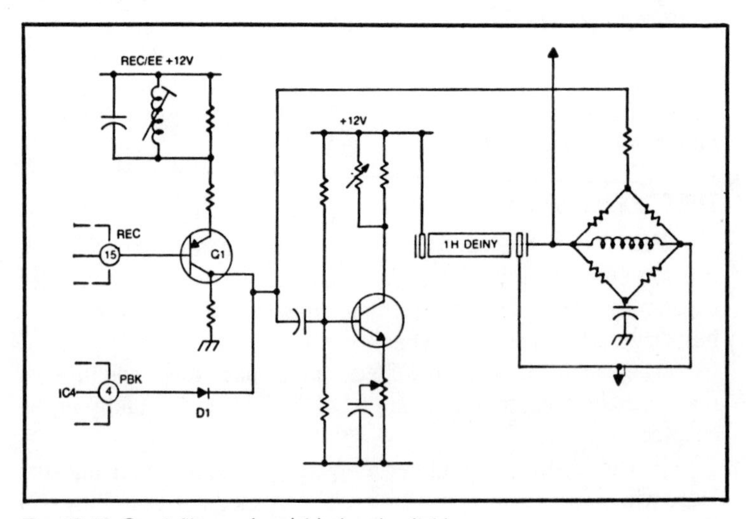

Fig. 12-13. Comb filter and rec/pbk signal switching.

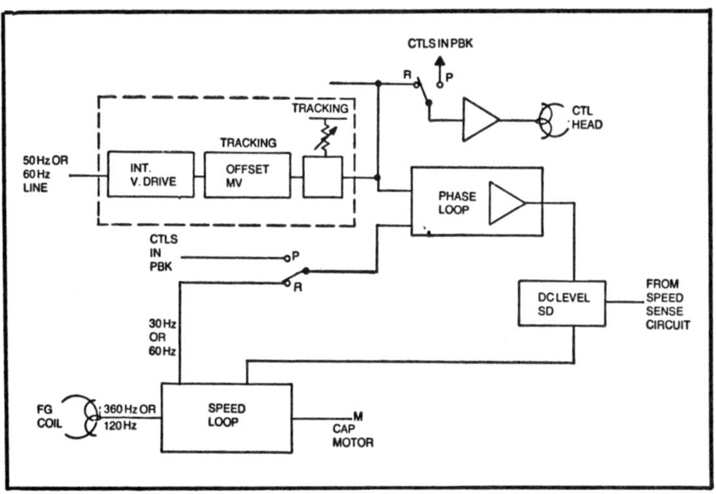

Fig. 12-14. Capstan servo block diagram.

an IC and the 30 Hz output is fed to the 'offset' multivibrator. This produces a narrow output pulse about 4 ms wide which is used to trigger the tracking multivibrator. The offset is used only in the 2 hour mode playback. In recording and in the 1 hour mode a signal from the auto speed sensing circuit allows the 30 Hz from the divider to pass directly to the tracking multivibrator. Note that the tracking control is in the reference circuit of the capstan servo, and not in the feedback circuit from the CTL pulses.

PLAYBACK TRACKING SHIFT

The playback tracking shift shown in Fig. 12-4 is easily accomplished. Instead of the heads following the same path used on record they are offset to one side so that the overlap part of the track is divided between the top and the bottom of the head. This offset is easily accomplished. All VTRs have a tracking control in either the head or the capstan servo. This is usually a multivibrator whose pulse width can be varied by the tracking control on the front panel. In the 2 hour machines an extra multivibrator with a narrow pulse width is used to delay the pulse input to the tracking multivibrator. The amount of offset required was initially determined experimentally.

The speed sensing circuit is shown in block diagram form in Fig. 12-16. When power is first applied it settles into the 1 hour speed. When a tape is first played back it is automatically played in the 1 hour mode. If this is correct then the CTL pulses arrive at a frequency of 30 Hz and nothing happens. If the tape is a 2 hour tape, then the frequency of the pulses will be 60 Hz and this change is sensed in the 2H detect circuit. This produces an output pulse which changes the state of the flip flop in the speed memory, and this output now changes the capstan speed and the equalizing circuits in the playback video and audio circuits. The CTLs now return to the 30 Hz rate and the circuit rests as before.

If, after a while the tape reverts to the 1 hour mode, then the CTL frequency falls to 15 Hz. This is sensed by the 1H detect circuit, which changes the speed memory flip flop again.

Both detect circuits rely on the time it takes for a capacitor to charge, and a different circuit is used for each detect function.

The mute circuit is activated by both detect circuits, and it emits a pulse about 3 seconds long. This mutes both the audio and the video until the speed change has settled down, thus preventing noise on the screen and on the speaker.

In the record mode a simple switch selects that state of the memory flip flop, and operates the mute via the 1 H detector.

THE SYSTEM CONTROL CIRCUIT

The 2 hour machine contains a fairly simple control circuit. It is very similar to the 1 hour machines, and uses the same IC. The

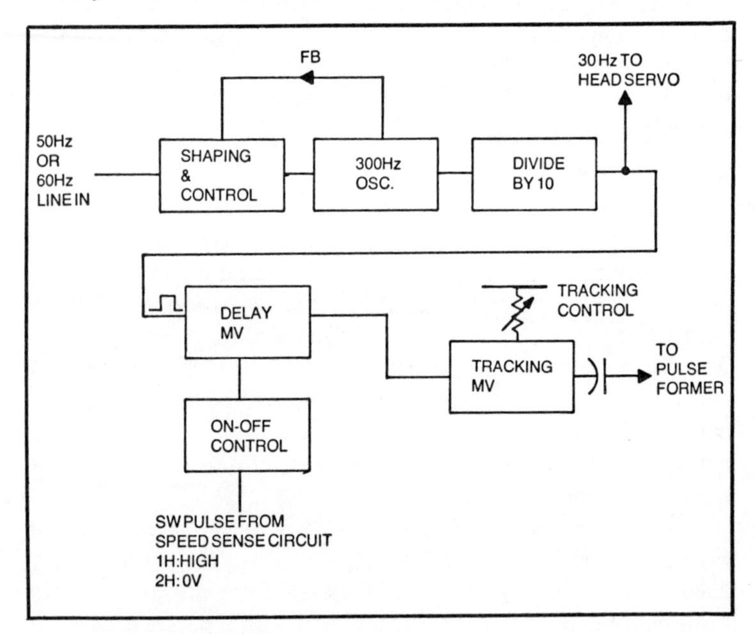

Fig. 12-15. Internal vertical drive and tracking.

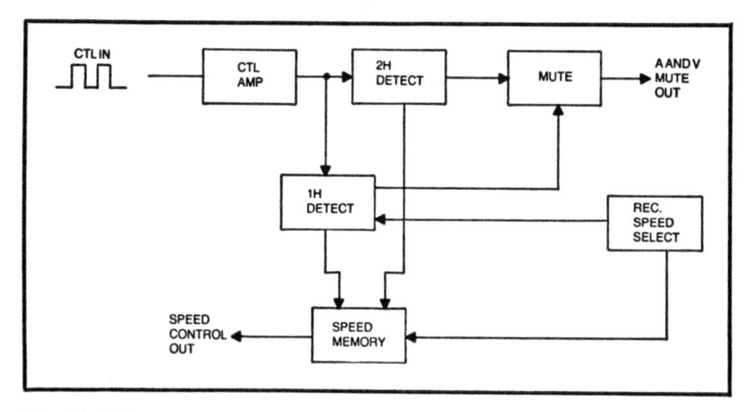

Fig. 12-16. Speed sensing circuit.

main difference is that the muting function has been moved to the speed sense circuit in the capstan servo.

AUDIO

The audio circuit is basically the same as that in the 1 hour machines, but the equalization needed in the 2 hour mode is different, due to the slower tape speed past the same head gap. The equalization in both record and playback is electronically switched by the speed change signal from the speed sense circuit.

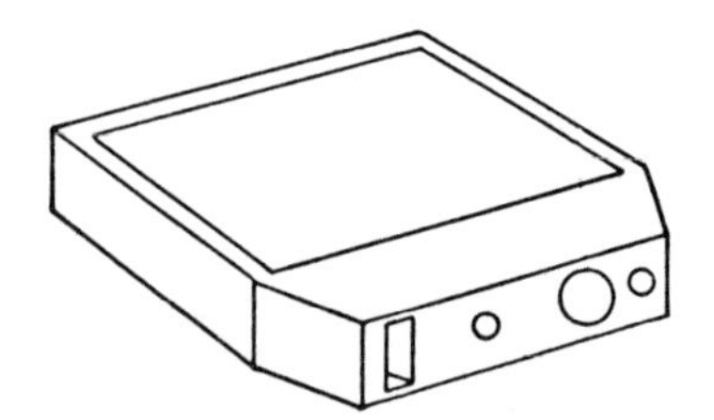

13 The VHS Machines

The VHS machines have some similarity to the Betamax in principle, but in practice they are different. The VHS machines are produced in two forms, the SP and the LP. The SP, or standard play, provides two hours of program time on a cassette, and the LP, or long play, provides four hours. The newest machines provide for even more program time.

The following sections will describe the VHS format and machines, concentrating on their principals rather than the individual model details.

FORMAT

The VHS format is based on the same ideas as the Betamax, but it is different in several respects. The VHS and Betamax tapes are not interchangeable since the format differences are too great. The VHS uses the same zero guard band idea as the Betamax but it is not quite so obvious. The LP mode has track, head and speed dimensions which produce an overlap of the tracks when recording takes place, just as in the Betamax. The dimensions are shown in Fig. 13-1. The head occupies a track 38 microns wide, the overlap is 9 microns and the clean track is 29 microns.

Because of the overlap of the tracks the azimuth recording system must be used, and the heads in the VHS are offset 6 degrees each way to give a 12 degree azimuth difference between the tracks.

When the SP mode, or two hour time, is used in dual speed machines, the tape speed is doubled and so guard bands are produced, as in Fig. 13-2. However, in the SP only machines, a thicker head is used to produce thicker tracks, and reduce the guard bands back to zero, as in Fig. 13-3. SP mode tapes are interchangeable between machines. However, the signal to noise ratio is slightly worse when tapes made on a dual speed machine are played on an SP machine, as the wider head covers an unrecorded area of guard band, as in Fig. 13-4.

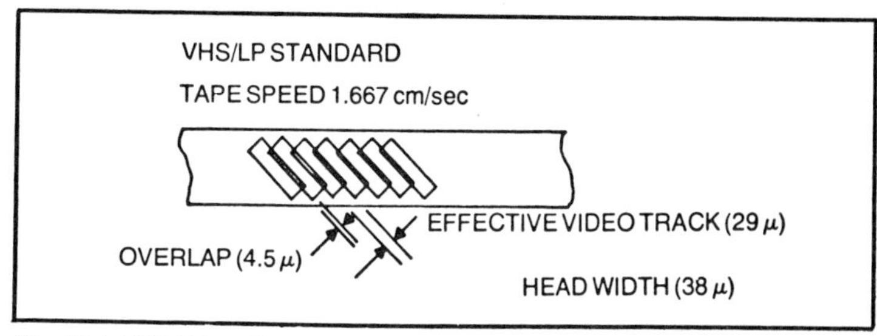

Fig. 13-1. VHS LP track layout.

VHS ELECTRONICS

The VHS system contains all the circuits previously described for the Umatics and the Betamax. The basic ideas are the same, but the individual circuits are slightly different and a few variations are used which are not found in the other machines. Much of the circuitry is contained in ICs. These are manufactured by Matsushita. See Table 9-2 for a list of their main functions and numbers.

There are some slight differences between the LP and the SP mode circuits. Here we will cover the LP circuits as these are more complex, and point out where the SP mode differs.

Luminance Recording

As with the Betamax, many of the main luminance recording functions are contained inside one IC. Figure 13-5 is a block diagram of the basic luminance record path, showing where the signal enters and leaves the ICs.

The video input is separated into the luminance and chroma components by filters, and the luminance enters the IC at Pin 11. Here the gain is automatically controlled in the AGC amplifier, then sent through an amplifier between Pins 9 and 7. At Pin 7 the signal splits 3 ways:

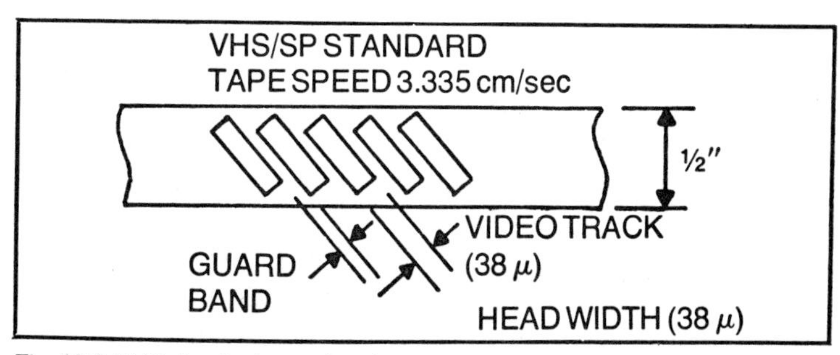

Fig. 13-2. VHS standard speed track layout.

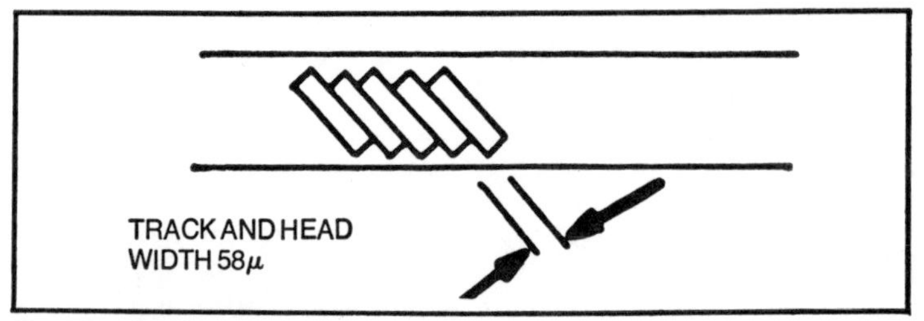

Fig. 13-3. VHS single speed track layout.

- The EE path.
- It enters Pin 6 and goes to the AGC detector.
- The main path.

The main path through the LPF rolls off above 3 Mhz. A level control adjusts the level of the signal and thus sets the FM modulator deviation. The signal now splits and feeds both the LP and the SP pre-emphasis circuits. The speed select switch feeds one part to pin 3 of the IC. This select switch is actually two emitter followers, only one of which is powered and will pass the signal. In the SP mode the LP pre-emphasis is omitted and only the SP path is used.

The signal is amplified and leaves the IC at pin 1 and re-enters at pin 15. Here it is clamped by the incoming horizontal sync pulses and a control sets the dc level at which the sync tips are clamped. This sets the unmodulated frequency of the FM modulator. At pin 14 the signal leaves the IC with about 4 v p-p amplitude, and white and dark clippings are applied.

At this point the ½H FM shift is added. This is used in the LP mode only. It is not necessary in the SP mode as the guard bands eliminate the interference.

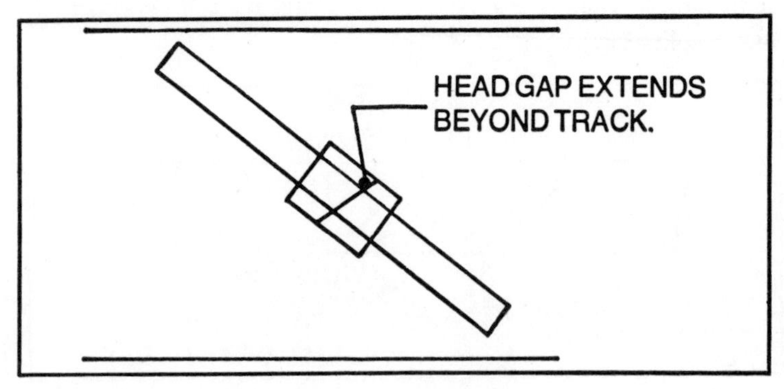

Fig. 13-4. Wide SP machine head on narrow LP track.

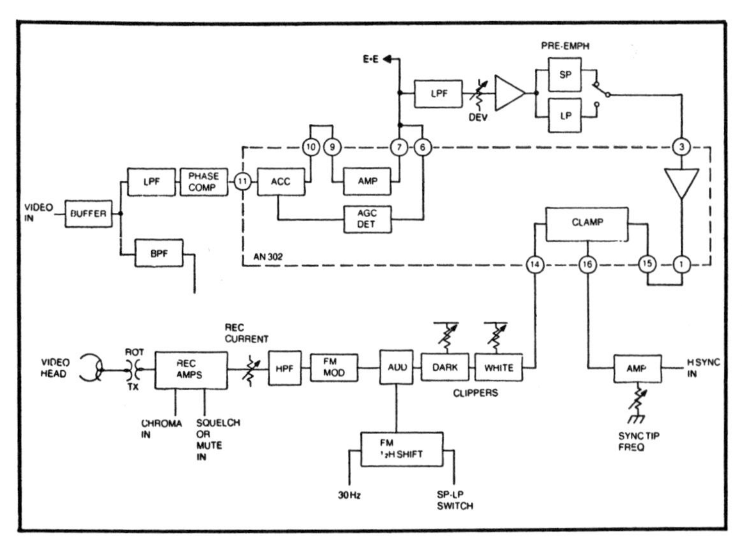

Fig. 13-5. Basic luminance path.

The modulator output is fed through a HPF which eliminates all frequencies below about 1.4 MHz, so that the spectrum is clean for the addition of the 629 kHz chroma. An emitter follower and level control feeds the FM to the record amplifiers and color is added by resistive mixing. A rotary transformer finally feeds the signal, via electronic switching, to the two heads. A squelch or mute circuit inhibits the record signal for about 3 seconds during threading. This prevents erasing the signal already on the tape during threading.

FM Modulator

Figure 13-6 is a circuit of the FM modulator used. It is basically two transistors used as a multivibrator, and this circuit is made with discrete components and is not inside an IC. The video acts as a varying dc level which alters the charge and discharge time of the 2 capacitors, and thus varies the frequency.

½H FM Shift

This circuit is also shown in Fig. 13-6. The 30 Hz input from the head servo switches all 3 transistors on and off. As Q3 goes partially on, its emitter voltage depends on the setting of the adjust control. This is set so that the dc level variation introduced into the video signal will cause the right amount of FM deviation. In the dual speed machines when the SP mode is chosen, the speed select input goes low and thus holds off Q1, thus disabling the circuit. In SP only machines, this circuit is simply omitted.

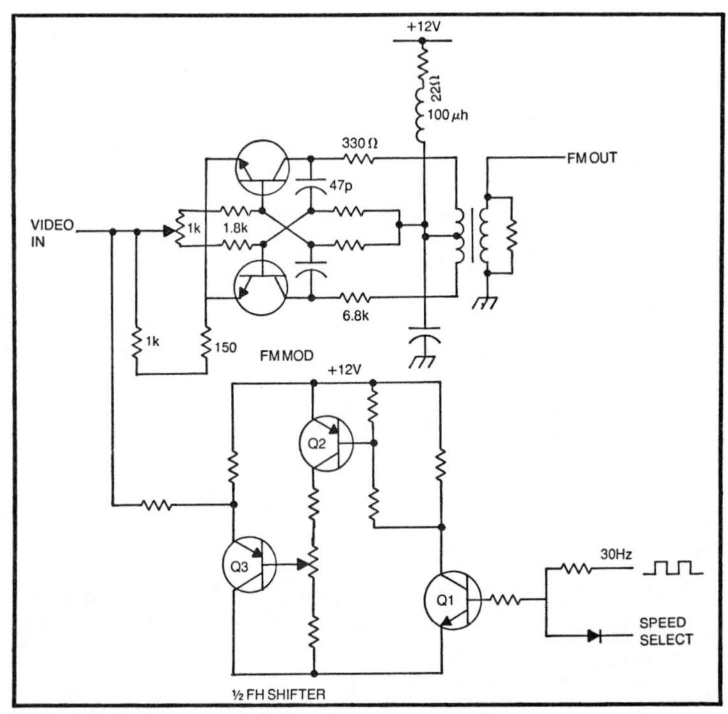

Fig. 13-6. FM modulator and ½H shifter.

Clip Circuits

These are usually external to the IC and are simply diodes connected between the video output line and a constant voltage point, provided by the emitter of a transistor. If the video peaks are excessive D1 in Fig. 13-7 conducts and they are clipped. Overshoot in the sync tips causes D2 to conduct and they are clipped.

The Record Amplifier and Head Switching

The luminance FM and chroma are resistively mixed and an emitter follower feeds the combined signal to the record amplifiers. This is a complementary symmetry amplifier which drives the low impedance heads via the rotary transformers. The current to each head is set by the individual record level controls. See Fig. 13-8.

In record, the record 12v turns Q3 on and grounds point B of the rotating transformer primary coil. In playback the playback 12v grounds point A of the primary and Q3 is off, so the playback feed is taken from point B.

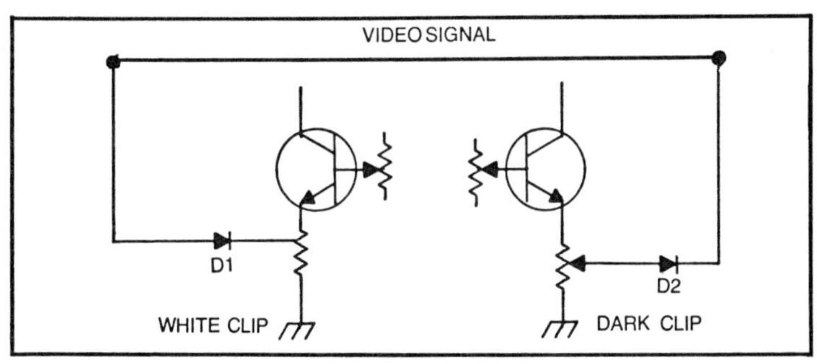

Fig. 13-7. Clip circuit.

Luminance Playback

The head pre-amplifiers are inside the IC shown in Fig. 13-9. The heads are impedance matched at the input with variable resistors and capacitors, and the 1-2 mv from the heads is amplified by the high gain, low noise pre-amplifiers. The outputs are switched by the 30 Hz head tach pulses and then resistively combined. A channel balance potentiometer balances the outputs which are then separated by filters to feed the luminance and chroma paths.

In the Luminance path an amplifier equalizes the signal and a HPF removes the low frequency amplitude variations caused by the 629 KHz chroma and the signal enters the luminance playback chain. Figure 13-10 is a block diagram of the system used in most VHS machines. Slight differences are found from manufacturer to manufacturer, but this general plan is followed.

The signal enters the IC(AN 316) at pin 2 and between pins 3 and 4 an HPF removes HF noise. After amplification a switch

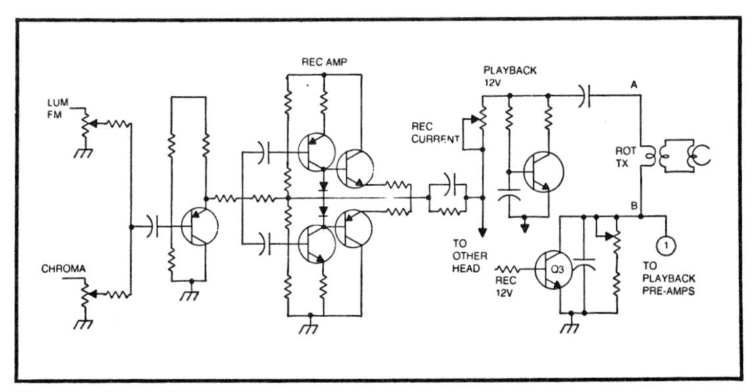

Fig. 13-8. Record amp and head switching.

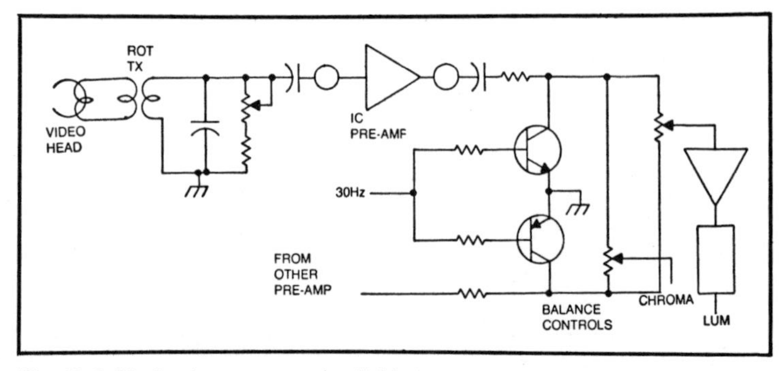

Fig. 13-9. Playback pre-amp and switching.

selects either the direct or delayed path and the output appears at pin 9. One path from here is fed back through the 1H delay line to form the delayed signal used in the dropout compensator. The main path feeds the limiter and the demodulator section. Two common circuits are used here, which are described later.

After demodulation the signal is now de-emphasized and then a 30 Hz signal restores the FM offset effects in the LP mode. In the SP mode, this is not used. Noise is removed by an LPF and a level control feeds the video signal into the final IC (AN 303). In this IC the chroma is added, and the output signal is clamped, muted when required, and then fed to the output of the machine. The way in which this IC is used is slightly different between different machines.

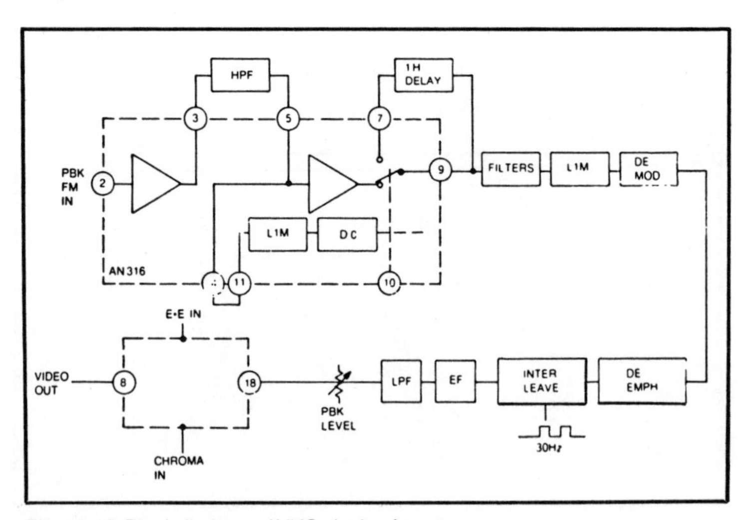

Fig. 13-10. Block diagram of VHS playback system.

The main variations and points of interest in this overall circuit are:

- The limiter circuits and the demodulation methods
- SP and LP de-emphasis.
- FM interleave removal.
- Output IC circuit.

Each of these will be covered in separate paragraphs.

The Limiter And Demodulation. The limiting and demodulation is performed inside an IC (SN 76670). Fig. 13-11 shows a simple block diagram of the process. The IC is a chain of differential amplifiers that are used as limiters, and an output differential amplifier which performs a comparison function.

The FM input to pin 14 is amplified repeatedly until all the amplitude variations are removed, and the square wave output appears at pin 6. Pin 13 provides balancing of the output stage of the limiters. A direct internal connection is made to one side of the output amplifier, and a delay line or phase shifter is used to connect the signal from pin 6 to the other side of the output amplifier. The two inputs to the final amplifier produce a pulse train output which is integrated into the video signal. This simple limiting and demodulation is used in several VHS models.

Some VHS models use a double limiter system. The slow tape speed, the slow writing speed, and the narrow tracks require greater pre-emphasis of the video signal to overcome the signal to noise ratio than that found in VTRs with a faster linear and writing speed. Because of this excessive pre-emphasis, the FM signal is apt to suffer from AM variations at the times it reaches its highest FM frequency.

At the times of high pre-emphasis the FM frequency reaches its highest level, and the amplitude of the FM carrier tends to fall. The signal then tends to settle about the common average level, which causes low frequencies to appear in the overall waveform.

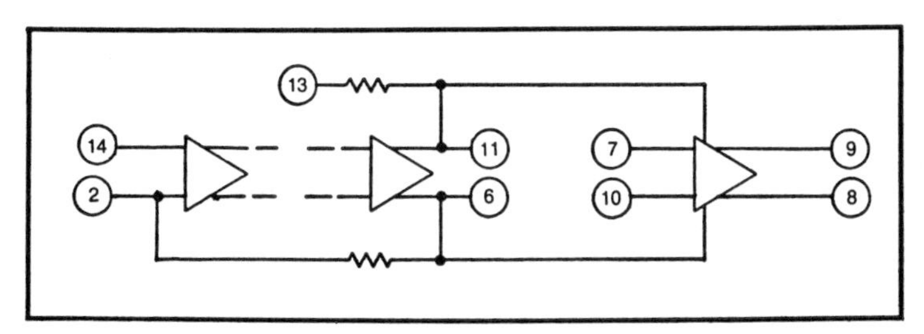

Fig. 13-11. A simple limiter.

A simple limiter will now cut off the small amplitude variations and produce a waveform as in Fig. 13-12. This now looks like a low FM frequency, and is demodulated as a black or blanking, while in fact it probably started out as a peak white. Also, as the high frequencies in the FM are lost, so also is the high frequency video they represented, so the picture loses detail. To prevent this, a double limiter system is used.

A HPF and a LPF separate the signal into the components shown in Fig. 13-13. The HPF output is amplified, phase corrected, limited, and added back to the LPF output. The combined signal is now amplified and limited to give a clean FM waveform suitable for demodulation. Figure 13-13 shows all the waveforms in the circuit. Figure 13-14 is a typical circuit from a VHS machine.

SP AND LP De-Emphasis. The simple RC circuit of Fig. 13-15 is used in many VTRs to provide de-emphasis of the demodulated video signal. Figure 13-16 shows a circuit used in some VHS machines. Q3 and Q4 are alternately switched on or off, so that either the SP or the LP path is selected. The video enters

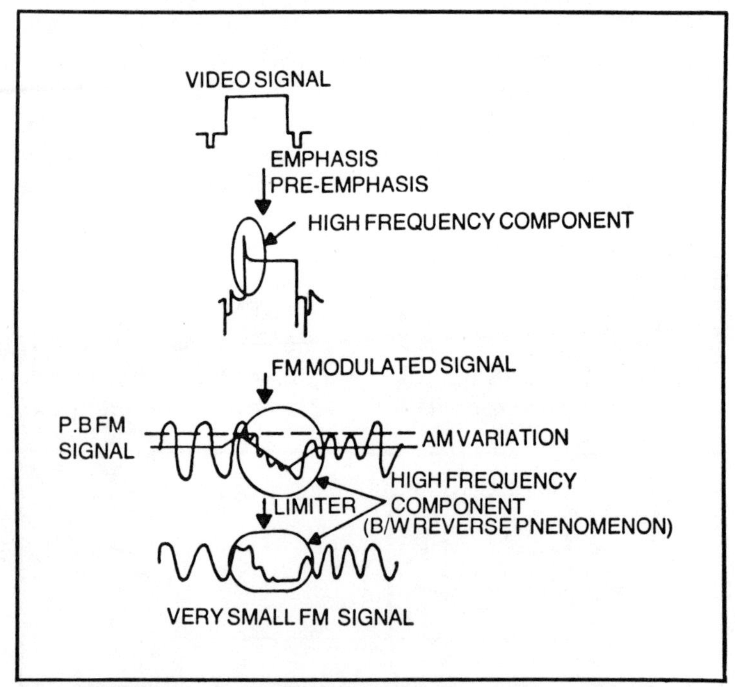

Fig. 13-12. FM level change due to excessive high frequency peaking.

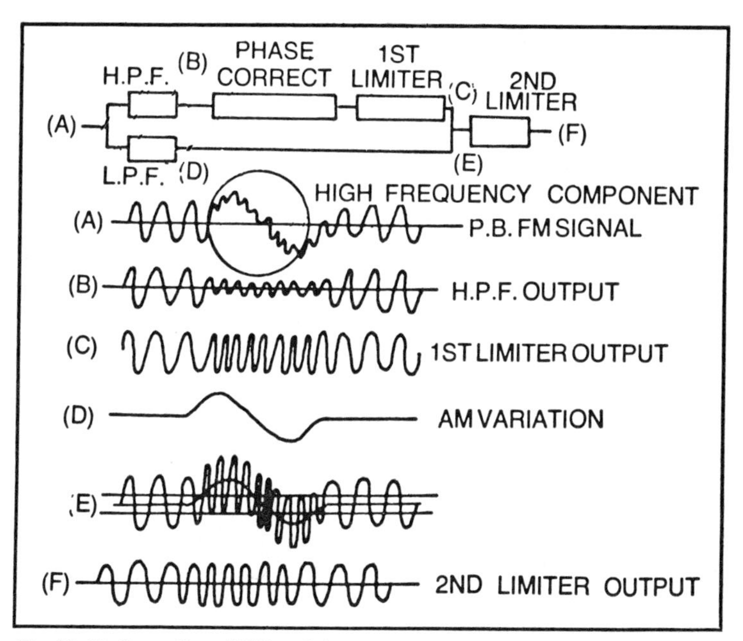

Fig. 13-13. Correction of FM level change.

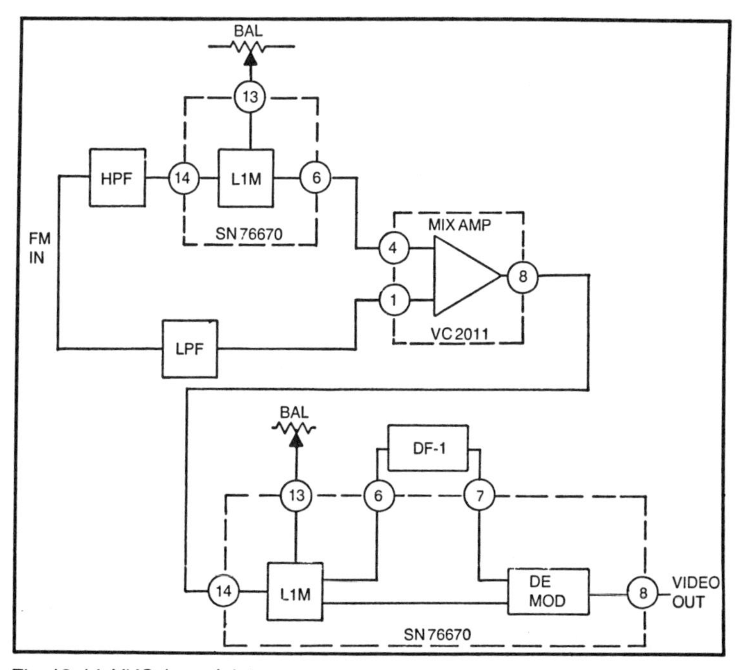

Fig. 13-14. VHS demodulator.

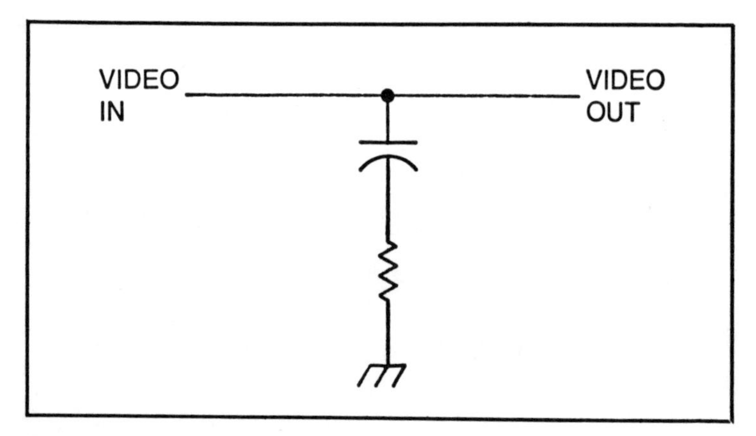

Fig. 13-15. Simple de-emphasis.

emitter follower Q1 and feeds both de-emphasis paths at all times. The SP path is very simple, and the two diodes provide a logarithmic de-emphasis at high frequencies and will actually clip excessive levels. The LP path again uses diodes to provide a level dependent non-linear de-emphasis. Q2 is an emitter follower which feeds the video to the next stage.

FM Interleave Removal. As with the Betamax, the FM ½H shift causes a slight change in the overall video level, which must

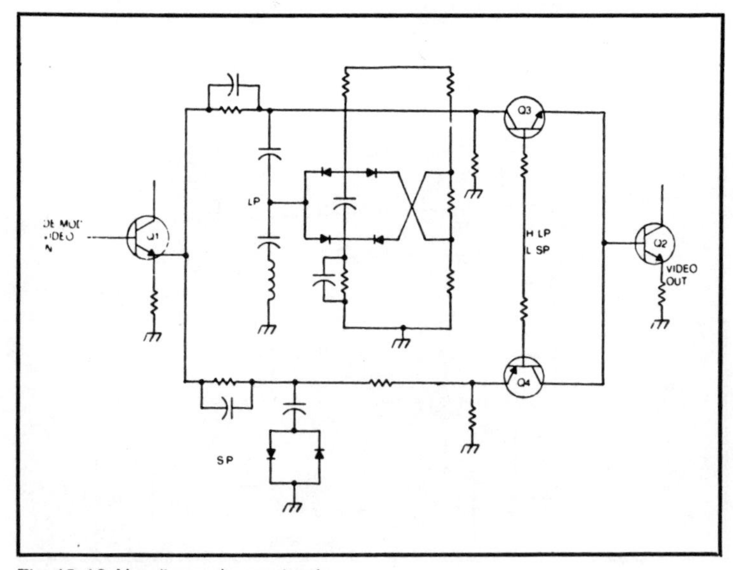

Fig. 13-16. Non-linear de-emphasis.

be corrected. Again the 30 Hz pulse is used to provide a dc level change at some point in the video circuit, usually just after the de-emphasis.

The Output Circuit. An IC (AN 303) is used for output purposes in most VHS machines, but the connections can vary. Two set ups are shown here.

In Fig. 13-17 the luminance video signal enters at pin 18, is amplified and appears at pin 1 and pin 3. The pin 1 output or the EE input in the record mode enter pin 16. The pin 3 path feeds through an aperture corrector for high frequency correction and enters pin 13. The chroma enters at pin 15. These three inputs are combined in the mix amp and the output appears at pin 11, and enters the clamp at pin 10. Muting is applied in the IC when required and a final amplifier feeds the output to pin 8. This is the video output of the machine, and is fed to the RF modulator and the output connectors.

Figure 13-18 shows an alternative set up. No aperture correction is used, and the inputs to pins 13 and 15 are changed.

Basic Chroma Circuit

The basic chroma processing circuit is contained in an AN 305 IC, as shown in Fig. 13-19. This circuit is used to convert the input 3.58 MHz down to 629 kHz in the record mode, and to convert the 629 kHz playback signal from the tape back up to 3.58 MHz. To do this the input signals are heterodyned with a 4.27 MHz from the AFC circuit. The VHS system uses a chroma phase change, but it is quite different from that use in the Betamax. It is covered later when the 4.27 MHz is discussed.

The input to this circuit can be either the record 3.58 MHz or the playback 629 kHz, and the correct signal is switched by a diode. The signal is automatic gain controlled in the ACC circuit. This is controlled by the burst input at pin 15 and a different input is used for record and playback. The output from the IC at pin 7 is a wide range of frequencies which result from the heterodyne process. The correct frequency for record or playback is separated out by a filter and sent to the appropriate signal path by an emitter follower switch.

The switching circuits to separate the signals use either emitter followers, FETs or MOSFETS, all of which are operated by the record or playback 12v, thus allowing signals to pass only in the desired mode.

The input circuit is shown in Fig. 13-20. This uses a diode switching system. The input 3.58 MHz passes through a 1.2K

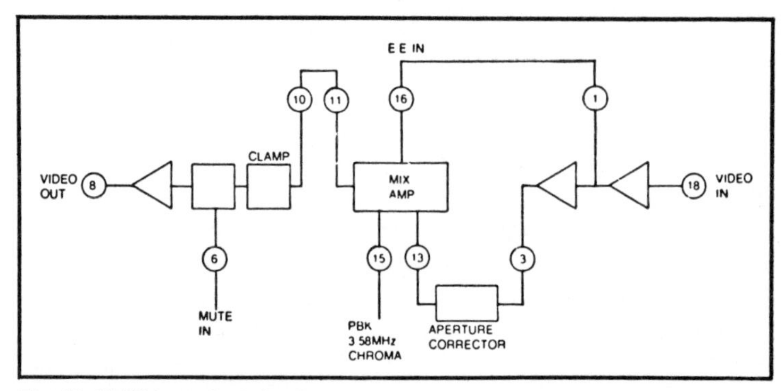

Fig. 13-17. Video output circuit.

resistor and enters the BPF. In playback the playback 12v forward biases the diode and places the 0.01 capacitor in the circuit. At 3.58 MHz this capacitor has a reactance of about 4 ohms and so it shorts the 3.58 MHz to ground. In record, the 629 kHz does not appear at the input as no 12v playback is applied to the IC pre-amplifiers.

The ACC circuit controls the gain of the chroma signal. The control voltage is developed from the input at pin 15. This input is always the 3.58 MHz signal. See Fig. 13-21. In record, the FET is turned on by the record 12v to pass the 3.58 MHz signal from pin 11 to 15. In playback this FET is off. The playback 629 khz cannot reach pin 15. During playback the playback output amplifier is powered by the playback 12v, and so this provides a feed to pin 15.

Inside the IC a burst gate allows only burst to pass, and this is rectified to dc which is used to control the ACC amplifier. The time

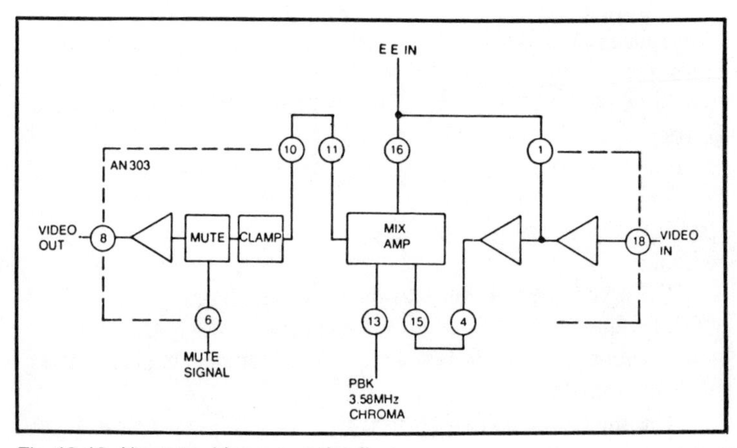

Fig. 13-18. Alternate video output circuit.

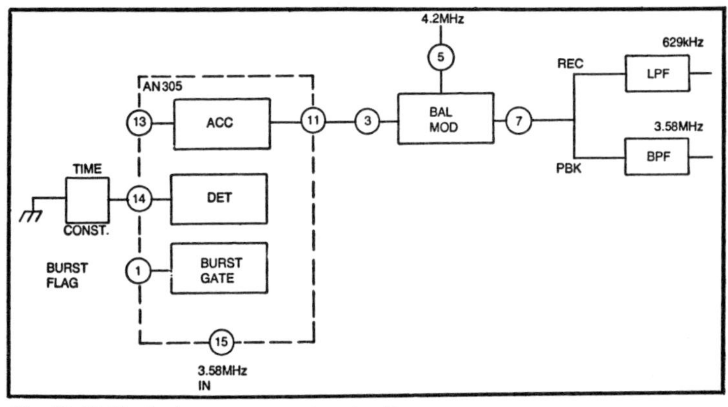

Fig. 13-19. Basic chroma processing circuit.

constant of the detector response is varied from record to playback by the components at pin 14, which are switched in and out by emitter followers powered by the appropriate 12v.

In the record mode the burst gate is not really necessary as the double amplitude burst is usually much larger in amplitude than the chroma signal. As the ACC is a peak type circuit, it works on the burst peaks only.

The output of the IC appears at pin 7, and this is a complete range of frequencies from the heterodyne process in the balanced modulator. The signal is sent down two paths. A LPF is used in record to separate out the 629 kHz, and a PBF separates out the playback 3.58 MHz. The outputs of the filters are passed or blocked by simple emitter followers powered by the correct 12v.

The next sections will look at the record and playback paths in a little more detail.

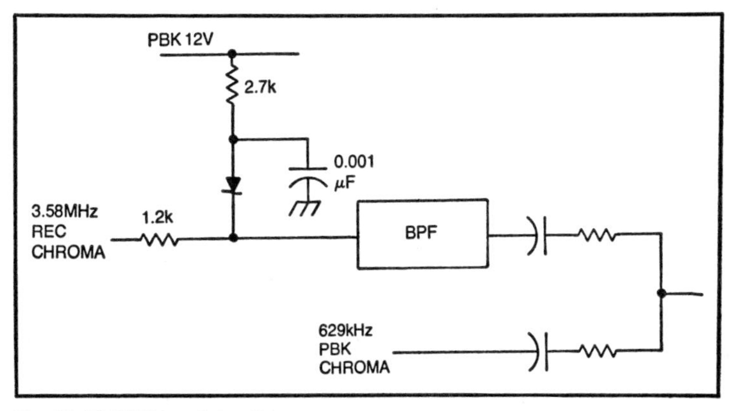

Fig. 13-20. VHS input circuit.

Chroma Recording

A more complete recording block diagram is shown in Fig. 13-22. The input signal is fed from the luminance circuits through a BPF to extract the 3.58 MHz. An amplifier, either a single transistor or in an IC feeds this signal to a circuit which increases the burst only by 6 db in the SP mode. This allows much better signal to noise control of the chroma. In the LP mode this circuit is not used. A switch selects this path and feeds the signal to pin 13 of the IC. At pin 11 an emitter follower or an FET feeds the 3.58 MHz to pin 15, where it is used for gain control. At the output at pin 7, another emitter follower after the LPF selects this path and feeds the signal to the record amplifier, via the killer amplifier. Another emitter follower blocks the playback path. The killer amplifier operates only if color is present. In the case of a monochrome input it inhibits the signal and thus prevents noise being added to the output.

The main circuit of interest here is the burst expander. This doubles the amplitude of the burst only. It is increased by 6 db. The

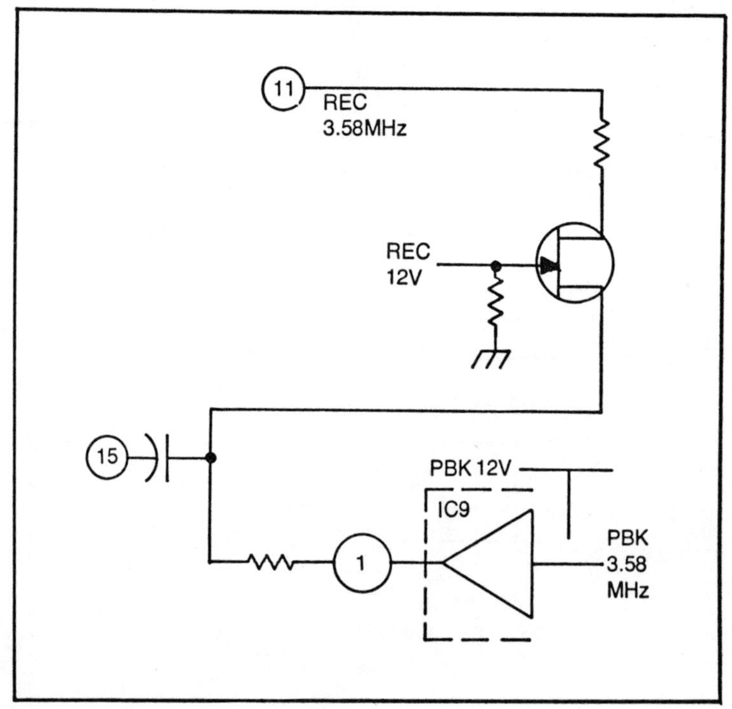

Fig. 13-21. Input signal to ACC circuit.

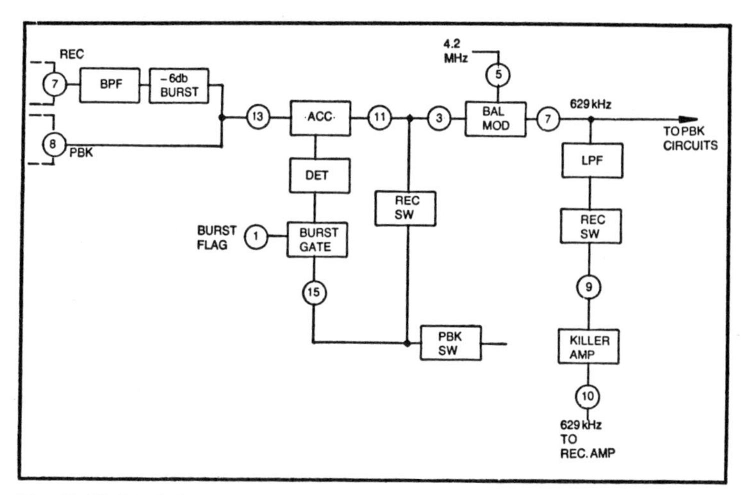

Fig. 13-22. Block diagram of recording system.

circuit is shown in Fig. 13-23. The two resistors form a voltage divider. In the play mode the playback 12v turns the MOSFET on and the signal from pin 8 is divided by two at all times. In the record mode the MOSFET is held on by the output of the burst gate, and so the signal is again divided by two. But when the negative burst gate pulse appears it cuts off the MOSFET and thus places the inductor in the circuit. This has a reactance of about 5 K at 3.58 MHz and so the voltage divider now produces no attenuation of the signal. So the signal entering the IC at pin 1 is now twice the level of what it is normally. As this occurs only during the time of burst, only burst is increased in amplitude.

CHROMA Playback

The playback signal from the heads is amplified and switched to form a continuous signal, as described in the luminance playback. The signal is then split and one path passes through an amplifier and then a LPF to extract the 629 kHz chroma. Here it passes through the same amplifiers as the record 3.58 Mhz, but the burst expander circuit is off, so the signal passes unchanged. It enters the IC at pin 8 and the circuit is now as shown in Fig. 13-24.

The signal is AGCd and leaves the IC at pin 11 and enters it at pin 3. The same balanced modulator is used and the 3.58 MHz is extracted from the output by a BPF. The input at pin 15 is the playback 3.58 MHz which is used to AGC control.

At pin 7 two slightly different circuits are used, as in Fig. 13-25 and Fig. 13-26, but the result is the same. A BPF separates

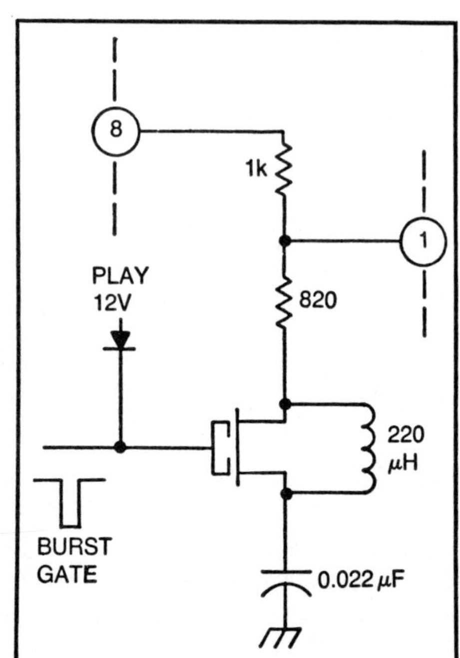

Fig. 13-23. Burst expander.

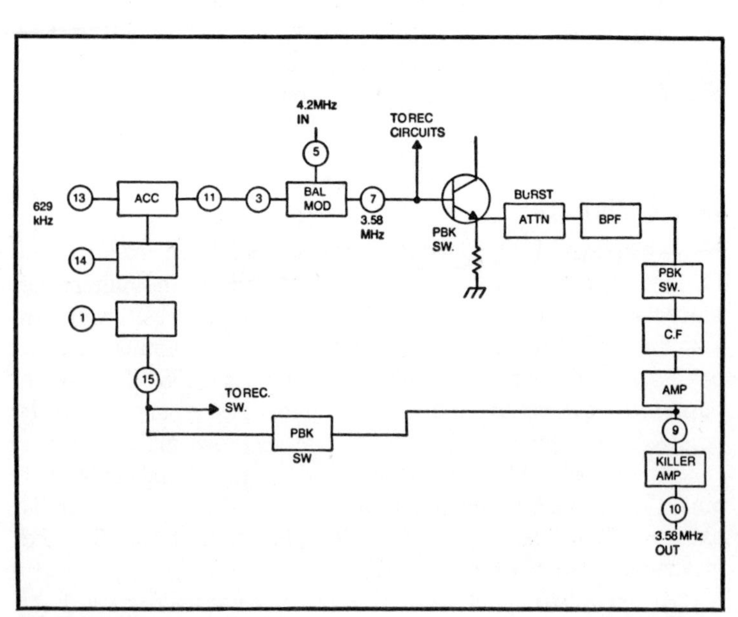

Fig. 13-24. Chroma playback.

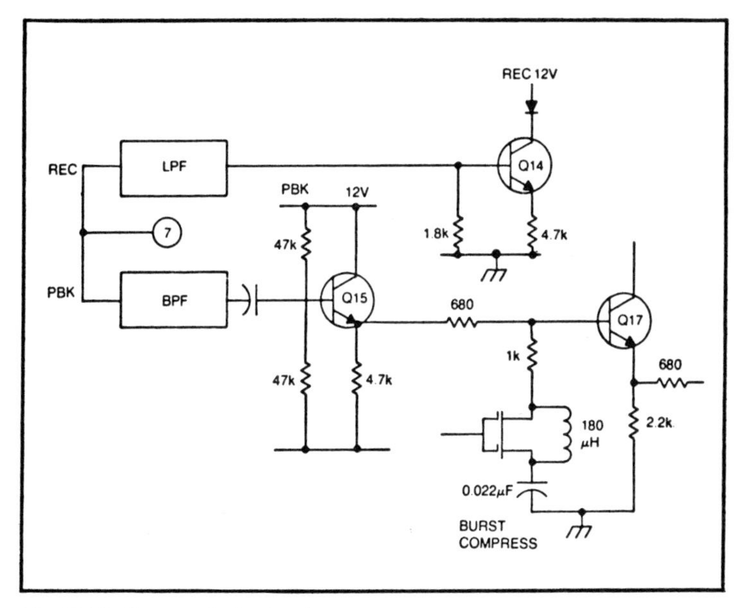

Fig. 13-25. Chroma playback path.

out the 3.58 MHz and passes it. In the SP mode the burst is restored to its proper level with respect to the rest of the chroma by the burst attenuator circuit. In the LP mode this is not used. The

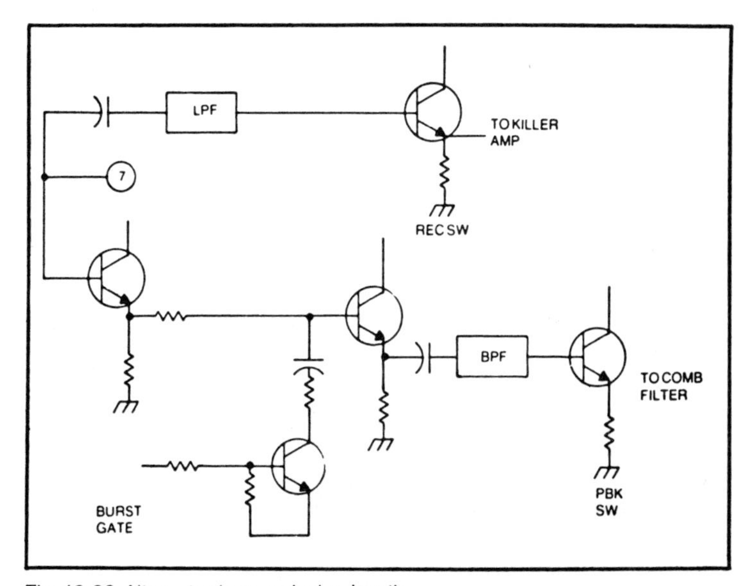

Fig. 13-26. Alternate chroma playback path.

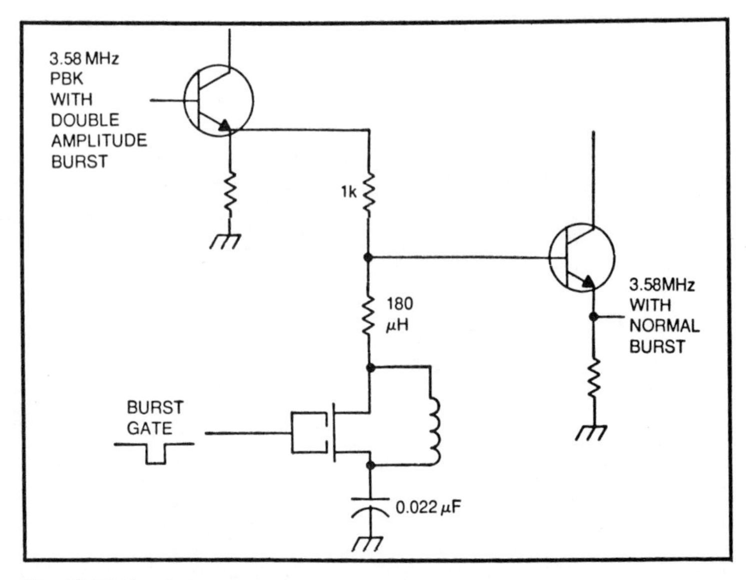

Fig. 13-27. Burst compressor.

signal is then passed through the comb filter where the chroma crosstalk from the adjacent tracks is removed. An amplifier now feeds the signal to the killer amplifier in the IC and to pin 15 for AGC control.

The burst compressor circuit is shown in Fig. 13-27. Normally the MOSFET is held off. The inductance has a reactance of about 4 K at 3.58 MHz and so very little signal attenuation occurs. The burst gate pulse turns on the MOSFET, and now the circuit looks like two 1K resistors, so that signal is effective halved during the time of the burst.

VHS MECHANICS

The main difference in the VHS from the other formats is the tape wrap. This is a very simple looking wrap compared to the other systems; but the actual mechanism is about the same in complexity as the other machines.

When the cassette is inserted it sits on the two reel tables, the protective door opens and the section of the tape between the reels drops over the two pull out pins (pins 2 and 3), and the capstan, as in Fig. 13-28. It also closes a microswitch to provide power to the system control part of the circuit. When the **play** button is pressed the pull out pins pull the tape out and thread it around the head drum and the guides. Pins 1 and 4 guide the tape, keeping it in an horizontal plane around the angled head drum. Two mechanical stoppers hold the guide pins, in the correct place at the completion of threading, and the two impedance rollers move into contact with the tape. These help to keep the tape moving smoothly and keep an even tension in the tape. The slanted head drum has a machined guide around its lower half to guide the tape. The pressure roller closes toward the capstan and the tape is driven from left to right. In some VHS machines the main erase head moves with the impedance roller to contact the tape. The audio and CTL heads are fixed.

The tape is not automatically threaded upon insertion of the cassette. Threading occurs only when the **play** button is pressed during playback and recording. When **stop** is pressed the tape is retracted back into the cassette. Fast forward and **rew** are conducted with the tape inside the cassette.

The ends of the tape are sensed by light shining through the clear leader which attaches the tape to the cassette reels. A single lamp is mounted in the machine and positioned at the center of the cassette, and a phototransistor is mounted at each end of the cassette. When light reaches the transistor it initiates the stop mode.

As with other machines each reel table has a brake, and the supply reel has a tension control band operated by the tape. The drive to the two reel tables is supplied by belts and movable idlers driven from the capstan flywheel.

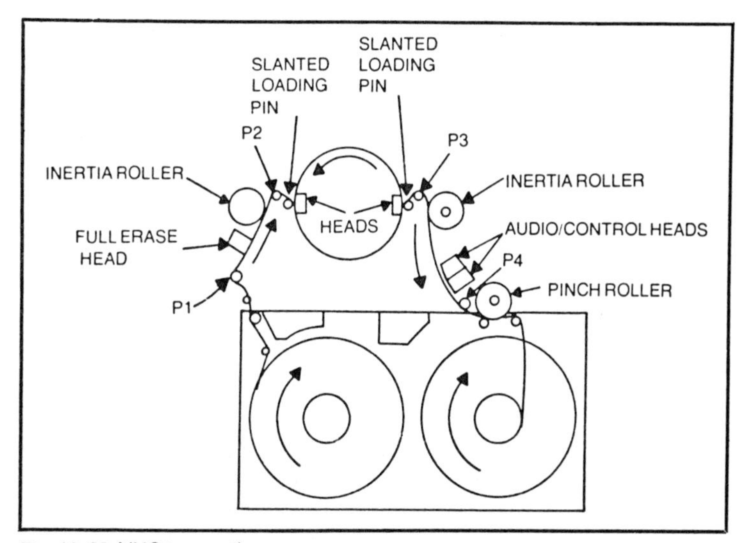

Fig. 13-28. VHS tape path.

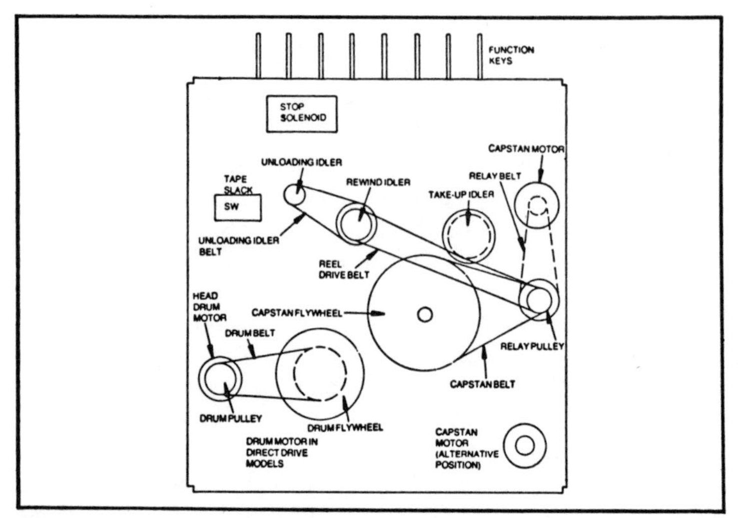

Fig. 13-29. VHS belt system.

Figure 13-29 shows the underside of the machines. The threading and unthreading is controlled by a gear wheel system which is moved into contact with the capstan flywheel when the **play** or **stop** button is pressed. It is a complex arrangement which is well covered in the service manuals of the individual models. For clarity it is omitted from Fig. 13-29.

The position of the capstan motor varies slightly from model to model, and so does the belt drive arrangement. Two types of head drum drive are used, a belt drive from a dc motor, and a direct drive dc motor mounted in the head drum.

Figure 13-30 shows the mechanical layout and positions of the most important mechanical parts. Each of the function buttons or levers operates a microswitch, which engages the correct electronics to provide the desired functions. When used, the buttons are mechanically locked down, and are released by the **stop** button or the stop solenoid.

The advantages are the very short tape path and the short time it takes to thread and unthread the tape. These give fast operation once a mode is selected, and allows the tape to be easily returned to the cassette. The short path reduces tape stresses and thus helps to extend tape life.

The main disadvantages levelled against it are the large number of sharp tape bends and the larger tape tension than in the Umatic or Betamax system. High tension is a definite disadvantage when using the super thin tapes of the long play cassettes.

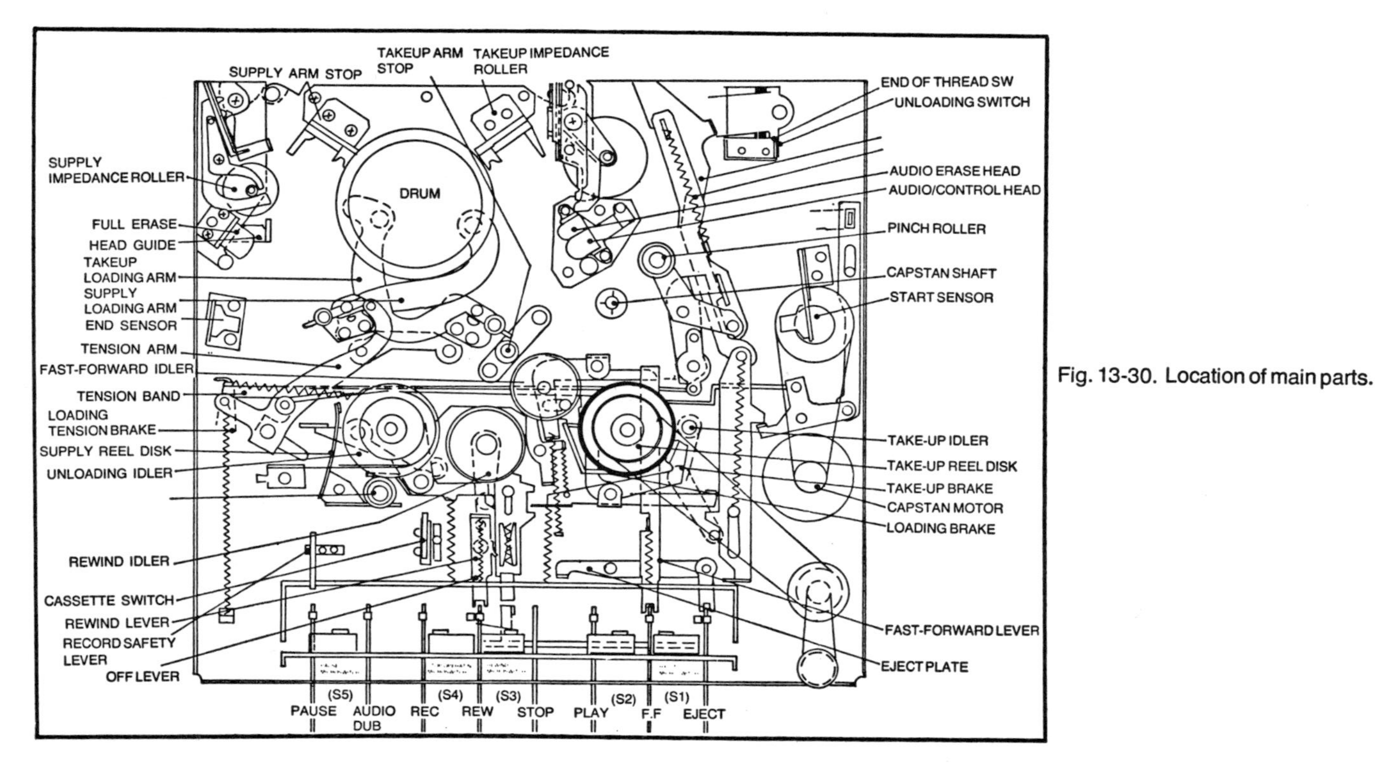

Fig. 13-30. Location of main parts.

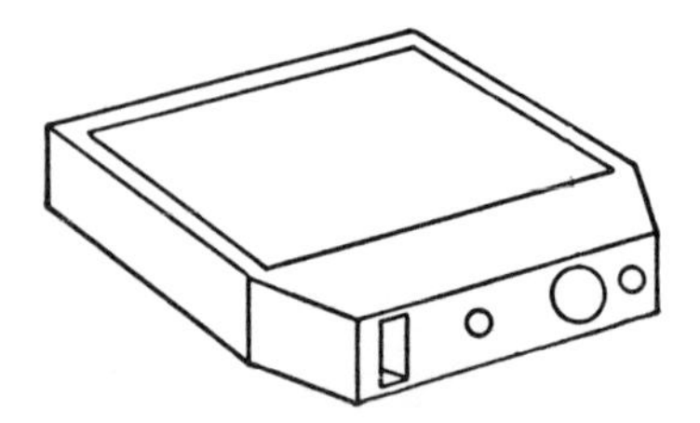

14 Color Correction

No VTR will playback a color signal directly, all need extensive correction of the color signal.

Most of the color problems are due to physical causes such as tape stretch, head wear, slight speed variations in the heads and the tape, etc. They all show up as timing errors and variations in the color signal. The slightest timing variations or errors will completely ruin the color.

In small VTRs the only satisfactory way of avoiding these problems and producing a good color playback signal is to heterodyne the 3.58 MHz down to a lower frequency and add it to the luminance FM signal. The same heterodyne frequency from the same circuit is then used in playback to reproduce the 3.58 MHz from the lower frequency. Switching the same circuit between the record and playback chains saves considerable circuit space and cost in the machines.

The correction to the color signal is applied during the heterodyne up process, where the timing errors in the signal are introduced into the heterodyne frequency.

The consumer VTRs have to be treated slightly differently, as with their zero guard bands and track overlap, the head azimuth offset will not eliminate the chroma crosstalk between the tracks. Consequently, the color correction in these machines must remove both the crosstalk and the timing errors. To do this a major modification is make to the color signal. Phase changes are introduced into the chroma signal during the record down conversion process, and are re-introduced during the playback up conversion.

The Betamax and the VHS use slightly different systems but the principles are the same, and are based on the basic color correction used in the Umatics.

During this discussion, approximate frequencies are used.

PRINCIPLES OF COLOR CORRECTION

The basic principle of color correction is very simple, and in all cases it involves the heterodyne frequency in the playback mode. The heterodyne frequency must be accurate to ensure the interchangeability of tapes between machines. In record it is produced by either a crystal oscillator or by a tightly controlled VCO.

When a tape is played back the timing errors must be removed from the color signal and this is performed in the up conversion process. The timing errors in the playback signal are sensed, either by comparing burst or H sync with a stable oscillator, and then introducing the timing variations into the heterodyne frequency. When the heterodyne frequency with these errors is now heterodyne with the playback chroma subcarrier, the sum and difference products contain no errors.

The reason why this works is best explained by the simple heterodyne arithmetic of the following equation:

$$(4.27\text{MHz} + \Delta) - (688\text{kHz} + \Delta) = 3.58\text{ MHz}$$

where Δ (Delta) means the frequency variations of errors. Notice that this correction is applied to the chroma only; the luminance is not corrected in this fashion. Hence, the original stable phase relationship between the two has not been recreated. This is the main disadvantage of this system. Although this produces a color signal good enough to be viewed on any monitor, it does cause some problems in editing. But these are somewhat alleviated by vertical interval switching and by the use of monitors with fast recovery horizontal circuits.

Although this correction circuitry removes most of the serious time base errors, it does not restore the phase relationship lost during the recording process. Hence, the resulting playback is still 'non phase color'.

The color correction circuit is the heart of the color playback system, and although the circuit details may differ the principle used is the same. It assumes that the errors are constant over the length of an horizontal line. This is not quite true but it does suffice in practice to produce fairly good correction.

The next section of this chapter will discuss the Umatic color correction, as this is quite straightforward and simple, and illustrates the color correction principles very well. After this the comb filter is briefly covered, as this is necessary to understand the more complex circuits used in the color correction in the Betamax and the VHS. After this section, the Betamax and the VHS are covered separately.

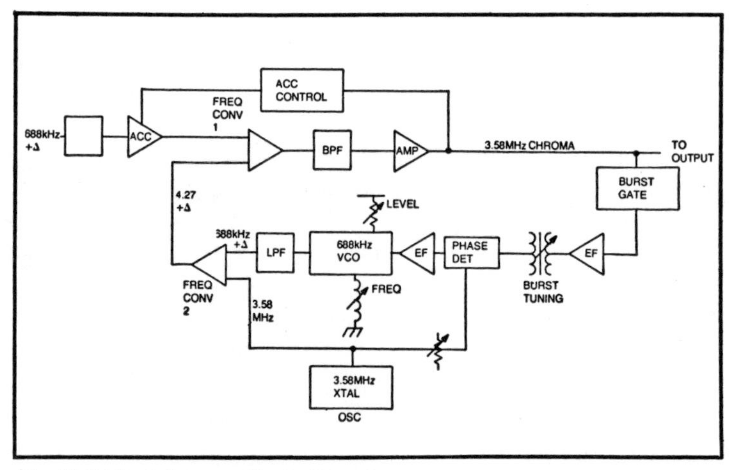

Fig. 14-1. Block diagram of typical Umatic.

UMATIC COLOR CORRECTION

The Umatics use a very simple and straightforward color correction system to remove the color timing errors.

In record the 3.58 MHz is heterodyned with 4.27 MHz and the difference frequency of 688 kHz is put onto the tape. The 4.27 MHz is produced directly by a crystal oscillator or in a voltage controlled oscillator (VCO) controlled by a crystal (usually 3.58 MHz). This makes it stable enough so that the interchange between tapes and machines is assured.

In playback the 688 kHz from the tape is again heterodyned with the 4.27 MHz and the difference frequency of 3.58 MHz is extracted. But now the 4.27 MHz is formed slightly differently and it has the timing errors introduced into it.

Figure 14-1 is a block diagram of a typical Umatic color correction system, showing the signal path in the playback mode. The main input is the 688 kHz chroma with errors (688 + Δ) from the tape. This is passed through an ACC amplifier and fed to the main frequency converter (Freq Conv 1). Here it is heterodyned with the 4.27 MHz and a BPF extracts the 3.58 MHz. Another amplifier buffers the signal and then feeds it to 3 main paths.

The first path is the main output, where the 3.58 MHz chroma is fed to the output mixing amplifiers and added to the luminance to form a non phased NTSC-type signal. This is not a true NTSC signal as it is non phased.

The second path is the feedback to the ACC amplifier, which is used to control the gain of the input.

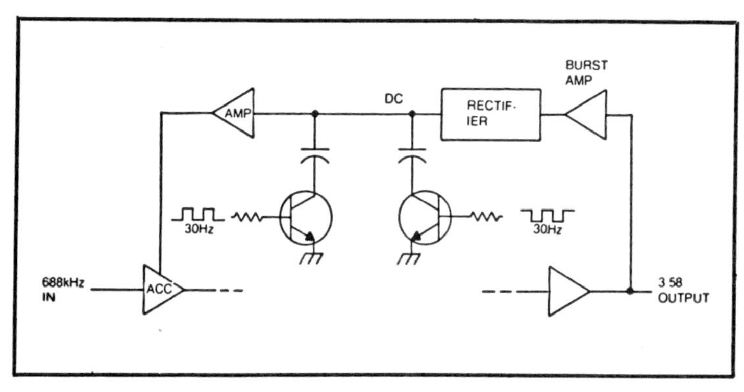

Fig. 14-2. ACC field memory.

The third path is through a burst gate and into the color correction circuitry.

Because of the diversity of models available, different parts of the circuit are likely to be found in IC or to be made with discrete components.

The 4.27 MHz is formed in a second frequency converter (freq Conv 2). The inputs to this are a reference 3.58 MHz and 688 kHz from a VCO. The reference 3.58 MHz is formed in a crystal oscillator, which feeds frequency converter 2 and a phase detector. The phase detector produces a dc output corresponding to the jitter from the tape by comparing the off tape burst with the stable reference. This dc is now used to control the phase of the 688 kHz produced in the VCO. Thus the 688 kHz has timing errors introduced into it which are the same as the timing errors from the tape. The same errors are now introduced into the 4.27 MHz and when this 4.27 MHz with errors ($4.27 + \Delta$) is heterodyned with the 688 kHz with errors from the tape, the errors cancel to produce a stable 3.58 MHz output.

Several refinements are often added to this basic circuit. The most important are:

- The ACC field memory
- Vertical interval drift compensation
- Still frame and slow speed compensation
- External subcarrier input.

ACC Field Memory

This is shown in Fig. 14-2. The burst output is rectified to a dc and fed via an amplifier to the ACC amplifier at the input. This way any changes in the output burst are used to control the input level, and they act to keep a constant output level. Because of differences which can develop between the heads as they wear, a 30 Hz color flicker can appear. To prevent this the dc level also charges a capacitor to an average level and this keeps the dc level from changing too abruptly. A separate capacitor is used for each head, or field, and they are switched by the 30 Hz head tach pulses. This way each head output is controlled by its own average dc level, thus preventing flicker.

Vertical Interval Drift Compensation

This is shown in Fig. 14-3. In the broadcasting machines the burst phase detector which feeds the 688 kHz VCO has two output lines, the VCO has 2 inputs, and a LPF is connected to each line. One line is now set as a reference by the color lock control and the other is controlled by the AFC error from the drum servo. This provides correction for drum and tape speed changes and for temperature variations etc. Also, the time constant of the circuit is changed during the vertical interval. This prevents the dc levels from drifting a long way off during the vertical interval when no burst is present; and prevents color flashes during the first few lines on the screen. The overall result of these changes is much more even control of the color.

Still Frame and Slow Speed

In still frame and slow speed the video tracks are slightly shorter, but they are still scanned in the same time. The output

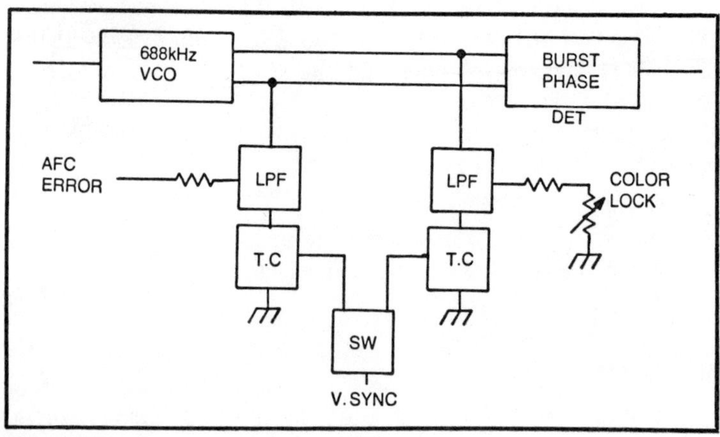

Fig. 14-3. Vertical interval drift compensation.

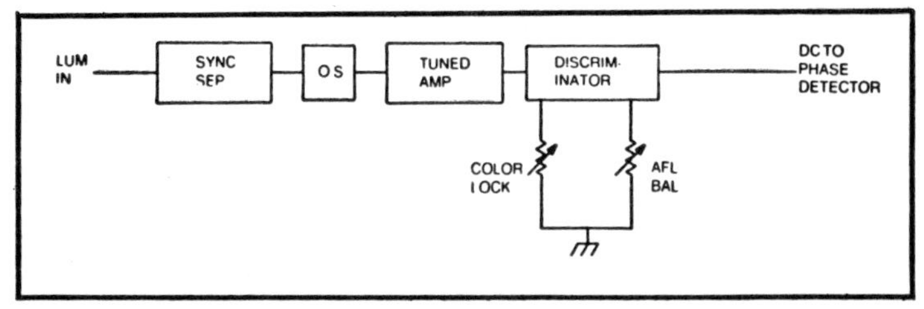

Fig. 14-4. Frequency correction for still frame and slow speed.

frequencies from the tape rise slightly. This rise in frequency is enough to upset the color correction. To obviate this the H sync pulses are converted to a square wave by a one-shot, as in Fig. 14-4. A tuned amplifier feeds this square wave to a discriminator circuit to produce a dc output. During slow speed and still frame the frequency rises and so the dc output changes slightly. This dc is used to pull the output of the phase detector slightly and raise the 688 kHz from the VCO to about 692 kHz. Thus the 4.27 MHz is raised a little, allowing a correct 3.58 MHz to be produced at the output.

External Subcarrier

The 3.58 MHz reference crystal oscillator can be locked in phase to an external 3.58 MHz subcarrier. Usually the crystal is disabled and the external subcarrier keeps the oscillator running and in phase. This allows the heterodyned color correction to be phased to a digital time base corrector.

The biggest fault of the heterodyne color correction is that the color signal is corrected but the luminance component is not. This 'half corrected' signal is quite unsuitable for a time base corrector, which is capable of correcting the total signal. To make the signal acceptable the color must contain the errors. The easiest way to do this is to re-introduce them via a subcarrier from the TBC. This is easier than modifying the color circuitry so that it does not correct the signal. Also, a modified machine would be unable to perform correction when no TBC was available.

The Comb Filter

A comb filter consists of an amplifier, a resistive bridge, and a 1H delay line (a delay line which delays the signal 63.5 microseconds). Normally it is used to separate the luminance and chroma components of a color video signal. Its great advantage over a purely passive filter is that it does not restrict the bandwidth of the two components of the video signal and it does not introduce severe phase errors.

In Fig. 14-5 the video input is split into two paths. One goes directly to the resistive bridge and the other path is inverted before passing through the delay line. Assume a signal arrives which corresponds to a sudden increase in video level. When this signal arrives at the output of the delay line 63.5 microseconds later, the same point in the next horizontal line will arrive at the input. The luminance signal will be the same, but the subcarrier will be out of phase. This is applied to the bridge at the same time the previous line arrives at the output of the delay line.

Figure 14-6 shows the current paths of each signal. The direct luminance signal causes current to flow in the direction of the solid arrows, and the delayed luminance signal causes current to flow in the direction of the dotted arrows. Across resistor R3 these currents add, and so an enhanced luminance signal can be taken from point B. Across R4 the luminance current paths cancel, so there is no luminance output from point C.

Figure 14-7 shows the current paths for the two chroma subcarriers signals. The direct subcarrier signal causes current to flow in the direction given by the solid arrows, and the delayed subcarrier signal current flows as shown by the dotted arrows. Across R3 these currents flow in opposite directions and thus cancel. Across R4 the chroma currents add to produce an enhanced chroma signal.

So one side of the bridge produces luminance only, and the other side chroma only. Note the principle by which this works. A signal is delayed by exactly one horizontal line (1H). The luminance signal from one line to the next changes very little. The

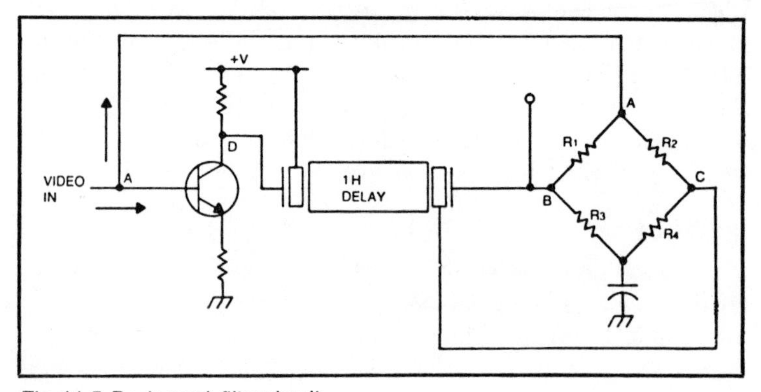

Fig. 14-5. Basic comb filter circuit.

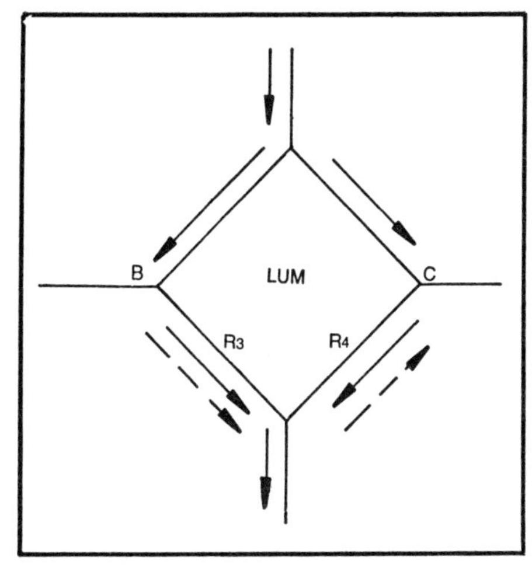

Fig. 14-6. Luminance current paths.

color also changes very little but the 3.58 MHz subcarrier representing the color has changed. It is 180 degrees out of phase. The comb filter makes use of this reversal of phase to separate the color signal from the luminance signal.

To remove chroma crosstalk, the comb filter is used slightly differently. However, its principle of operation is the same. Also, in the Betamax and the VHS, the recorded chroma signal must be modified in phase slightly. This is explained shortly.

BETAMAX AND VHS PHASE CHANGING AND CROSSTALK REMOVAL

The Betamax and the VHS differ from the Umatics and other VTRs as they either have no guard bands or they have a track overlap. These two conditions introduce the problem of chroma crosstalk, which cannot be removed by the azimuth offset of the heads. Color correction in these machines involves both time base error removal and crosstalk removal.

The time base errors removal is similar in principle to that of the Umatics, but the off-tape timing errors are taken from the playback H sync pulses and not the burst.

The crosstalk removal uses a phase changing process in the chroma signal which is not found in other VTRs; and a comb filter to remove the crosstalk. The phase changing method used in each of these formats is slightly different, and will be covered first, as it is essential to understand how the color correction works. Both machines make use of the heterodyne frequency to introduce the phase changes, and the same circuits are used on record and playback. The circuits are covered later in the chapter.

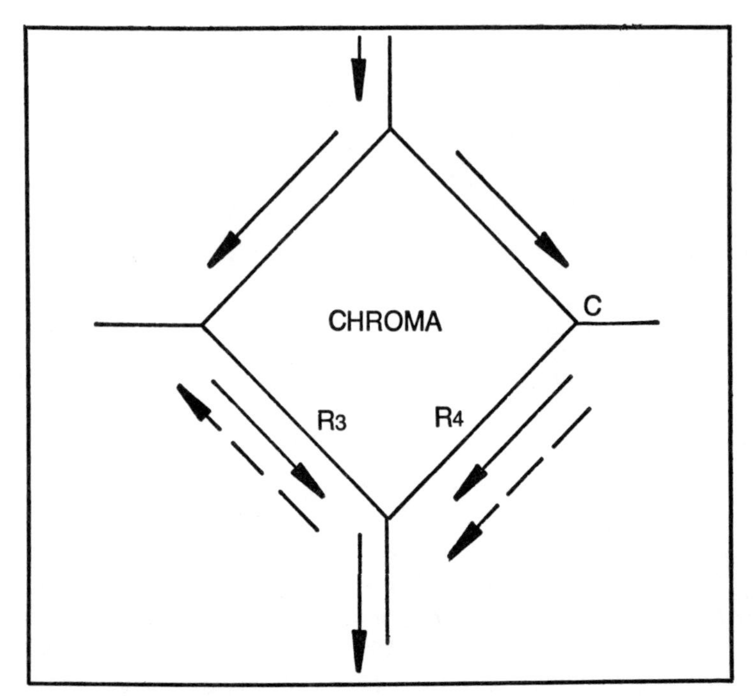

Fig. 14-7. Chroma current paths.

Betamax System

The Betamax uses the simplest of the two systems of phase changing, so it will be covered first.

In the NTSC signal the subcarrier changes phase in each successive line. Any one line is 180 deg different from those adjacent to it. It also changes on a field to field basis, so that corresponding lines in successive fields have a subcarrier of opposite phase.

In Fig. 14-8 the arrows in line A represent the subcarrier phase for the lines in one NTSC field. Only 8 lines are shown for illustration purposes. Line B shows the phase of both the previous

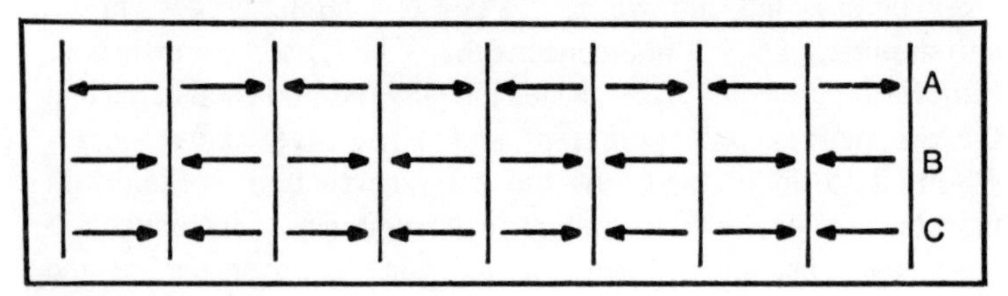

Fig. 14-8. NTSC field subcarrier field.

and following lines, and Line C shows the corresponding line for the other field.

When a video signal is laid down on a helical VTR, the full fields are laid down on the long slanted tracks across the tape. The recorded horizontal lines, adjacent to each other on the screen, are close to each other on the tape. These will be of opposite phase, as shown in Fig. 14-9.

In the Betamax, the signal applied to the B head is recorded as an unaltered NTSC field, as in Fig. 14-10. The signal applied to the A head, which is the alternate NTSC field, has the chroma phase changed on alternate lines. Instead of recording the pattern on line B, it records that of line C.

In the playback mode the B head plays back the track it recorded, but due to slight mistracking and the proximity of the A head tracks, the B head output contains some cross talk from those tracks. So the B head output can be represented by Fig. 14-11, line A. This signal is fed to the comb filter.

The same is true for the A head output; it also will contain cross talk components from the adjacent B tracks. This output is shown in line B. During the process of the chroma playback, the A head signal is phase inverted every other line and becomes as shown in line C. This is fed to the comb filter. In both cases the comb filter removes the cross talk.

Note how in both cases the required signal alternates in phase, while the cross talk does not. It is because of this that the comb filter can remove the cross talk and increase the required signal.

The Comb Filter. The comb filter consists of a delay line of one horizontal line duration, a resistor bridge, and an amplifier. The circuit is shown in Fig. 14-12.

The input chroma signal follows two paths. One is directed to the resistance bridge and the other is directed through the delay line and into the bridge. The bridge combines the two signals and produces an output which is fed to the amplifier and then to the luminance circuits.

In the bridge, the two signals combined are the horizontal line currently arriving at the input and the previous line of the same field which has been retarded by the delay line. The solid arrows show the current path of the undelayed signal, and the dotted arrows show the delayed signal of the previous line. The output of the bridge is taken across resistor R4 and here the two current paths are flowing in opposite directions. The chroma signals of the

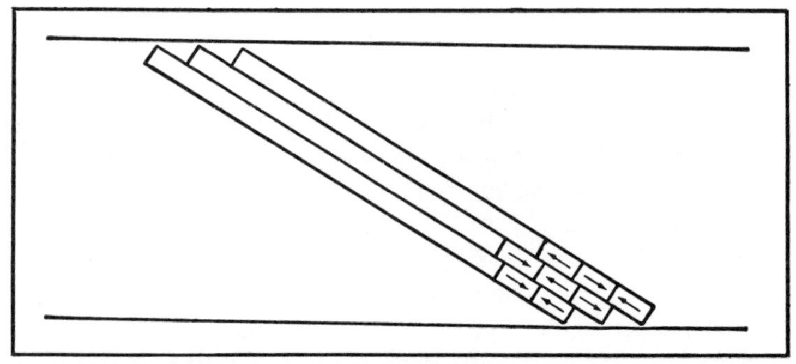

Fig. 14-9. Video signal laid down on a helical VTR.

two lines from the same head will add, but the cross talk from the adjacent tracks is differenced. Figure 14-13 helps to explain this for the B head.

Line A shows the undelayed B head signal, and line B shows the delayed B head signal. Figure 14-14 shows the effective reversal of the delayed B signal when it flows through R4. It can be seen from this that the main B signal chroma components add and the cross talk cancels. Figure 14-15 and 14-16 show the same thing for the A head signal.

The balance control in the undelayed path is adjusted to eliminate the cross talk as much as possible. The output from the amplifier has a very strong corrected chroma signal component, with almost all of the cross talk completely eliminated.

BETAMAX COLOR CORRECTION CIRCUIT

The color circuits are the same in the record and playback modes, but their use in the record mode is much simpler, so this will be covered first. Figure 14-17 shows the basic heterodyne circuit. This is only part of the total color circuit, and is sometime called the AFC loop.

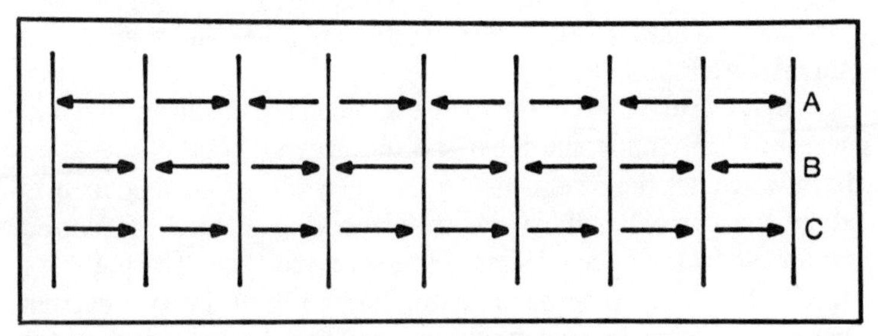

Fig. 14-10. The signal as recorded as an unaltered NTSC field.

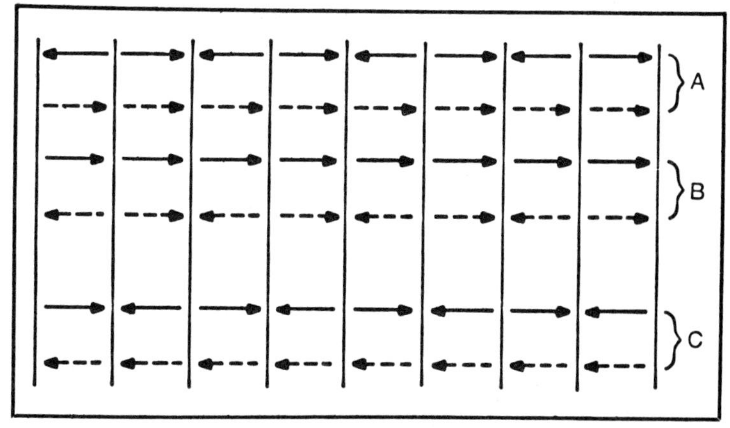

Fig. 14-11. Playback mode characteristics.

Instead of using a crystal to form the 4.27 MHz, a phase locked loop (PLL) type of circuit is used. This produces a 692 kHz output which is exactly 44 times the horizontal frequency, and is phase locked to the incoming H sync pulses. This is heterodyned with 3.57 MHz (not 3.58 MHz) from a crystal oscillator and the resulting 4.27 MHz extracted by a filter. This now passes through the phase

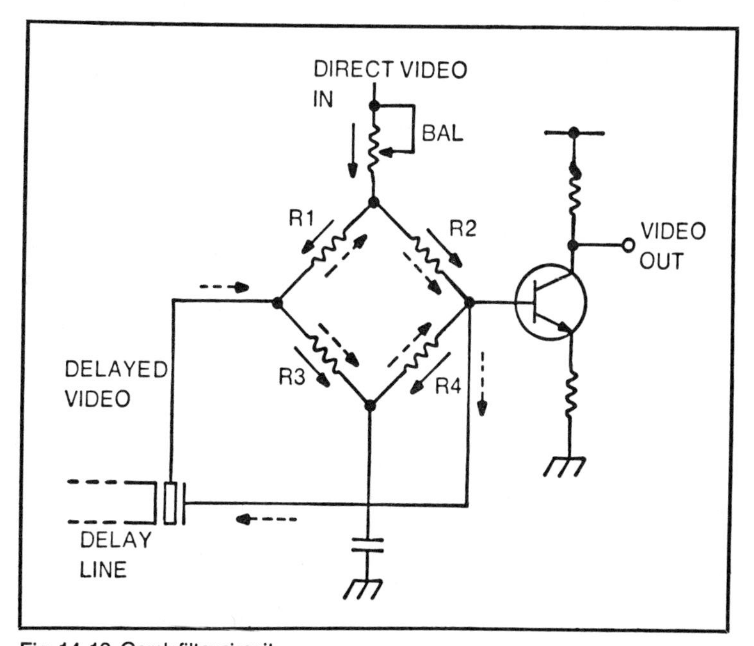

Fig. 14-12. Comb filter circuit.

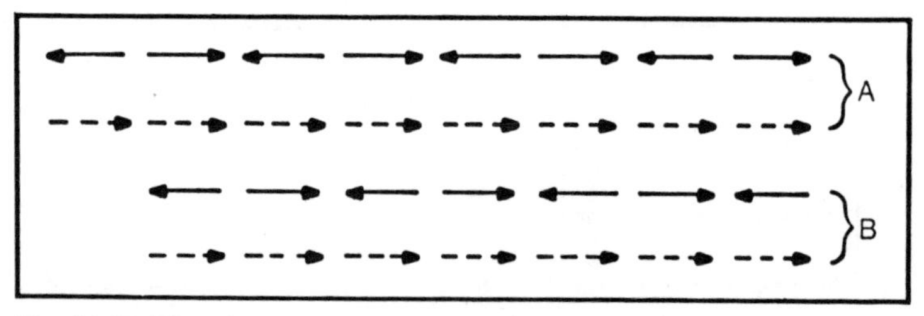

Fig. 14-13. B head signals. (A) Undelayed B head signal (B) Delayed B head signal.

changing circuit and enters the main frequency converter in the chroma record chain. As the 692 kHz is locked in phase to the incoming H sync, then so is the 4.27 MHz and so also is the 688 kHz formed in the main frequency converter.

The phase changing of the 4.27 MHz is performed in the circuit shown in Fig. 14-18. The 4.27 MHz is fed to the primary of a transformer, and each end of the secondary is alternately selected by a switch in the IC. When the A head is recording, H sync pulses alternate the switch position to invert the phase of 4.27 MHz by 180 degrees on alternate lines. When the B head is recording the 30 Hz head tach pulses lock the flip flop and the switch in one position so the 4.27 MHz phase remains constant.

Figure 14-19 is the full chroma correction circuit used during the playback mode. It is basically the same as that used in record, but with a few addition. The main addition is the APC loop.

The 688 kHz is heterodyned in the main frequency converter with the 4.27 MHz to form the 3.58 MHz which will be added to the output signal. The 4.27 MHz is again formed in the AFC circuit. The input to the AFC circuit is now the playback H sync pulses. As these contain all the timing errors and jitter, these errors will be introduced into the 4.27 MHz. So when this is heterodyned with the 688 kHz with the same timing errors, the resulting 3.58 MHz

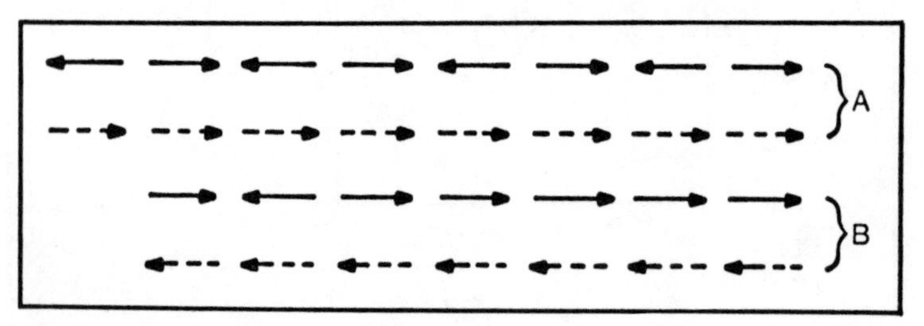

Fig. 14-14. B head signals. (A) Delayed. (B) Reversal of delayed signal when it flows through R4.

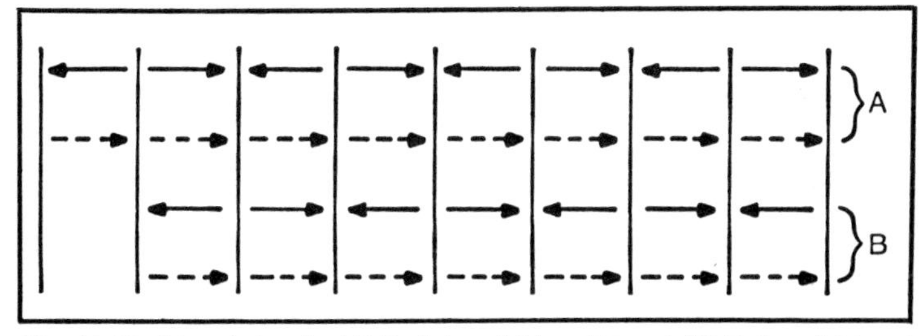

Fig. 14-15. A head signals. (A) Undelayed A head signal. (B) Delayed A head signal.

will be error free. The 4.27 MHz is phase changed in the same way as in the record mode, and the same circuit is used.

However, in the playback mode the 3.58 MHz is not entirely free of errors. The 692 kHz produced in the PLL is 44 times the horizontal frequency, and 688 kHz chroma is only 43 ¾ times the horizontal frequency. As the timing errors in the H sync are multiplied by 44, along with the H sync, then the timing errors introduced into the 4.27 MHz are also 44 times the horizontal frequency. This causes a slight overcorrection to occur in the final output. To offset this, the output burst is separated from the output 3.58 MHz and phase compared to a 3.58 MHz from a crystal oscillator. This is in the APC loop. The dc from the phase comparator is now used to pull the frequency of the 3.57 MHz oscillator - which is now used as a VCO instead of a crystal. This slight frequency pull adds a correction into the 4.27 MHz as it is formed, and the result is the removal of the remaining slight timing errors.

During playback it is possible that the phase changing can occur on the wrong line or the wrong field. This can be caused at power-on, at the start of a tape, or after a power surge or a drop-out. If this occurs then the playback burst will be 180 degree

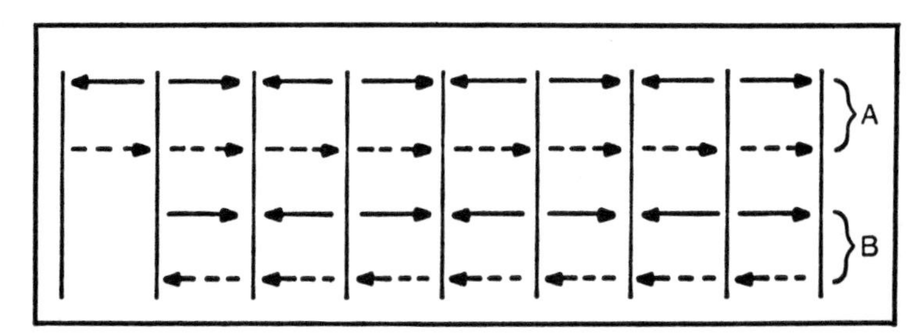

Fig. 14-16. A head signals. (A) Delayed. (B) Reversal of delayed signal when it flows through R4.

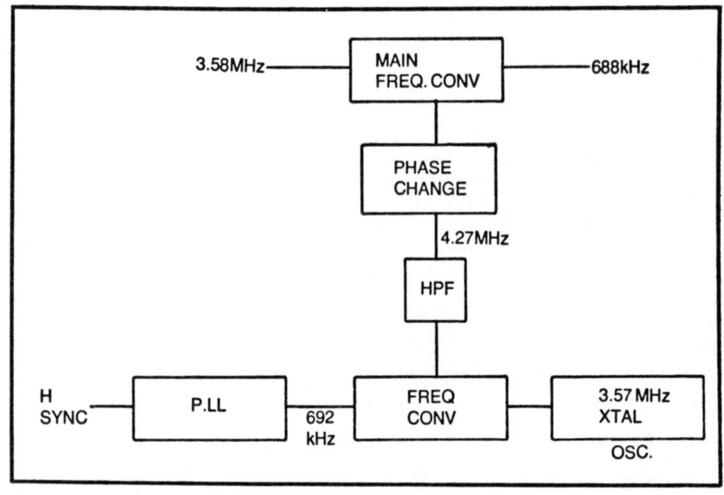

Fig. 14-17. Betamax hetrodyne correction circuit.

out of phase. To correct this the burst ID circuit is used. This is shown in Fig. 14-20. The phase comparator shown here will now produce a large voltage output error, which will switch the flip flop and thus correct the phase changing of the 4.27 MHz. When the burst phase returns to normal, the comparator output changes back to normal.

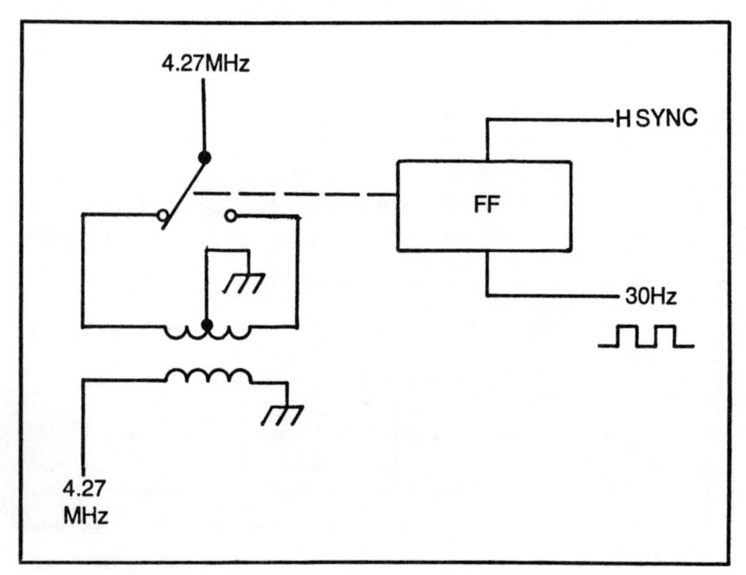

Fig. 14-18. 4.27 Mhz phase changing.

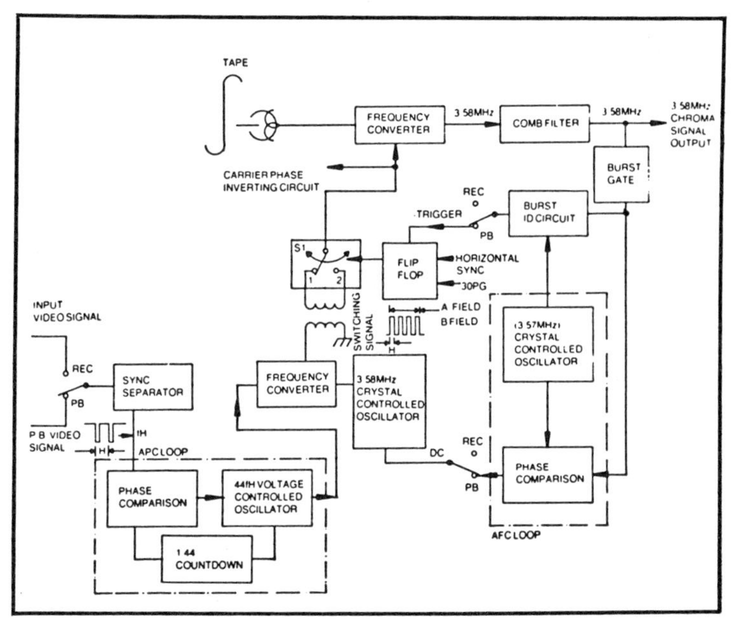

Fig. 14-19. Betamax chroma correction circuit.

THE VHS SYSTEM

The VHS system is a little more complex than the Betamax system. The diagrams will help in following the explanation. Line 1 in Fig. 14-21 shows a signal with normal NTSC phasing. Line 2 shows the phasing used when the A head is recording. Here the chroma subcarrier is phase advanced 90 degrees on each successive line. Line 3 shows the NTSC phasing for the next field, which will be recorded by the B head. Line 4 shows the chroma phase retarded by 90 degrees each successive line. Note how in each case the first line of the field retains its normal NTSC phase. Line 5 shows the A head playback signal. The solid arrows are the main signal from the A track, and the dotted arrows show the crosstalk picked up from the adjacent B track. The playback chroma process now restores the correct NTSC phasing to the main signal by reversing the phase 90 degrees on each successive line. This also makes all the crosstalk have the same phase, as in Line 6. It is this signal which is fed through the comb filter, where the delay causes each line in the A head to be compared with the next one. The directions of the current in the resistive bridge are now arranged so that the main components add and the crosstalk components cancel (lines 7 and 8).

Line 1 of Fig. 14-22 shows the B head playback with the A head crosstalk in dotted arrows. Line 2 is the phase corrected signal. Again the crosstalk is of the same phase and the main signal has NTSC phasing.

VHS Color Correction Circuits

The VHS circuits are similar in principle to the Betamax, as they must perform exactly the same job, but the details are quite different. Figure 14-23 is a block diagram of the chroma record and playback heterodyne circuit, and the frequencies shown are for playback and record.

The subconverter in IC 14 accepts the 629 kHz from the AFC circuit and heterodynes it with 3.58 MHz from a VCO. The resulting 4.2 MHz is extracted by a BPF and used in the main converter in IC 8. The 629 kHz formed in the AFC circuit is phase locked to the incoming sync, and also has a phase rotation applied. This is different from the Betamax, as here the phase rotation is applied to the 629 kHz before the 4.2 MHz is formed.

The AFC Circuit

The incoming H sync pulses trigger a monostable to form a square wave. This functions as the phase reference for a phase detector, and it is also divided to a ½H signal. The VCO in IC 16 produces a 160 H signal which is counted down to provide 80, 40

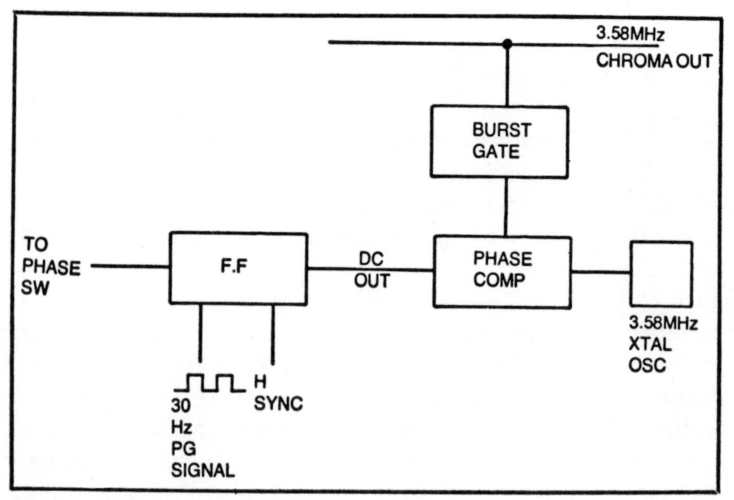

Fig. 14-20. Burst I.D. circuit.

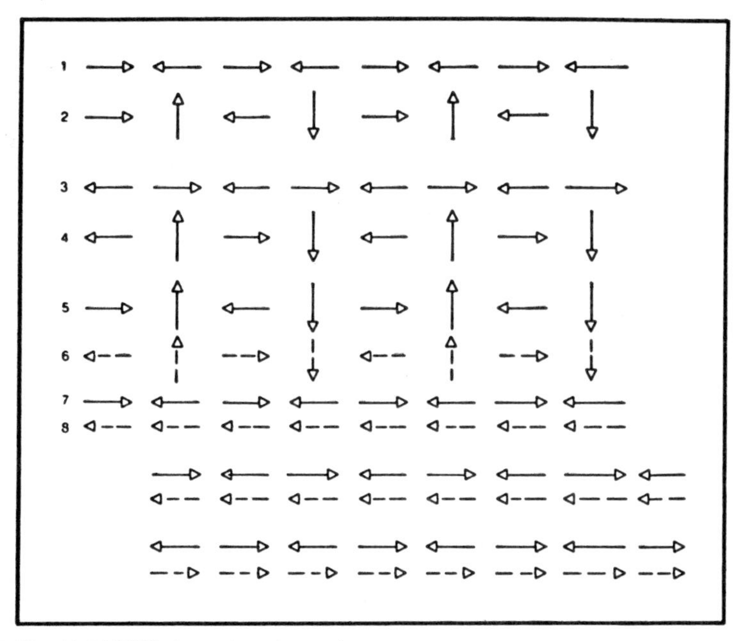

Fig. 14-21. VHS phase changing system.

and 10 H signals. The 10 H is now divided by 10 to provide a feedback signal at the H rate for the phase detector. The output of the phase detector is a dc control for the VCO. All the outputs at the multiples of the H frequency are now applied to the data selector IC. See Fig. 14-24.

The data selector IC 13 accepts the various H rate signals and these act as codes to select which of the 'D' inputs will appear at the Q and $\overline{Q}$ outputs. Table 14-1 shows the code. The outputs are combined resistively, and are controlled by the outputs of the flip

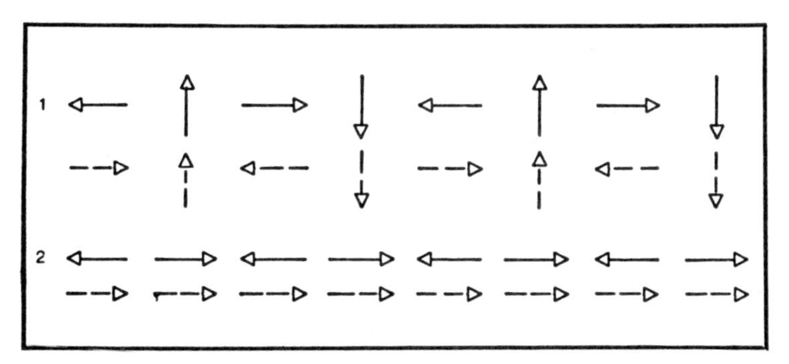

Fig. 14-22. The phase corrected signal.

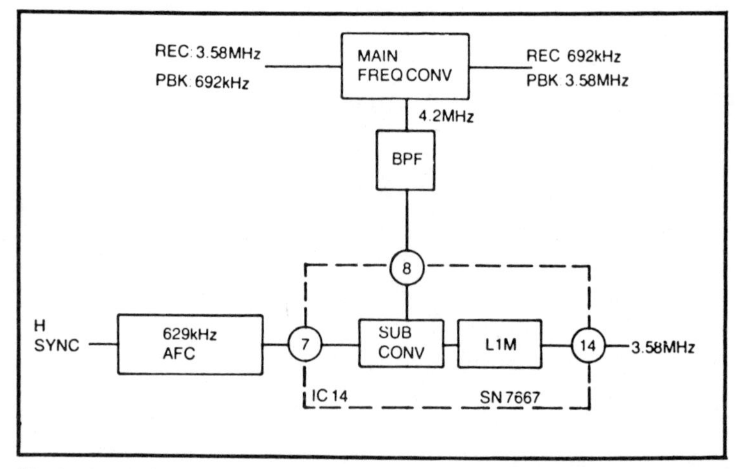

Fig. 14-23. VHS chroma circuit.

flop in IC/12. See Fig. 14-25. This toggles at the ½H rate and effectively selects the Q or $\overline{Q}$ from the data selector. Both these outputs are square waves at the 40 H rate - which is 629 kHz, but with the characteristic that at the beginning of each line the relative phase of the square wave to H sync pulse has changed by 90 degrees. See Fig. 14-26.

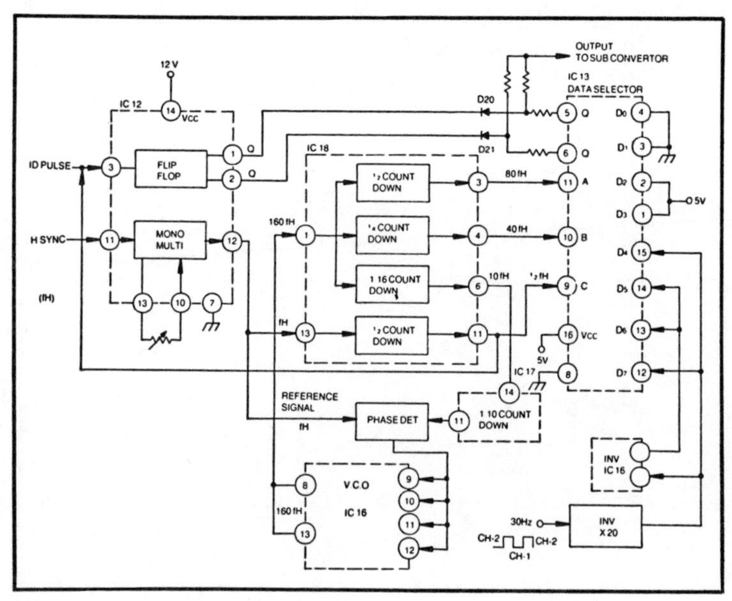

Fig. 14-24. Block diagram of AFC system.

Data Selector IC Inputs and Outputs.					
Input			Output		"L": Low potential "H": High potential
C	B	A	Q	$\overline{Q}$	
L	L	L	D_0	$\overline{D_0}$	
L	L	H	D_1	$\overline{D_1}$	
L	H	L	D_2	$\overline{D_2}$	
L	H	H	D_3	$\overline{D_3}$	
H	L	L	D_4	$\overline{D_4}$	
H	L	H	D_5	$\overline{D_5}$	
H	H	L	D_6	$\overline{D_5}$	
H	H	H	D_7	$\overline{D_7}$	

Table 14-1. Data Selector IC Inputs and Outputs.

It is this square wave which is heterodyned with the 3.58 MHz from the VCO to produce the 4.2 MHz with the 90 degree phase shift, which is now used in the main converter. A square wave is basically a sine wave of the same fundamental frequency with many harmonics added. The heterodyne products of the harmonics are easily filtered out, leaving only the heterodyne product of the fundamental with the 3.68 MHz.

The APC Circuit

This section of the circuit is used only during playback. It is shown in Fig. 14-27 and it has 3 main functions:

- Phase control of the color signal.
- Color killer operation by phase detection.
- AFC control, with an ID pulse.

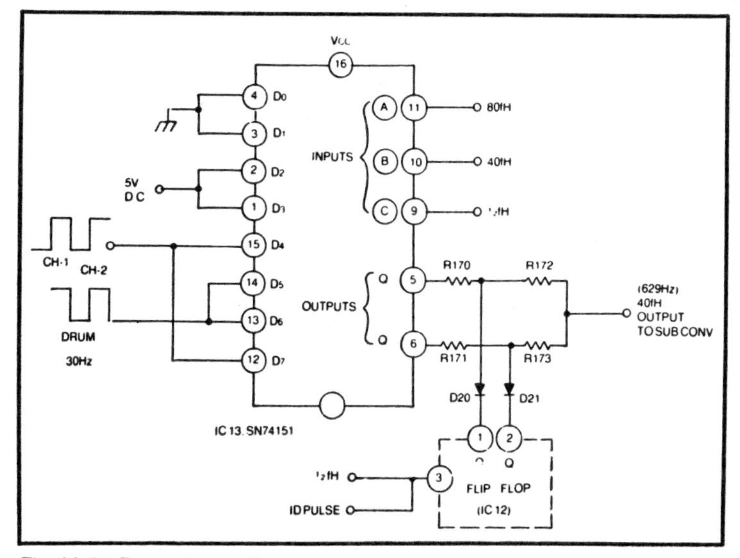

Fig. 14-25. Data selector IC.

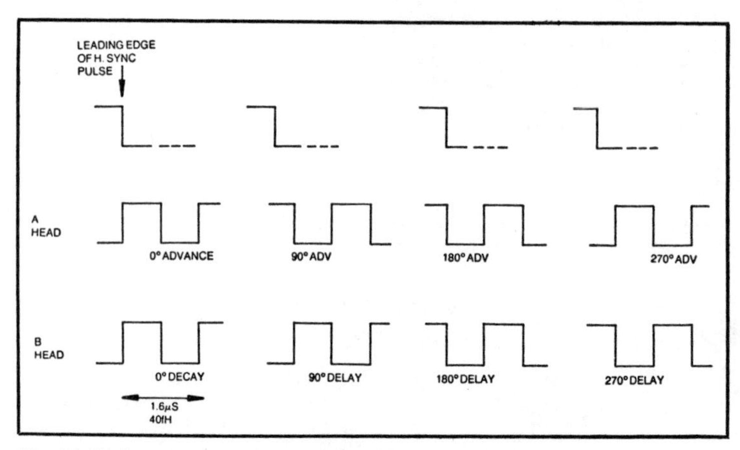

Fig. 14-26. Square wave phase relationships.

Burst is separated from the 3.58 MHz output color signal and is phase compared to a direct 3.58 MHz and a 90 degree phase shifted 3.58 MHz signal from a crystal. Phase detector 1 produces a dc output which controls the 3.58 MHz VCO used to form the 4.2 MHz. This removes residual timing errors from the output signal. Phase detector 2 produces 2 outputs. One produces the burst ID pulse. If switching occurs on the wrong phase during playback, then this sends a positive pulse to the flip flop in the AFC circuit which corrects the phase of the 629 kHz to the subconverter. Both

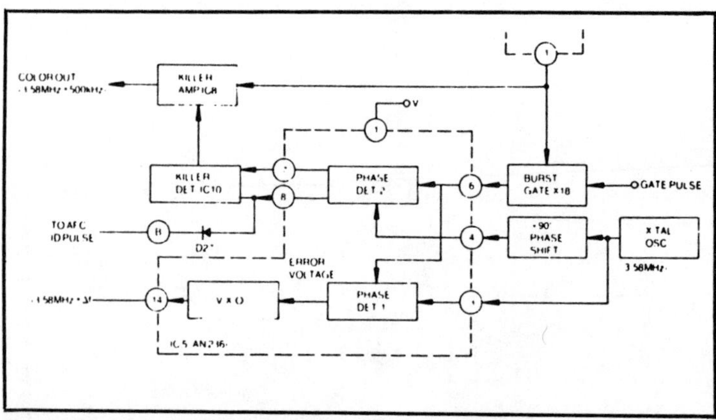

Fig. 14-27. Block diagram of automatic phase control (APC) system.

outputs from phase detector 2 enter the killer detector in IC 10. The output of this is dc which controls the final color amplifier. It inhibits the output until the color phase is correct.

15 General Care and Simple Maintenance

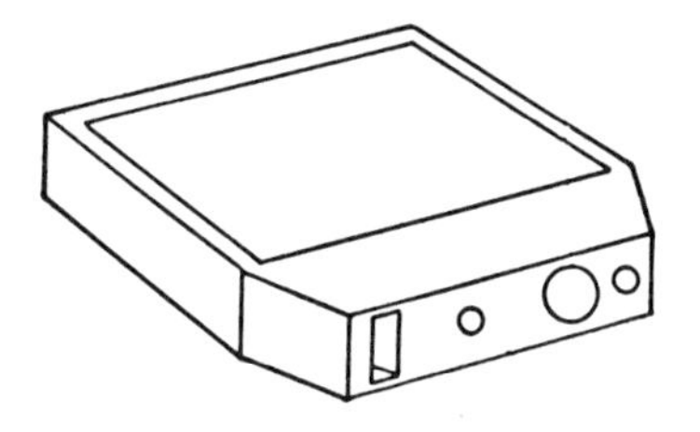

This chapter covers the things an operator can do to look after the tapes and machines and to keep them in good running order. It is divided into three parts.

The first covers general care and looking after the machines and the cassettes. It outlines procedures which can help avoid trouble and aid in curing the simplest problems which might arise.

The second part covers the initial steps an operator should take when a malfunction appears to occur and progresses into the simplest of the maintenance jobs an operator can do.

The final section delineates the jobs an operator should learn to do and those which are best left to an experienced technician or engineer. It concludes with a list of tools and the spares that should be on hand at all times.

Reference is made in the following to a *test tape* and an *alignment tape*.There is a difference between these two items which should be understood before any fault tracing or maintenance is attempted.

TEST TAPE

This is a tape which has been recorded locally by the operator and is know to contain good video and good audio on both tracks. This tape serves a definite purpose in any studio or taping setup. It provides a known good tape which can be played back through any system suspected of having problems. It removes a variable from a doubtful situation and thus provides a known good source of video and audio. It should be used in all situations involving a defective machine when a tape is needed to be put onto that machine. Should it become damaged, lost, or destroyed, another can easily be generated.

All systems should keep a tape as a test tape. Ideally, the video should be color bars or multiburst, but anything will suffice if these are not available. The best audio is piano music, next to this any music, and after this a simple tone. This is a work tape and should be treated as such.

ALIGNMENT TAPE

It should be fully understood that this is not a maintenance item.

All manufacturers produce an alignment tape for their machines, which should be used only for final checks in a known working system with no faults. Alignment tapes are expensive and not always readily available.

The alignment tape should never be used to check out a defective or suspect machine, because such a machine can easily ruin the tape. When a machine is being repaired, the test tape should be used to check it out and get it into working condition, after which the alignment tape is used for the final adjustments of the electronics.

This tape should be kept in a safe place and its box marked with red tape to distinguish it from others.

GENERAL CARE OF THE TAPE AND MACHINES

There are certain simple regular procedures that can obviate and limit many of the trivial problems which plague the tapes and machines. Most of these involve nothing more than general care in using the machines and cleanliness around the tapes and machines. Each of these are outlined briefly, but their importance should not be overlooked or minimized.

The Work Area

The cleaner the work area in which the cassettes are used, the less chance there is of trouble from extraneous dust and dirt. Although a "clean room" atmosphere is not necessary, things like eating, smoking, drinking, and having hands that are dirty or greasy should be discouraged around any VTR or studio. Ashes and food particles can ruin heads and tape, and spilled coffee or other drinks are a disaster for electrical equipment and are dangerous to personnel.

Grease from fingers can adhere to the tape and the machine parts, and this will pick up dust and dirt. If this gets onto the tape, it can become embedded to the detriment of the tape and heads, and can actually later transfer to another machine and continue its bad work.

Fortunately, most problems of this type are lessened by the enclosed format of the cassettes, but these things cannot be totally ignored:

- Do not allow excess moisture to enter the machine or tapes.
- Do not cover the air holes in the machines by putting things over them.
- Always handle the machine very gently; do not bang it around.
- Do not drop cables and plugs onto the floor. This can bend and shatter them.
- Do not pass cables through doorways which are allowed to fully close. This invites broken cables and damaged plugs. For the same reasons do not step on them or allow heavy equipment to roll over them.
- Never use damaged tape in a machine. It must be removed and discarded.

Cleaning the Machine

Cleaning is essential to any VTR. The two main things which cause cleaning to be necessary are the oxide shed from tape and the environmental dust, dirt, and grease. Most modern tapes do not shed as readily as the older types, and the enclosed nature of the machines helps to keep much of the extraneous material from the critical parts.

The buildup of oxide shed from the tapes occurs over a period of time and must be cleaned off. It will gradually affect the tape path, and the interchange of the tapes will be ruined. In serious cases the heads can completely clog, and no recording or playback is possible until they are cleaned.

With the cassette machines, a special cleaning cassette is available. This contains a mildly abrasive tape which will remove the excess oxide and dirt from all the places the tape touches. It is simply inserted and then played for about 30 seconds, which is about ten numbers on the tape counter. It is then removed without being wound to the beginning. Next time it is inserted and played again for the same period of time and again removed without being rewound. In this way it is gradually played to its end. Only then is it rewound to the beginning. Now it can be used all over again, and this can be repeated up to ten times. Excess use will wear out the heads and the guides, so its use should be severely limited.

It is possible to clean the machine with more conventional techniques, and this should be encouraged. For cleaning, the top of the machine is removed to expose the interior. A lint-free cloth or cotton swab should be used for cleaning and should be wet with a suitable cleaning solvent such as denatured alcohol, Freon, or any of the cleaning fluids offered by the manufacturers. The audio and control head, the erase head, and all the tape guides should be carefully and completely cleaned. So should the capstan, but care is needed with the pressure roller. The wrong cleaning agent will either dissolve it or cause it to harden and crack.

Cleanliness with any VTR is very important. The normal buildup of dirt and oxide can ruin both playbacks and recordings and eventually destroy the video heads. It is easy enough to do, and there is no excuse for ignoring it.

Cleaning the video heads is a special task which is covered later.

Cassette and Tape Care

There is rarely a fault with a cassette or a tape. They are so simple that very little can go wrong with them. The problems which do beset them usually are caused by outside factors, and these can range from a faulty machine to bad environmental conditions. In a good machine in good surroundings, a cassette will behave perfectly every time, provided there is no operator misuse.

Much has been written about tape and tape care, and manufacturers will readily give out pamphlets concerning this important subject. One of the reasons for the development of the cassette format was to alleviate much of the handling care required with the open reel tapes, because handling is a cause of much trouble in nontechnical areas and nonstudio use. All of the handling problems have been largely offset by the enclosed format, but certain problems still can arise, and there are areas in which caution is required. One of the most important of these is the environmental conditions.

Environmental Conditions

Apart from the general cleanliness of the surroundings, there are other factors which will affect the operation of a videocassette machine. Temperature and humidity have a great affect on both the machine and the tape. A high ambient temperature will allow the tape to stretch, and if the tape tension is too great, it can be disastrous. The length of the video tracks will be altered, and thus further use of the tape can be impaired. With too high a temperature and stretch the tape may never recover its original dimension and will be useless. Although a high temperature will cause a slight expansion of the metal parts, this will not permanently harm the machine, but it can affect tape tracking at the time.

The main problem with a low temperature is moisture condensation which causes tape "sticktion" and a reluctance of the mechanical parts to move easily. The combination of high humidity with high temperature causes the tape to stick, and also can be the cause of water particles or vapor being entrapped in the layers of the tape. This will result in layer to layer sticking, with its attendant problems. A low humidity has static electricity associated with it, which is always a problem around sensitive electronic equipment.

For these reasons most manufacturers recommend that the machine and the tape be kept in as constant an atmosphere as possible at all times, and in ideal room conditions. In fact one machine manufacturer states quite definitely that the machine should only be used at room temperature. If it is cold or a cold tape is used, then it should be allowed to acclimitize so it does not cause the machine to jam.

If environmental changes are made, then as long as possible should be allowed for the tape and machine to readjust. As little as 4 hours and as much as 24 hours have been suggested.

Tape Disentangling

Occasionally the tape will become entangled in the insides of the machine and will have to be extricated. This must be undertaken with the utmost caution and care so that the video heads are not damaged. Usually the tapes which suffer this become unusable and must be sacrificed. The biggest problem now is not getting the tape out of the works but removing the cassette itself. This cannot be accomplished until the **stop** button has been used and full retraction has occurred.

The best way to untangle the tape is: Depower the machine. Then the tape is freed from the mechanism and is held by hand so that it forms a loop with its ends disappearing back into the cassette. The **stop** button is pressed, and power is reapplied. The unthreading process will complete, and the tape will then try to retract back into the cassette as the rewind period occurs. This will usually pull all the tape into the cassette. After this the eject should be used and the cassette removed from the machine.

With care, this can be accomplished without damage to the machine or further damage to the tape.

End Damage

Cassettes often suffer damage to the end few feet of the tape; it becomes bent, warped, and rippled along the edges. The main reasons for this are too many stops and starts with their attendant threading and retractions, operator misuse, and playing of the tape on a faulty machine.

When this occurs, future threading and unthreading can be seriously impaired, playback becomes impossible, and often the tape will end up entangled in the insides of the machine.

A damaged cassette should never be put into a machine, and all damaged tape should be removed immediately to prevent this possibility.

Tape Repair

Damaged tape should be pulled from the cassette, cut off, and then discarded. The remaining good end then should be attached to the plastic leader and the tape rewound into the cassette. The tape should be spliced using an approved aluminum-backed sticky jointing tape, which should be applied to the tape backing and not to the oxide side of the tape. Joints should not be made in the tape itself. If a joint is improperly made, then the tape will not track correctly and a good recording or playback will never be possible. But more important is the fact that the joint will now attract and retain dust and other debris. This will then get deposited as a sticky blob somewhere around the tape path. When a good tape is then played, it can become contaminated and ruined. A further problem is that the sticky tape can pull at the video head chips and ruin them.

In cases of severely damaged tape being pulled back into the cassette, it may be necessary to dismantle the cassette to extricate the tape. Dismantling a cassette is very easy. All the screws on the underside are removed, and the cassette is separated into two parts. This can be hindered by the label, which may have to be cut along the joint. When separating the two parts, care must be taken to prevent further damage to the tape or the reels, so it is best conducted on a flat tabletop. At this time note carefully how the tape is guided in and out of the cassette by the metal posts, and ensure that it is rethreaded in exactly the same path.

When the tape is spliced it should be rethreaded carefully and the cassette assembled. The screws should be carefully replaced and tightened. Turn the reels by hand to check that they are not binding and that the tape moves freely before using the cassette in a machine.

SIMPLE MAINTENANCE

This section covers the simple front line maintenance procedures which an operator can do in the event of some problem or trouble. Most of these are not serious and can be determined and corrected by an operator with a minimum of effort and knowledge. Many of the apparent problems do not involve the cassette machine but are due to external items, such as those covered in this section.

A series of initial checks are described, which should be worked through in the event of a problem. These are followed by some simple maintenance jobs which will account for the majority of the operational trouble.

Initial Checks in the Event of Trouble

At first sight there are three observable faults which arise to impair the use of the videocassette machine. These are as follows:

- Nothing happens at all to the machine or the tape. There is no tape motion and no picture in any mode.
- The tape is run out and is transported, but no picture is seen in the **playback, e-e** or **record** modes and no audio is heard.
- The tape begins to transport, but it malfunctions.

In the first two cases, the simple lists below should be checked first. This procedure will not necessarily cure the problem, but in many cases one of these simple things will be the sole cause of the fault. The order is not important.

List No. 1. Check that the machine is turned on. Check to see that it is plugged into the wall power point. Check to see if the wall point is delivering power.

(Use a monitor or lamp for this.) Check the machine fuses. Check the fuse or circuit breaker for the wall point. Is the videocassette in? Has the cassette dropped into position? Has a main function key or button jammed down, thus preventing any other function from being selected? Is the power cable good? If a power distribution strip is being used, is it turned on? Has its fuse blown? Is it any good?

List No. 2. Some switch on the cassette machine or monitor is in the wrong position. Check them all. A plug or a cable is faulty. Check by changing them. A cable is plugged into the wrong place. Check it. The TV monitor is not plugged in or is not turned on.

If these simple initial checks do not produce the picture, then try the list of periphery items later in this chapter. If these do not help, then the procedures and tables in Chapter 16 should be consulted.

In the case of reason 3 above, this indicates that there is some problem within the machine itself. In this case the tables and advice of Chapter 17 should be followed, and the service manual should be consulted.

Although these may seem obvious, it is surprising how often one of these simple things turns out to be the sole cause of the trouble. If none of these produce the desired mode of operation, then changing cables is a good place to start. Changing the TV monitor and dusting the test tape are also good checks to make. Each of these actions helps to isolate which of the items is causing the trouble and malfunction. If none of these eliminate the fault, then it could be the machine itself or some other piece of equipment. If this appears to be the case, then these should be checked according to their service manuals.

Cables and Plugs

With continual use and flexing, cables will both break and pull loose from the internal pins in the plugs. The most common breaking point is close to the plug where the most stress is applied. This is one of the most common faults in a system and is one of the easiest to fix.

To fix a broken cable and plug, first get the correct tools and carefully disassemble the plug. Once you are inside the plug, carefully check and note to which pins the cables are attached. Often the wires are color coded, and this should be noted in the margin of the service manual or on a card. This ensures correct placement. Removing the old wire and attaching the new is mainly a matter of care and experience and is a case of practice making perfect. Once reassembled, continuity should be checked with a meter before the cable is reintroduced into service.

Cables should never be tied into knots or bent into tight loops. This is especially true with video cables, because they have a solid center conductor which breaks easily, and the insulating dielectric can affect the signal if it is deformed.

With the special video plugs, such as the BNC, F, and PL 259 types, special tools are available. It is advisable to use them at all time.

Switches and Knobs

All switches will give trouble after long use, and all knobs will work loose in time.

The two main problems with switches are that they become dirty and defective. When they are dirty, they often can be cleaned with a spray cleaner. If this does not cure the problem, then replacement is advised. Dirty switches cause video flashes, audio noise, intermittent operation, or a combination of all these. Defective switches are dangerous, especially if they carry power; the only way to deal with these is to replace them.

In the videocassette machines, three types of switches are in common use.

- **Microswitches.** Figure 15-1 shows a microswitch, and these are used extensively in the cassette machines, both in the operator keys and in the internal mechanics. Usually they will operate for a long time without trouble. When they fail replacement is necessary because repair is impossible due to their molded nature.

They are easy to replace: the mounting bolts or screws are removed and the wires unsoldered. The new one is put in the exact place of the old one and bolted down and the wires are reattached. Often their position is critical and needs careful adjustment, especially if they are actuated by some moving part within the machine. At all times consult the service manual when changing such a switch.

- **Slide switches.** A few slide switches are used. Most machines do not have these mounted on the printed circuit boards or operated by the internal mechanics. When they give trouble they should be replaced.

- **Toggle switches.** Toggle switches are almost always hand operated, and so they tend to be exposed and obvious. Again, they should be replaced if troublesome, because often they carry power.

Switch replacement is not hard and is well within the capabilities of the average operator. A minimum of tools is required. Always replace a switch with an exact replacement. Although a substitute will work electrically, often it will not fit mechanically, and this may be the most important factor governing its use. Depower the machine every time a switch is serviced.

Knobs simply need to be tightened when they work loose. The correctly sized screwdriver or wrench should be used; otherwise damage can result. A knob shaft should never be turned with pliers or a wrench in the absence of a knob. This will gradually ruin the shaft and make the replacement of a knob difficult or impossible. Many of the knobs on the cassette machines are the pull-off, push-on type and need no tightening. Always replace these with the correct item if they become lost or defective.

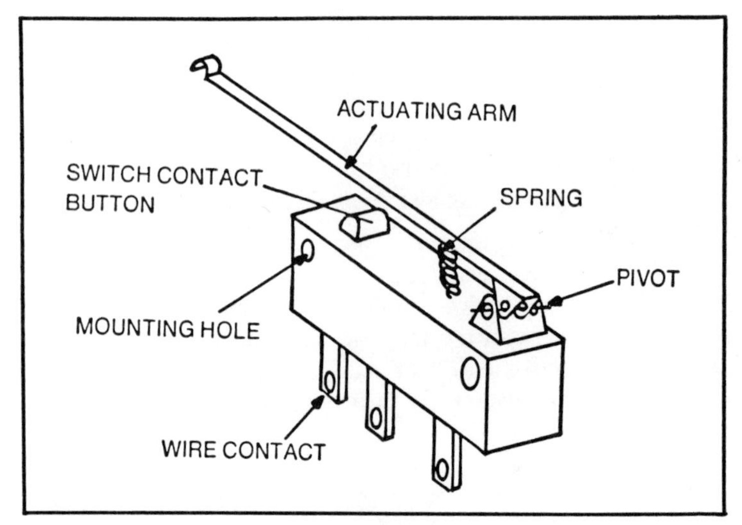

Fig. 15-1. Microswitch

Peripheral Equipment

Many of the operational problems that occur when using videocassettes are due to the other items of equipment which must be used in conjunction with the cassette machine. An operator should be able to determine if any trouble lies with these and not with the cassette machine.

The easiest way to locate a source of trouble is to change the most likely offending item with some similar item. For example, if there is no playback picture on the monitor, then try another monitor—perferably one that is known to be working. If a camera appears not to be producing a picture, try another camera with the same cassette machine and monitor. Or try another camera with a different machine and monitor.

A few moments of careful switching around like this will soon locate the offending item and enable it to be removed from the system for later servicing. (This servicing is not covered here.)

With peripheral equipment, a list of the typical things to check can be given as below.

● **Monitors.** Check the on-off switch, brightness, contrast, horizontal and vertical hold, and volume control. Also check that the TV-VTR switch is in the correct position. Check the plugs at the rear.

● **Cameras.** After the power switch, check the beam and target controls; remove the lens cap and check that the iris is open.

See if internal or external sync is being used, and check to find out if a switch for this is available and in the correct position.

● **Video switchers.** After checking the power switch, check to see if the noncomp or comp inputs and outputs are correctly used, and find out if sync is available.

● **Audio mixers.** After checking the power, check that the main and individual faders are not at zero. The inputs may have level and impedance switches which, if in the wrong position, could affect the audio. Check the output in the same way. Also check the plug wiring; it may be different from the audio sources and the equipment used for recording.

Operator's Maintenance

The videocassette machines have been reduced to a level of simplicity suitable for the nontechnical person, but this does not relieve the operator from learning and doing a few necessary simple tasks. None of these requires a knowledge of advanced electronics, and all can be handled by someone who has done nothing more complicated than change a light bulb or look after a hi-fi set.

If possible, practical advice and a few minutes instruction should be sought, especially if the operator is a newcomer to the video field. Most engineers and technicians in this industry will be willing to help in this. The learning will be repaid in less downtime and the speed with which a troubled piece of equipment will be made operational again, and the knowledge can only be to the operator's advantage.

The following is a list of the jobs an operator should learn to do as part of the normal operational function:

The Cassette Machine

Check the video output, using a monitor or an oscilloscope. Check the audio output, using an amplifier or headphones. Measure the tape tension as described in the service manual. Check the video input with a monitor. Check the audio input. Run the alignment tape to check performance. Clean and possibly change the heads. See Chapter 16.

Monitors and TV Sets

Check the video input.
Check and adjust scan sizes.
Adjust all external controls.

Cameras

Check the video output with either a monitor or an oscilloscope.

Change the lenses and set up zoom lenses.

Replace the vidicon tubes.

Adjust the target, beam, and electrical focus.

Adjust scan sizes.

Check to see if the video is composite or noncomposite.

Adjust the amplifier for optimum response using a test pattern.

Other Peripheral Items

Change the fuses and plugs.

Fix loose nuts, bolts, screws, etc.

Change the indicator and pilot lamps.

Replace defective plugs and switches.

Check and repair the cables.

Make adaptor cables.

Jobs Best Left for a Technician

There are some jobs best left to a trained technician to perform. These all presuppose some education and experience in electronics, plus familiarity with videotape and cassette machines. The following list is given as a guide for the operator; it represents typical jobs which require the above training:

Changing and adjusting the video heads. See Chapter 16.

Changing and aligning the audio control and erase heads.

Adjusting and replacing damaged tape guides.

Making adjustments to the threading ring and mechanism.

Making tape tension adjustment.

Replacing reel tables.

Changing drive belts inside the machine.

Repairing printed circuit boards and changing soldered-in transistors and integrated circuits.

Making video and servo circuit alignments.

Making equipment modifications.

Tools and Spares

Any establishment which has more than one or two machines or has a small TV studio should back them up with a supply of spares and small hand tools. The following lists are given because they represent good starting points for items required which should be on hand continually.

Because so many of the nonbroadcast equipment is of Japanese origin, metric tools and hardware are advised. Although most standard American tools will suffice, there are a few instances where they will not. With regard to the hardware, almost all American standard nuts and bolts will not fit any metric counterpart.

A set of tools also can be a great help to a visiting serviceperson who does not have all of his or her own tools. Those in the house can save a further expensive service call or the loss of a machine which is returned to the shop for servicing.

The cost of the tools shown here is minimal, but the savings they can effect are enormous.

With regard to the spare parts, individual items obviously depend upon the models and the manufacturers, so only general advice can be given here. Most service manuals have a list of all the electronic and mechanical parts required to build another machine, and obviously not all of these should be kept on hand. The parts list is thus a general list of the more important items which are most often needed.

Tools List

- **General tools**

Screwdriver set—both flat blade and Phillips.
Nut driver set.
Allen wrenches—American sizes, especially 3/32 in.
Allen wrenches—metric sizes.
Adjustable wrench—½ in. maximum jaw size.
Dikes or wire cutters with side bite and end bite.
Needle nose pliers.
Watchmakers' files.
Stripping knife and spare blades.
Wire strippers.
Hemostats—1 in. jaws.
Can of contact cleaner.

As far as possible these should all be top quality tools and should have insulated handles.

- **Soldering tools**

One or more soldering irons in holders or bench cages. Use 25 to 45 watt elements.
Spare tips and elements for above.
Soldering aid with a brush at one end.
Solder tip cleaner.
Flux cleaner and brush.

Solder sucker.
22 gauge wire—insulated and not insulated.
Flux cleaner and brush.
Small hand drill.
Pin vice.
Several No. 56 drill bits.

● **Head cleaning tools**

Head cleaning fluid—perferrably in a spray bottle.
Cotton-tipped swabs.
Special head cleaning pads.
Head demagnetizer.

● **Other items**

Cleaning cassette tape.
Alignment tape.
Test tape.
Tension tape reels
Tension gauges.
Multimeter or test meter.
Frequency counter.
Oscilloscope—this must be a good quality professional item; hobbyist types are unsuitable for several reasons.
All service manuals.

● **Spare parts**

Hardware: About 50 of the most common nuts, bolts, and screws should be on hand at all times. These are best determined from the list in the service manual. They will all be metric. The standard American sizes also should be stocked in the same quantities.

Semiconductors: A list should be compiled of the most used transistors, diodes, and integrated circuits used. About 10 should be kept at all times.

Mechanical parts: All the common small items should be on hand. A typical list would include knobs, reel tables, belts, idler wheels, pulleys, microswitches, and indicator lamps. Spare audio and video plugs also should be available.

Expensive items. The advisability of keeping spare video and other heads must be judged by the size of the establishment. These do not often become defective, but when they do the machine is totally out of service until a replacement is installed. This is a decision which must be governed by the size of the studio, its budget, and the availability of local personnel to install these items.

16 Advanced Maintenance and Alignment

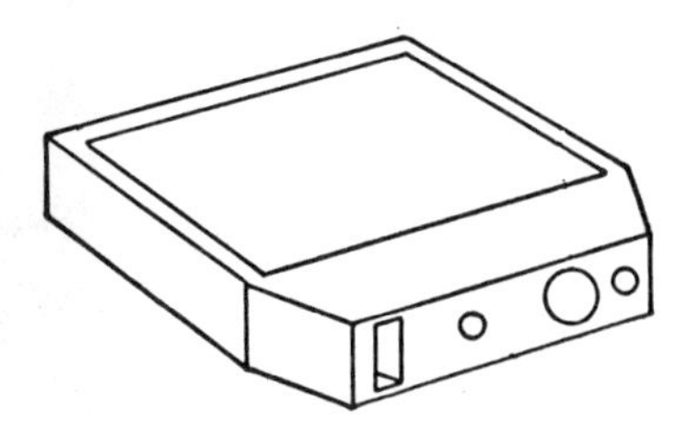

There are certain tasks which should be performed on a regular basis for the purpose of avoiding trouble and ensuring that the machines are continually in the best working condition. They are mainly mechanical adjustments and electrical alignments, which also are needed after a repair or when components have been replaced. They are distinct from specific fault tracing but often are part of the overall procedure.

The procedures included here are those which are the most important to the correct running and operation and are common to all the machines. Only the principles are described, because the actual details differ from model to model and are too diverse to be covered fully. Many of the mechanical procedures require special jigs and fixtures for their correct completion, and the electronic procedures require the use of advanced test equipment. No maintenance at this level should be attempted without the service manual, which contains the exact details for the particular model, and anyone who is unfamiliar with such advanced equipment and procedures is strongly advised to leave most of these jobs to a trained technician or engineer.

Five basic areas of interest will be covered: mechanical adjustments, video alignment, servo alignment, audio tests, and power supply adjustments.

Because all of these must be performed with the machine out of its case, this subject is covered first.

REMOVAL FROM THE CASE

Videocassette machines are built in two basic sections: the main mechanical chassis and playback and servo electronics, and the record electronics on the smaller chassis. This is explained by Fig. 16-1. A common top and front operator panel stretches the length of the machine, as in Fig. 16-2.

Removal of the internal chassis is easy, and a simple procedure can be followed.

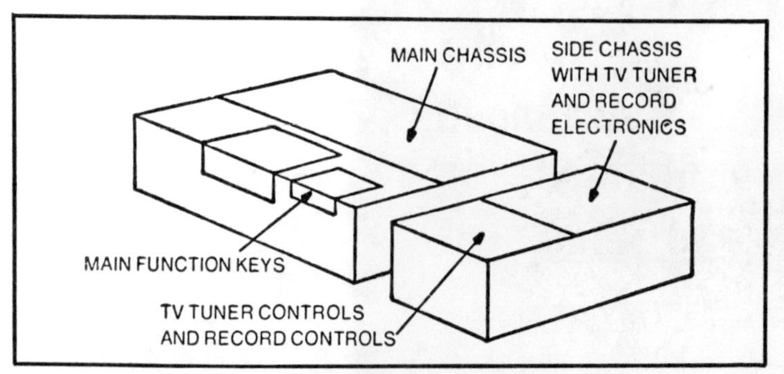

Fig. 16-1. Electronic chassis layout.

- Use a 3/32 in. Allen wrench to remove the bolts at the rear which hold the top in place.
- Remove the bolts holding the front panel in place and pull off all the operator knobs. The front panel now has to be jiggled over the cassette compartment. Do not lose the felt washers which sit over the control knobs.
- Remove the two large screws in the rear which hold the main deck in place.
- Remove all the screws in the base of the case. These hold the main chassis and the record chassis separate.
- Remove the plastic connecting plugs which connect the two sections. Also remove any plugs near the front of the machine and the ground lead near the plastic plugs. The plugs are either numbered or color coded to facilitate replacement. They also are indexed. See Fig. 16-3.
- Remove the record section first because it is easier.
- Lift out the main section by the handles shown in Fig. 16-4.
- The main section can stand flat in its normal operating position, or it can be stood on end, using the flanges provided as legs. See Fig. 16-4.

Once they are out of the case, the machine sections can be reconnected and put into operation. When the machine is in operation make sure it is not standing on end but is laying flat as in the case.

All of the PC (printed circuit) boards and test points are accessible, and none need to be removed for normal service and alignment checks. None of the boards are plugged into a socket, so all the leads are either directly wired or use push-on pins to make the connections. Figure 16-5 shows the board layout for a Sony machine.

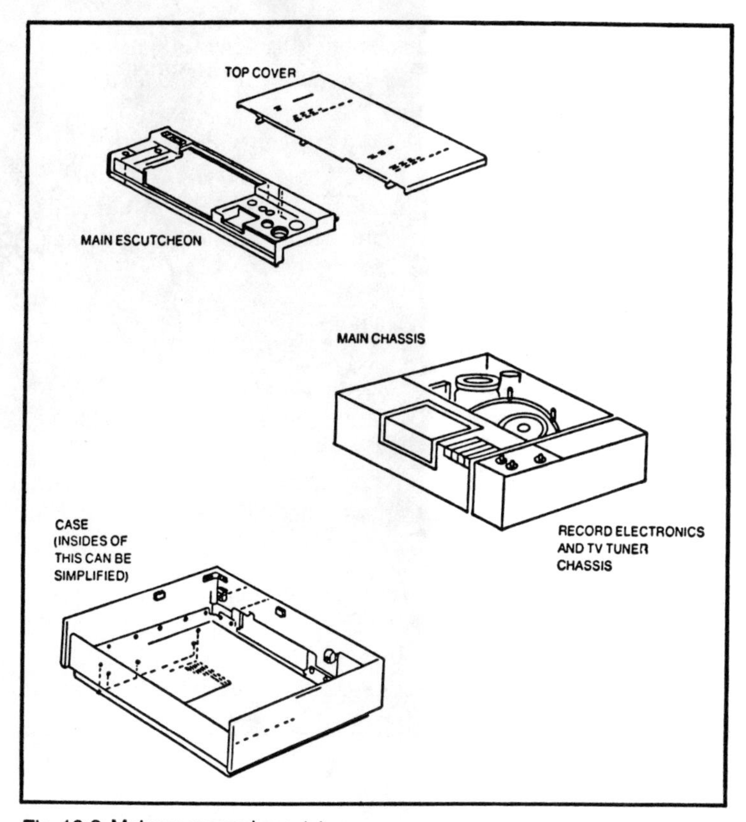

Fig. 16-2. Main covers and escutcheon.

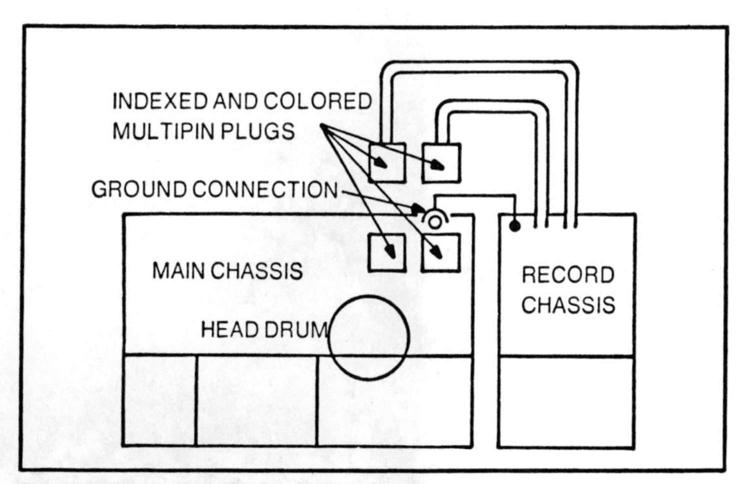

Fig. 16-3. Chassis interconnect plugs.

Many of the tension checks and adjustments must be performed with the cassette compartment removed. This is easily accomplished in most machines.

There is no power interlock to contend with, as with TV sets when the back is off, and the highest voltage found is the 120V ac supply. All dc voltages are quite low. The deck should not be placed on a metal table for servicing; use a nonconducting surface.

MECHANICAL ADJUSTMENTS

Much of the mechanical alignment and parts replacement requires the use of special jigs and fixtures for its correct implementation. If a repair of this magnitude is needed, then the machine is best returned to an authorized factory service center. If the user intends to tackle this type of job himself, then he is strongly advised that the correct items be available and that he have the requisite mechanical aptitude and experience.

Video Head Cleaning

The rotating video heads are the most delicate and expensive parts of the cassette machine. They must be treated with the utmost care and respect and be kept clean. Cleaning is really an operator's job, but due to the difficulties of getting at the heads in

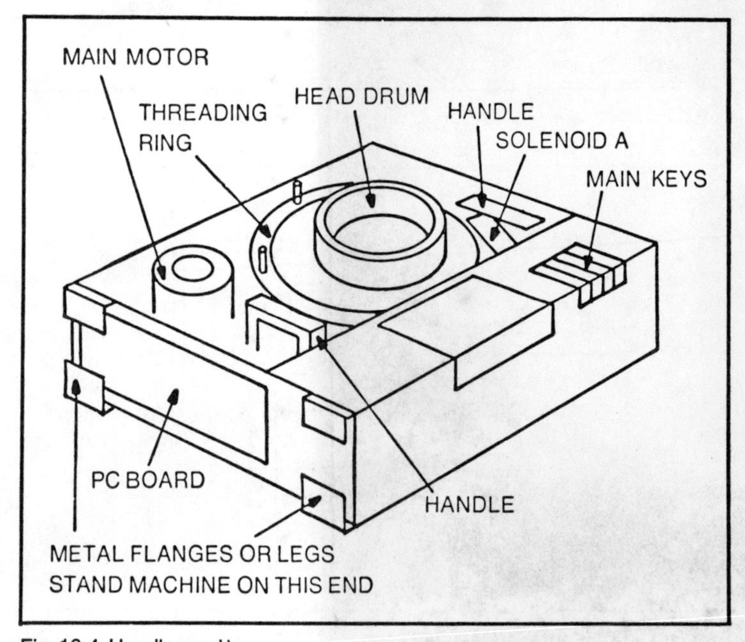

Fig. 16-4. Handles and legs.

TV LISTINGS

SATURDAY AFTERNOON

April 21, 1990

	12:00	12:30	1:00	1:30	2:00	2:30	3:00	3:30	4:00	4:30	5:00	5:30
3	ALF-tales □	ALF □	**Better Your Home** Topic: adding a family room.	**Runaway With the Rich and Famous**	**Senior PGA Golf**: Legends of Golf. Third round, from Barton Creek Country Club in Austin, Texas. (Live)				**Greatest Fights Ever** The first installment of this occasional series profiles the 1974 "Rumble in the Jungle" -- George Foreman vs. Muhammad Ali. Hosts: Mary Albert and Ferdie Pacheco.			
5	**Flintstone Kids** □	**Crime Stoppers 800**	**Movie:** ★★½ *"The Stooge"* (1953, Comedy) Dean Martin, Jerry Lewis. A singer decides that he needs his partner to be a success.				**PBA Bowling**: Greater Hartford Open. (Live) □			**Wide World of Sports** Featured: "The Mystery and Magic of Nadia," the story of Nadia Comaneci; Wood Memorial horse race (Live); Paris-Roubaix bike race from France. □		
7/4	Woman to Woman		News/ Weather	Register Locals	News/ Weather	Register Locals	Register Classifieds		News/	News/	News/	News/

What's happening

Music

AKRON RUBBER BOWL, on University of Akron Campus. For more information call (216) 247-2722. Tickets available at

to 8 p.m. May 29, 31 and June 1 and 2.

LAKESIDE HOOVER AUDITORIUM, corner of Walnut Ave. and Third, Lakeside, for more information call 798-4461.

MANSFIELD RENAIS-

Savage Hall, to order tickets by phone call 537-4231.

Theater

CLEVELAND BALLET, One

tion. Tickets available at box office or Ticketron.

1989-90 Broadway Show Series, the option of "Nunsense" for subscription holders.

• "Nunsence," Nov. 24-26, Palace Theatre.

TOLEDO BALLET, Masonic

6199 Wilson Mills Road, Highland Heights, Ohio. For ticket information call 216-524-0000 or 1-800-225-7337.

- Isley Brothers, April 7, rescheduled for 7 p.m. and 11 p.m. May 26.
- Ricky Van Shelton with Wild Rose, 8 p.m. April 20.
- Jay Leno, April 21.
- Bob Newhart, 7:30 p.m. April 28.
- Good Friends II, featuring Najee, Randy Crawford and Alex Bugnon, May 5.
- Ronnie Milsap with the Forrester Sisters, 7:30 p.m. May 6.
- Anita Baker, date changed

Buddy De Franco and Herb Ellis, 8 p.m. April 20, at Tri-C Metro Campus Auditorium.

- James Williams, 8 p.m. April 21, Tri-C Metro Campus Auditorium.
- Gunther Schuller directs Charles Mingus 'Epitaph,' 8 p.m. April 22, Severance Hall.
- Sandra Reaves-Phillips, 8 p.m. April 25, State Theatre.
- George Benson and Jimmy Smith, 8 p.m. April 27, State Theatre.
- Tony Bennett with special guest, pianist Ahmad Jamal, 8 p.m. April 28, State Theatre.

UNIVERSITY OF TOLEDO

TER, box office, 1519 Euclid Ave., for ticket information call 1-800-492-6048. Tickets available at box office and all Ticketron locations.

PLAYMAKERS CIVIC THEATRE, 604 E. 6th St., Port Clinton. For reservations call 734-5044, 11:30 a.m. - 1 p.m. and 3:30 p.m. - 5:30 p.m., Monday - Saturday.

- Lie, Cheat and Genuflect, 8:15 p.m. nightly, May 10 - 13 and May 17 - 20 with additional 2 p.m. matinee on May 20.

RITZ THEATRE, 30 S. Washington St., Tiffin, phone 448-8544.

STATE THEATRE, 107-109 Columbus Ave. For ticket information call 626-4319.

- Firelands Chorus Spring Concert, 8 p.m. April 21.
- Jerry Herman's Broadway,

- South Pacific, July 2[illegible] through Oct. 7.
- Anything Goes, Oct. 1[illegible] through Dec. 31.

Museums

BELLEVUE HERITAGE MUSEUM, 115 S. Sandusky St. Bellevue. March and April, Sat & Sun. 1 to 4 p.m.

- "Bellevue Whiz Kids 194[illegible] Class A Basketball Champs."

CLEVELAND CHILDREN'S MUSEUM, 10730 Euclid Ave. Cleveland, 216-791-KIDS. Hours: 1-5 p.m., Sun.-Fri.; 1[illegible] a.m.-5 p.m., Sat.

CLEVELAND HEALTH EDUCATION MUSEUM, 891[illegible] Euclid Ave., Cleveland, 216-231-5010. Hours: 9 a.m.-4:30 p.m. Mon.-Fri.; 10 a.m.-5 p.m., Sat.;

				Gulla	2)							predator.	
ESPN	Auto Racing: American	Indy 500: A Race for Heroes	Auto Racing: NASCAR Modifieds. From North Wilkesboro, N.C. (Live)			Indy 500: A Race for Heroes	Diving: USA Invitational. From Ft. Lauderdale, Fla.			Yachting: Congres-sional Cup.	Horse Racing: Arkansas Derby. (Live)		33
FAM	The Virginian "The Captive"			Rifleman	Wagon Train		Big Valley "Explosion" (Part 1 of 2)		Gunsmoke		Bonanza: The Lost Episodes "The Wish"		22
LIFE	Supermarket Sweep	Rodeo Drive	Days and Nights of Molly Dodd	She's the Sheriff "The Hero"	Moonlighting "The Dream Sequence Always Rings Twice"		MacGruder & Loud "The Price of Junk"		Spenser: For Hire "I Confess"		Movie: ★★½ *"Rags to Riches"* (1987, Comedy-Drama) Joseph Bologna.		31
NASH	Country Music	Side by Side (In Stereo)	Celebrity Outdoors	Remodeling Today	Country Kitchen	Gospel Jubilee	Tommy Hunter (In Stereo)		Church Street	On Stage (In Stereo)	Country Beat (In Stereo)		26
NICK	Count Duckula	Inspector Gadget	Lassie	Heathcliff	Count of Monte Cristo		Black Beauty		Can't on TV	Out of Control	Mr. Wizard's World □	Dennis the Menace	34
SC	SEC Weekly	Motor Racing	Sports Exchange		College Baseball: Alabama at Auburn. (Live)						College Baseball: Cleveland State at Akron.		
TNT	Movie: ★★ *"The Treasure of Pancho Villa"* Cont'd		Movie: ★★½ *"Mister Cory"* (1957, Drama) Tony Curtis, Martha Hyer.				Movie: ★★ *"No Blade of Grass"* (1970, Drama) Nigel Davenport, Jean Wallace.				Man From U.N.C.L.E.		09
USA	Movie: ★★ *"This Wife for Hire"* (1985, Drama) Pam Dawber, Robert Klein.				Movie: *"The Kissing Place"* (1990, Adventure) Meredith Baxter-Birney, David Ogden Stiers.				Double Trouble	My Sister Sam	Murder, She Wrote "Tough Guys Don't Die"		35
WTBS	Movie: ★½ *"Eat My Dust!"* (1976, Adventure) Ron Howard, Christopher Norris. The teen-age son of a California sheriff steals the best stock cars from a race track.				Movie: ★★★ *"Grand Theft Auto"* (1977, Adventure) Ron Howard, Nancy Morgan.			Movie: ★★ *"The Great Texas Dynamite Chase"* (1976, Adventure) Claudia Jennings, Jocelyn Jones.			Fishing With Roland Martin	Fishing With Orlando Wilson	17

these machines, the decision about undertaking it is perhaps not easy to make.

To clean the heads a special head cleaning pad, such as the one available from Sony, should be used. If this is not available a lintless cloth or muslin gauze can be used. The popular cotton-tipped swabs should not be used here—their fibers can catch on the head and pull and break it. The cleaning pad should be liberally soaked in the cleaning fluid and then gently and firmly rubbed *sideways* across the head. (Never rub it up and down; this may break the head.) The whole head drum should be cleaned with this sideways motion, as shown in Fig. 16-6. Cleaning should be undertaken at least once a day in a heavily used system, and oxide buildup never should be allowed to occur. The main ingredient in head cleaning is care. The more often it is done, the easier it is, with attendant safety for the heads.

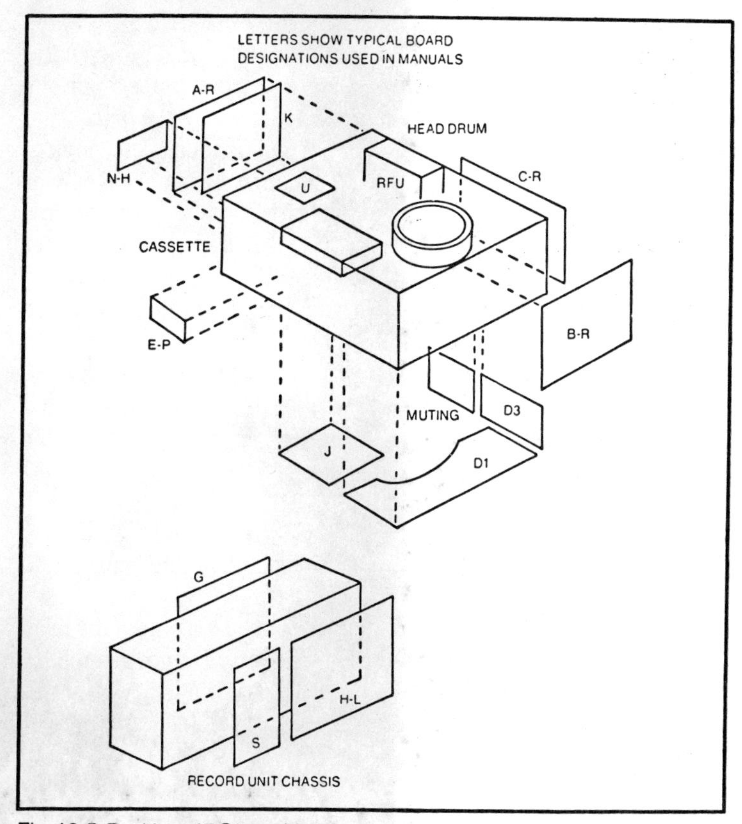

Fig. 16-5. Position of PC boards.

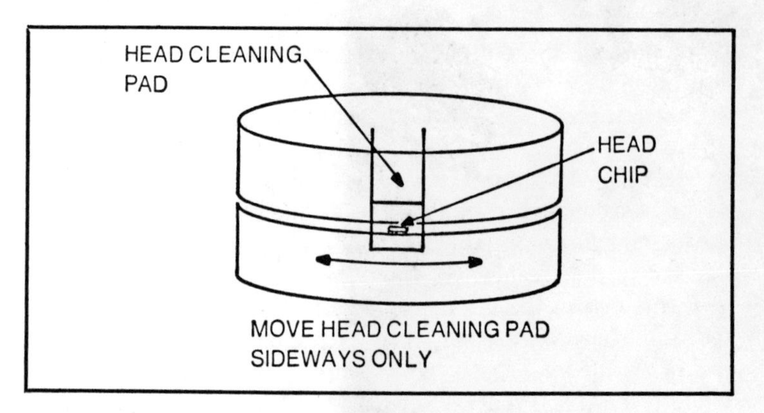

Fig. 16-6. Head cleaning.

Head Demagnetizing

Every metal part of the machine which comes into contact with the tape will gradually become slightly magnetized. This is most true of the video and audio heads, which should be regularly demagnetized. Magnetization of the tape from these spurious sources will cause loss of color and partial erasure of the tape.

It is very simple to perform demagnetization. The demagnetizing tool is plugged in, turned on, and then slowly brought near to the part to be demagnetized. It is moved slowly up and down a few times, and then the tool is slowly retracted to about 3 feet away. If the tool is turned off while close to a head it will have the opposite effect and leave it magnetized. So long as this is avoided the machine will be suitably demagnetized.

One word of caution: Use only the approved tool for this. The video head demagnetizers are weaker than most of those used for audio machines. Using one made for an audio machine can actually shatter the video head chips.

Head Changing and Adjustments

This is not a simple procedure and should not be approached lightly. It is best left to a qualified technician.

The heads are permanently mounted in the top part of the head drum, and this must be changed in its entirety. Do not attempt to change the individual head chips. The utmost care is needed at all times.

● Head Drum Removal

a. Unsolder the leads as shown in Fig. 16-7.

b. Remove the two Phillips screws shown.

c. Lift off the head drum. Do not remove the shim. Do not touch the central hex nut.

● Replacing the New Head Drum

First clean the new drum and the area into which it will be placed.

a. Hold the new drum over the lower portion so that it matches the position of the old drum top.

b. Position it so that the Phillips screws can be installed.

c. Install the screws, but do not tighten them.

d. Resolder the leads.

e. Check the eccentricity as described in the service manual.

f. Now tighten in place.

Associated with the head drum replacement are checks for the eccentricity and the dihedral. The head must be exactly concentric with the center of the drum, and this is what the eccentricity check and adjust accomplishes. The dihedral ensures that the heads are exactly 180 deg apart, but this is an adjustment seldom needed. The top penetration is set during manufacture and does not need to be checked.

Tape Tension

Tape tension is critically important in a videocassette machine. Too much tension will cause excessive head and guide wear, will cause tracking errors due to tape stretch, and can even permanently stretch the tape. Too much tension in the threading

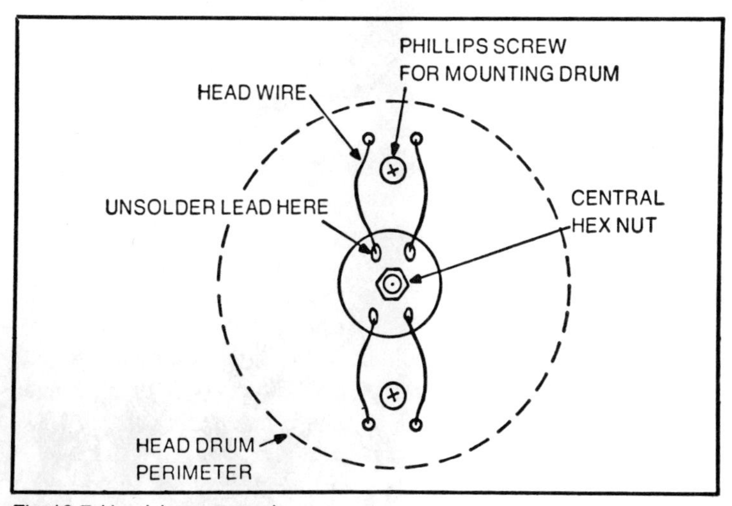

Fig. 16-7. Head drum removal.

loop can actually stop the tape from traversing its path and even pull it out of its path. Too low tension will allow the tape to fall out if its true path and not make proper contact with the heads, with the resulting misplacement of the video tracks and many dropouts.

Many tape transport problems are due to the wrong tension in the tape, and this is something which should be checked regularly. Checks also should be made when belts and other mechanical parts are changed or adjusted. A full tension check must be made in all of the following modes and conditions: forward and back tension in the **rewind** mode, forward and back tension in the **fast forward** mode, forward and back tension in the **play** or **forward** mode, and the back tension in the **threading** mode.

The back tension in the **play** mode is dependent upon the setting of the skew control. This is provided to allow operator adjustments so that tapes from other machines can be played with minimum difficulty and so that environmental conditions can be offset slightly. In the **record** mode this control is automatically flipped to a preset position, which provides a constant back tension throughout the recording.

Tension problems become apparent in **playback** by producing a hooking or bending of the picture at the top. Tension problems in **playback** are good indicators of tension problems in the other modes and suggest that a check and adjustment should be made.

Tape tension is provided and maintained by several different mechanical means, which are mainly brake bands and pads, and felt washers in or around the reel tables. Tension problems are caused by the wearing or maladjustment of these. Further causes lie in the drive belts and rubber pulley wheels which drive the reel tables. If these become worn or stretched, then insufficient drive torque is applied to the reel tables.

The basic method for checking tape tension is to wrap a few winds of tape around the hub of a spool and attach the free end to a spring gauge, as shown in Fig. 16-8. The machine is then put into the appropriate mode and the gauge held still. The tape will pull on the gauge, and the tension can be read directly from the scale. With the cassette machines the tension is checked with a nearly full reel of tape. The values read should lie within a certain range, which will be given in the service manual, and if it is outside of this range then corrective measures must be taken immediately. Table 16-1 shows the ranges for the videocassettes.

The approved tension adjustment depends upon the model and the mode of operation, and in general a different control is used for each of these modes. Checking tension is often a lengthy and tedious business, especially if adjustments must be made. Most of the adjustments are in the form of making slight bends in the metal linkage arm, and a few involve screw adjustments. If the approved corrective methods do not produce the correct tension, then it is likely the tension-producing element is faulty. Because these usually work by felt rubbing against a metal or plastic surface, the coefficient of friction is most important, and this cannot be altered in any way. Roughing up the felt or cleaning the surface sometimes helps but is usually a waste of time. The only safe method of alleviating tension problems is to change the brake pad or brake band, and often to change the entire reel table.

Often tension troubles can be traced directly to reel tables. Figure 16-9 is a simplified cross section of a reel table, and the tape tension is maintained by the felt washer between the two sections.

Reel Table Replacement

This requires the removal of the videocassette loading compartment, which exposes the two tables. The exact nuts and bolts to achieve this varies with each model, so the service manual should first be consulted.

Only the correct reel table should be used as a replacement, and the whole unit should be replaced. Do not attempt to replace

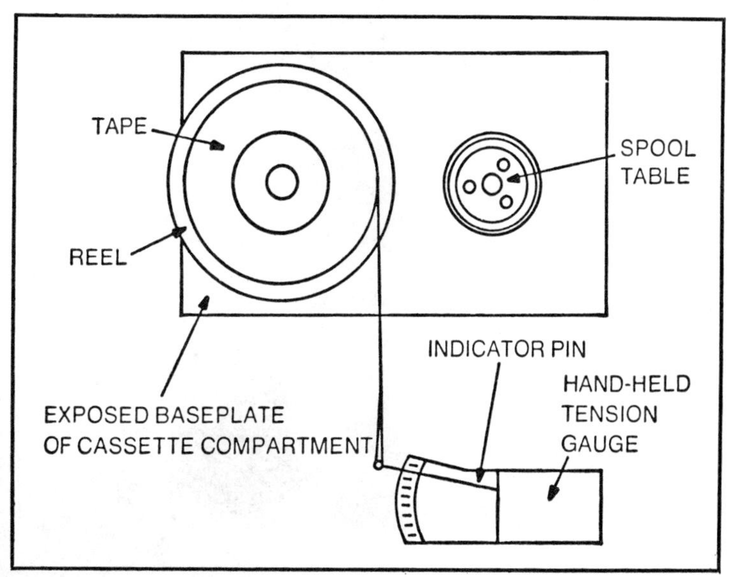

Fig. 16-8. Tension check.

Table 16-1. Tape Tension Ranges.

	Tension Measurement	
Mode	Supply Reel	Takeup Reel
Threading back tension	8 – 9 g	
Forward drive	70 – 75 g full reel 90 – 96 g small reel	84 – 104 g
Rewind	200 – 340 g	8 – 10 g back tension
Fast forward	8 – 10 g back tension	200 – 340 g
Brake	200 g before slipping in clockwise direction Less than 100 g to cause slipping in counterclockwise direction	

only the top or the bottom or to install a new washer. These seldom produce satisfactory results, and the cost of the complete item is small compared to the cost of the productions the machine must handle. If the top and lower parts are undamaged, then they can be cleaned and a new washer tried between them. This sometimes produces satisfactory results.

Belts

All the important and major drive functions in the cassette machines use belts to transmit the power. Service manual illustrations show the positions of the major belts.

Belts provide a good positive drive when they are new, but with age they become brittle and wear smooth and tend to stretch.

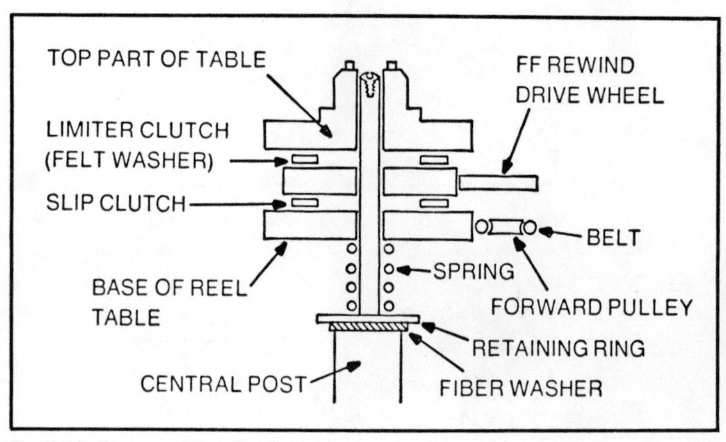

Fig. 16-9. Cross section of reel table.

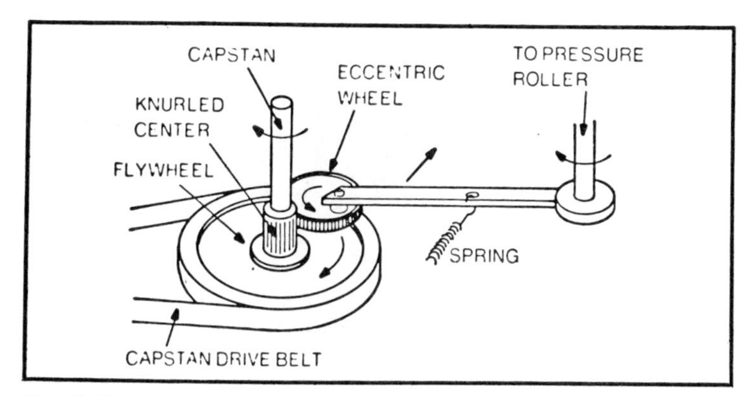

Fig. 16-10. Eccentric roller.

When this happens they cannot perform correctly and must be changed. Belt changing is very easy, but care must be taken not to stretch the belts while slipping them over the pulley wheels. Too much taking off and replacing will cause excessive stretch.

Belts usually have one dull and one shiny side. The correct side must be used. In some cases a positive tight drive is needed, so the dull side must be used. But in cases where a controlled slippage is required, the shiny side is required. As belts wear the servos can adjust to correct for the drive problems, but there comes a point where this is no longer possible. Hence, many drive and servo problems can be resolved by changing the belts. Anytime a belt is changed, it usually means that a servo alignment becomes necessary for optimum operation.

All belts and surfaces should be kept clean and dry. The pulley wheels should be cleaned before the new belts are put on, and no oil, grease particles, or moisture should be allowed to remain. Fingers should be kept off the belt drive surface as much as possible. A good stock of spare belts should always be on hand.

Pressure Roller

If this fails to close and drive the tape, then the whole mechanism should be checked. This roller is closed by a complex mechanical arrangement involving an eccentric wheel which is underneath the main deck. It is not an easy adjustment, and it occurs in connection with the capstan rotation and the pull in of solenoid A. See Fig. 16-10. The adjustments are given in the manuals and should be followed very carefully.

Changing the pressure roller is also covered in the manuals, and this occasionally is necessary.

Linkage Adjustments

These are very seldom required in a machine which has functioned well for a period of time. But after very long usage or a part replacement, the linkages may require some adjustments. It is best to read the manual carefully before attempting these and to observe the action of the linkages in operation before attempting anything. Always use the utmost care, and if in doubt call a serviceman.

Main Function Keys

In most machines these are mechanical locking keys, which operate slide and microswitches mounted on printed circuit boards. A typical trouble is that the keys will not lock down or release, or will not even go down and initiate the selected function. Most of these problems are within the key unit itself, and this whole unit is easily removed for working on.

A common problem is that the sliding bar which locks the keys will become stuck and will not slide sideways to release a key or hold the newly pressed one. Usually this can be cured by cleaning and a little lubrication. Sometimes the metal flanges on the key bars and slider became bent, or one of the microswitches will become loose and interfere with the movement. Although this unit looks horrendously complex it is amenable to careful servicing, and with care it can be fixed.

If one part of it continues to give trouble, then it should be replaced. Remounting of new parts is quite easy.

Replacing the unit in the machine is fairly easy, but it is essential to use care when engaging the linkages near and under the cassette compartment. It is essential that these be correct, because they are vital to the automatic operations as well as to the operator selected functions.

Other Mechanical Adjustments

Most service manuals contain information about the following items. These need attention much less often than the previous items and are best left to a service technician. They are lubrication, replacement of audio and erase heads, solenoid replacement and adjustment, tape transport adjustment, motor replacement, and threading ring problems.

VIDEO ALIGNMENT

These can conveniently be divided into definite categories and subsections.

Playback Checks

● **Head resonance.** This is performed by playing back an RF sweep signal on the alignment tape and adjusting specific components to get a given RF envelope pattern at a given test point. Although the individual procedures vary, the basic RF pattern is similar to that shown in Fig. 16-11. This ensures the heads are matched to the preamplifiers and are delivering the optimum signal across the frequency spectrum of interest. This is usually needed only when the heads are changed or when preamplifier components are changed.

● **Preamplifier equalization.** The process is similar to the above, but the adjustments are to correct the frequency response of the preamplifier.

● **Preamplifier balance.** The heads and the preamplifiers are likely to have slightly different outputs when optimized. Because these are fed to a common limiter and demodulator chain, it is best to start with the same level of signal; this avoids 60 Hz fluctuations in the chroma of the final signal caused by the AGC action. The adjustment is made with a stairstep signal from the tape, and it merely picks the same level from each preamplifier. Note: This is often referred to as preamplifier equalization.

● **Carrier balance.** This is also referred to as demod or limiter balance. The stairstep section of the alignment tape is used, and the video output picture is viewed on a monitor. The indicated control is adjusted to minimize the RF and moire beats in the picture.

● **Video level output and AGC.** These are not usually the same adjustment, but they must be performed in conjunction with each other. They ensure a standard 1 V peak-to-peak video output at all times.

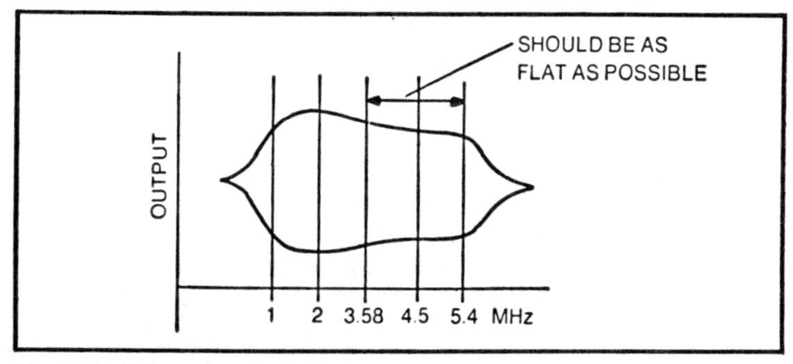

Fig. 16-11. RF envelope in playback.

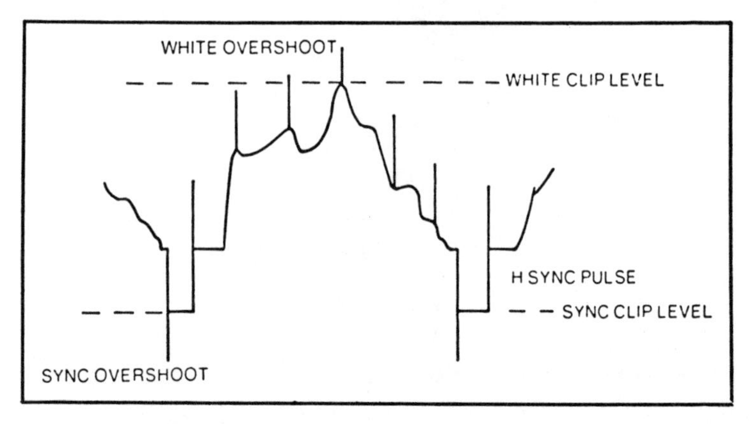

Fig. 16-12. White and svnc overshoots.

● **Dropout compensator sensitivity.** This ensures that annoying dropouts on bad tape will be masked and thus not be seen on the monitor or copied onto another tape. If an external dropout compensator is used, then possibly the "tag time" will need to be changed. For this, seek the advice of a service engineer.

Record Checks

● **Video input level and AGC.** The video input signal is specified to be within certain limits. Over these limits the video AGC will work and present a definite level of video to the FM modulator. This is an important video adjustment and should be done carefully. Note that the video AGC does not have the wide range of the audio AGC.

● **Clip levels.** Before the AGC video signal is fed to the FM modulator it is preemphasised. This causes overshoot spikes on the high whites and the sync tips. See Fig. 16-12. These must be clipped at the correct level. It is a simple adjustment.

● **FM frequency.** The peak white and sync tips correspond to a definite frequency from the FM modulator. This must be correct, otherwise the interchange of the tapes will be affected. It is set by feeding a standard stairstep signal into the video input and then injecting the output of an oscillator into some point in the electronics. The oscillator is set to the FM frequency, which corresponds to sync tips and the E-E picture is viewed on a monitor and an oscilloscope. The FM deviation is adjusted for minimum interference. The injection frequency is now set to the peak, while frequency and the adjustment are repeated using the appropriate control in the circuit. See Fig. 16-13.

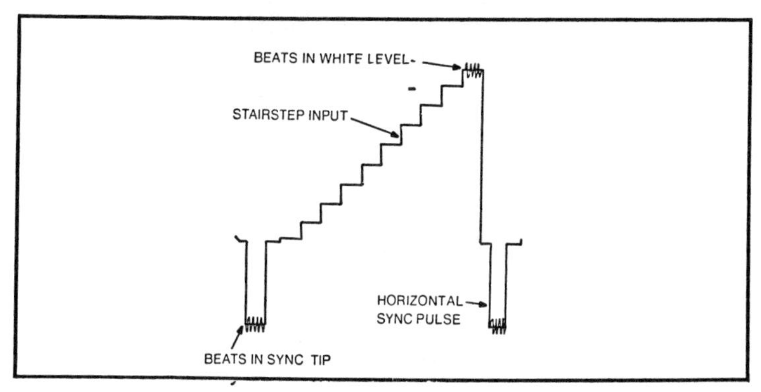

Fig. 16-13. Interference beats in stairstep signal.

In the U-matic videocassette system, peak white frequency equals 5.4 MHz and sync tip frequency equals 3.8 MHz.

● **Record amplifier level.** The heads respond to the record current as shown in Fig. 16-14. The optimum condition is at the peak point A. This peak is found by recording a signal on the tape and increasing the record amplifier level in stages from minimum to maximum. The exact level setting is indicated by speaking into a microphone and recording it on the audio track. This is listened to on playback so that optimum setting can be determined by observing the output of the playback preamplifiers. This setting is then used for the record amplifier. Each amplifier and head must be treated individually. It is important to realize that this adjustment has nothing to do with the video input level for the FM deviation settings.

Color Circuits

● **3.58 MHz oscillator.** Although most of these are crystal controlled, some require minor adjustments. Use a frequency counter, and set the output to 3.579545 MHz ±5 Hz.

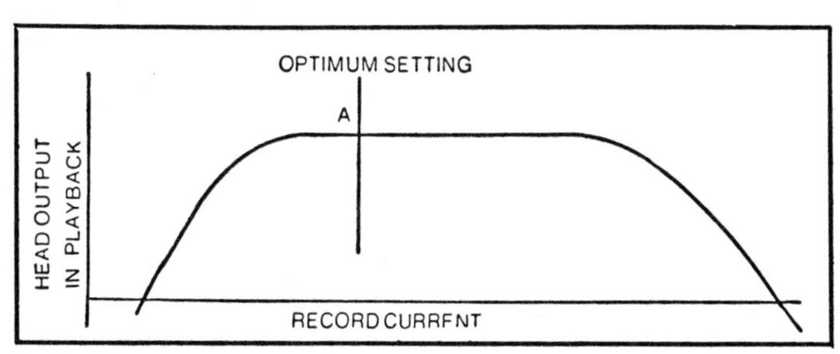

Fig. 16-14. Head record current setting.

● **3.58 MHz traps.** These must be set as described in the manual. They are used to shape amplifier responses and to keep signals from interfering.

● **4.48 MHz oscillator.** This is essential to correct color correction and is set as described in the manual.

● **Burst level.** The burst is used for several purposes in the cassette machines but mainly for the color correction circuits. The level must be properly set.

● **ACC (automatic color control).** If this is provided, follow the instructions in the manual.

● **Luminance to chroma ratio.** The relative levels of the luminance and the chroma are very important in the final output. Adjustments are provided to control the level of chroma added to the output.

SERVO ALIGNMENT

Periodic checking and adjustment of the servo is essential. The servo is an electromechanical arrangement in which the mechanical parts are controlled by the electronic circuits, but are actually driven by the belts. It is difficult to separate the electronic and mechanical problems due to their interdependence. Most of the mechanical problems can be fixed by checking tape tension and changing belts, but invariably this must be accompanied by electronic alignment. The individual checkpoints must be taken from the service manuals, but the principles in all cases are very simple and are the same.

Free Run Speed

The speed of the rotating heads and the capstan are controlled by allowing them to run slightly fast and then slowing them to the desired speed with controlled electronic braking. For this to operate properly the free run speed with no braking must be set to run about 1% fast. This is done by counting the frequency of a small generator in the head or by measuring the frequency of a multivibrator circuit. It must be within narrow limits before the rest of the servo can work correctly. If the adjustment controls will not pull it within range, then belt changing is indicated.

Sampling Waveform

All VTR servos work on the principle of a pulse sampling a controlled ramp waveform. A convenient point for observing this is always given and so is a convenient control.

● **Head switching.** The two heads must switch over at the correct time. This is controlled by the head pulse coils and some simple electronics. Easy controls are provided to monitor and set this.

● **Tracking.** This is easy to set (refer to manual). In many cases servo problems are caused by worn belts and faulty mechanics, each of which provide loads that are too much for the circuits to control. If the simple checks and adjustments do not produce satisfactory results, then the mechanics of the machine should be checked throughly—especially tape tension and drive belts.

AUDIO TESTS

Audio problems are very rare within the cassette machines. Most are due to normal operational misuse rather than internal problems. The checks provided are very simple and are akin to those of a normal small audio recorder. These are the playback level, the playback **agc**, the record level and **agc**, and the bias and erase levels.

Occasionally the output switch will develop problems, and so will the plugs. Because these machines do not have speaker level outputs, their audio output circuits are not the type which need protection and seldom give trouble.

POWER SUPPLY ADJUSTMENTS

All the power supplies are of the regulated output type, usually with voltages of 24 and 13V. Adjustment merely involves metering at the correct point and adjusting the indicated control. The conditions under which the adjustment is made should be checked and adhered to, because different operating conditions place different loads on the supplies, which can alter the output levels slightly.

The major problems are fuses blowing, the control transistor burning out, and some element in the machine causing a short.

CONTROL CIRCUITS

These do not need checking and adjusting in the same way the other circuits do. Just about all of them are mechanical switches in the system and transistors used as switches and current amplifiers to drive relays and solenoids. The only real problems encountered are the mechanical switches not opening or closing and the transistors becoming defective. These circuits either work or they do not; there is no half measure about their performance, and this makes troubleshooting quite easy.

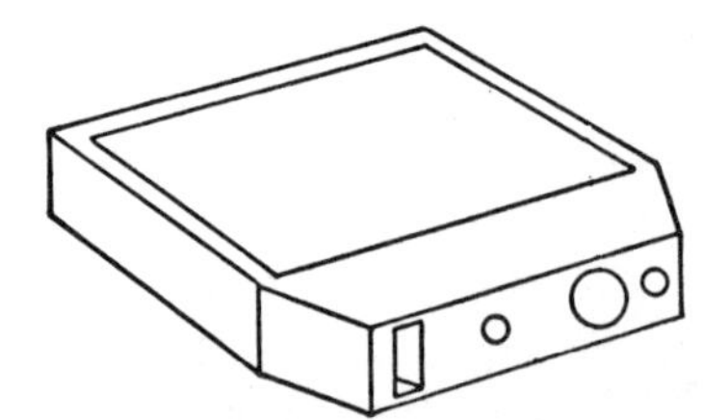

17 Troubleshooting and Further Maintenance

This chapter covers the common faults and problems likely to occur with the cassette machine itself. It is a maintenance guide to enable the user to determine the probable cause of the fault and which corrective measure to take.

Most of the problems which beset the machine generally manifest themselves in several common ways. The machine will not play back or record either the video or the audio, transport the tape, etc. Because any of these apparently simple occurrences can be caused by a number of nonrelated reasons, it is difficult to set out exact faults against exact causes. Many internal problems can exhibit the same outward symptoms, and the interdependence of the parts and circuits can further cloud the issue.

Many times the problem will fall into the category of nothing happening when levers and buttons are operated, and often these faults are due to the simple reasons previously discussed. When checking these problems produces no answer, further investigation is required. When this point is reached the first thing to do is to remove the top cover and expose the insides. This can now give a good visual clue to what is going on and is often enough to solve the problem.

Some problems occur with sufficient regularity to be called common, and in most cases these are due to the previously discussed causes. If a serious and obscure fault does occur, then a trained technician or engineer should be consulted and the service manual used.

Service manuals contain occasional mistakes, but all manufacturers maintain an updating service to correct these and to inform the user of recent improvements and changes. The various models are too complex to rely upon memory alone in servicing, especially with changes, so the service manual is indispensable.

Once the top has been removed, the first things to look for are these:

Faulty key switches and function buttons.

Faulty microswitches or those which may be jammed by the mechanism or pushed out of position.

Relays and solenoids not operating.

Linkages bent or not moving.

The end-of-tape lamp not illuminated.

Belts off or broken.

Tape tension looks wrong—it is sometimes possible to see this if it is very much awry.

Reel table problems.

If none of these appear to be the cause of the trouble, then these less common points should be checked:

Plugs fallen apart or not put together.

Plug pins out of the plug.

Wiring problems such as pinched wires and shorted wires in between metal parts.

The dc motor may not work.

The ac motor may not work.

Printed circuit board problems.

Also look for smoke or burned components, especially if there is a burning odor.

If none of these can be observed, then a logical investigation of the fault should be undertaken. Most problems can be loosely classified into the categories in Table 17-1. Each of these are briefly described in the following sections, with tables and trees showing the most likely faults, their causes, and their cures. It should be emphasized that often no clear division of faults into exact categories is possible, but a start at tracing must be made somewhere, and these tables provide a reasonable starting place. The most likely and simplest events are described first wherever possible.

MECHANICAL PROBLEMS

The open reel VTRs and audio tape recorders are fairly simple mechanisms, and their mechanical faults tend to be obvious and easy to fix. This is not so with the cassette machines, where the complex links and levers needed to handle the tape transport and threading present new and unique problems. The lack of video on playback and an inability to record are often symptoms of mechanical trouble and have nothing to do with electronic problems. Fortunately it is easy to determine when mechanical problems are occuring, and they tend to be easier to diagnose than the electronic faults because they are more exposed and obvious.

Table 17-1. Electronic and Mechanical Troubleshooting Categories.

Electronic						Mechanical		
Video	Audio	Servo	Controls	RF	Power Supply	Tape Transport	Functions and Controls	Mechanism Faults
Playback Record	Playback Record			Tuner VHF modulator Antenna switching				

The mechanical problems can be divided into three simple categories: tape transport, function buttons and keys and other controls, and mechanism.

In many cases an exact division is not always possible, and all will in some way affect the quality of the playback or recording. A list of simple causes and effects serves to illustrate them best, and often this is sufficient to point out obvious corrective measures.

Tape Transport Problems

In many cases these are the result of some simple fault which is easily determined by inspection. For the purposes of distinguishing these from the other categories it is assumed that all the function buttons work and no definite machine malfunction has been established. In all cases some mechanical action is observed inside the machine.

If the previously covered obvious checks do not rectify the fault, then the following checks should be made:

Check if the operating-key interlocks release.

Check if the key or button operates its associated electrical switch.

Ensure the cassette itself is not defective.

Check that no wiring or plugs are obviously defective.

It is possible that a circuit is not working. This should be kept in mind at this stage but not yet acted upon.

If one of these appears to be the cause, then usually the action to be taken will be obvious. If these do not seem the likely cause, then the following individual faults are perhaps the most common, and they can be attributed to the appended causes.

Play button pressed but no picture is produced. This fault is best traced with the use of the tree shown in Fig. 17-1.

Stop button produces no action. Refer to the simple tree of Fig. 16-2.

No rewind or fast forward. In most machines these two modes occur with the tape inside the cassette and not run out around its threading path. The most likely causes are as follows:

A microswitch has not been actuated by the mechanism.

The mechanical interlock and its associated switches may be defective.

Relays are not closing.

Solenoids are defective.

Tape tension is wrong.

Belt has come off somewhere.

Faulty wiring.

AC motor problems.

Circuit problem on a (printed circuit) board.

Stop occurs at the wrong time. Check the following:

Check the position of the autorepeat switch.

Check the setting of the auto timer or counter.

The ambient or incident light may be too high and is causing the photosensitive circuit to work. (Keep the lid on.)

A scratch in the tape may allow the lamp to shine onto the photo element.

No auto stop at the end of the tape. The end of tape is not being sensed due to a defective photosensing circuit or element, the solenoid is not releasing, the eccentric is not releasing, the tape guides are in wrong position or the tape is pulled out of position. In each case check tape tension. Check for broken or bent linkages.

No auto repeat at end of tape. Check for auto repeat switch and timer settings. Check all items in previous fault.

No instant stop or pause. Check the **pause** switch, check to see if the pressure roller releases, and check the tape brakes. Observe the mechanism for linkage problems.

Tape entangles in mechanism. The tape tension is wrong or the reel tables are defective. These may need oil where indicated in the manual or may need replacing. Check the reel brake bands. The tape at the pinch roller may be out of position. This may be due to roller misalignment or to faulty tape guides. A pinch roller protector may be required. The cassette compartment may be misaligned. The threading ring may be bent and so may the threading arm.

Tape damaged by the machine. Caused mainly by improper threading. This is often due to the ring being faulty and in need or adjustment. Check the mechanism with an expendable tape. Both of these can be caused by defective tape being put into the machine. Do not use bad tape.

Tape moves too fast. This is usually due to nothing on the tape, especially if the control track is missing. This is now a servo problem.

Function Buttons, Keys, and Other Controls

Two main types of problem occur here: the mechanical part jams and will not operate, and the electrical switches actuated by the control become defective.

Because the controls are always associated with other mechanical levers or electrical circuits, the result of these malfunctions is usually major and requires immediate attention.

The main function buttons will not work. This is usually due to internal mechanical problems, either within the key block itself or in the interlocks associated with it. Force should never be used to make these operate. Often the whole block will have to be removed and examined to determine the fault. Usually it is a worn or bent part causing the trouble.

The record button will not depress. This is interlocked with the main buttons, and the procedures just described are usually necessary.

Fuses blow when the play or forward button is used. Several problems can cause this. Check these:

Diode in the main solenoid is shorted.
Main solenoid is shorted.

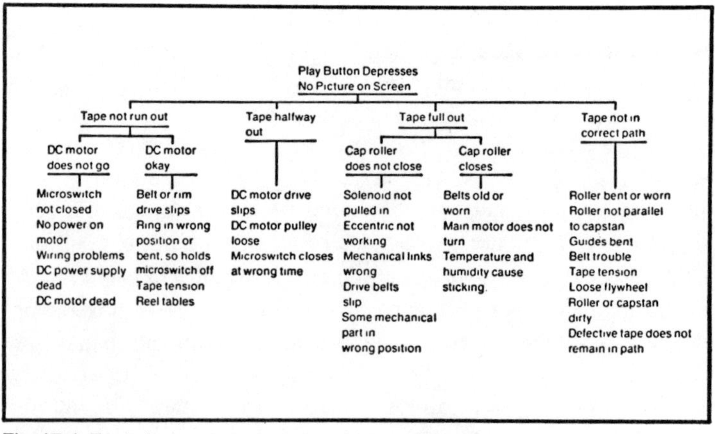

Fig. 17-1. Fault diagnosis- play button depressed, no picture.

DC motor is defective.
DC motor filter is shorted.
Wiring problems and circuits on boards.

Cassette does not drop into position when inserted. Check that main function buttons are not depressed. The cassette compartment may be defective—remove it and check.

Cassette does not pop up and eject. Tape may still be threaded. Check that main function buttons are not depressed. The main solenoid has not reset. The eject lever is faulty.

No auto stop after rewind. Check the repeat switch position. The linkages may not be working properly—inspect them. The wiring or switches may be defective.

Mechanism Problems

Occasionally the insides of the machine will go wrong and exhibit problems ranging from severe tape damage to no motion of any parts. This section limits these to actual problems with the mechanism itself as distinct from the previously discussed items.

Also included is a list of isolated problems which sometimes occur and do not appear to fit the previous classifications.

- **The main AC motor does not turn.** No power is supplied to the motor—check the power switch and relays, and microswitches. Thermal overload may have tripped. Motor may be dead.
- **The main motor does not stop.** Check the microswitches. Check the main function button and operate the switches. The relays may not be operating.
- **DC threading motor does not turn.** No power—check the microswitch and function switches. The filter circuit may be open. The dc power supply may be dead. The fuses may be blown.
- **Capstan does not turn.** Check the main motor because this drives the capstan. Check the drive belts.
- **Video heads do not rotate.** These are often belt driven by the main motor, so check this first, and then check the drive belts.
- **Belts come off.** The main motor is marked with an arrow to indicate the turning direction. If it is turned the other way it will throw the belts. The other cause of belts coming off is they stretch or break or have been tampered with.
- **Internal plug and socket problems.** These are usually indexed and are tight fitting. The main problem is that they have not been properly mated after servicing, the pins do not mate and push out or the wiring to them breaks.

Mechanical Noises

All the cassette machines operate silently under normal conditions. Any noise is indicative of some trouble. The main causes are as follows:

Old belts are squeaking.
Tape tension is too tight.
Temperature and humidity will cause the tape to squeal.
Wire may be rubbing against a moving mechanical part.
Wear in the mechanics.
Bent arms and levers rubbing against belts or motors.
Tape spool tables have worn too tight.

VIDEO PROBLEMS

Video problems can occur in both the **playback** and the **record** modes. Often they are caused by operational errors, but other than this the main causes are as follows:

Cables and plugs are faulty.
Mechanical problems as discussed in the previous sections.
Servo problems, which usually cause a bad picture rather than no picture.
The tape itself can be defective.
Circuit problems within the machine.

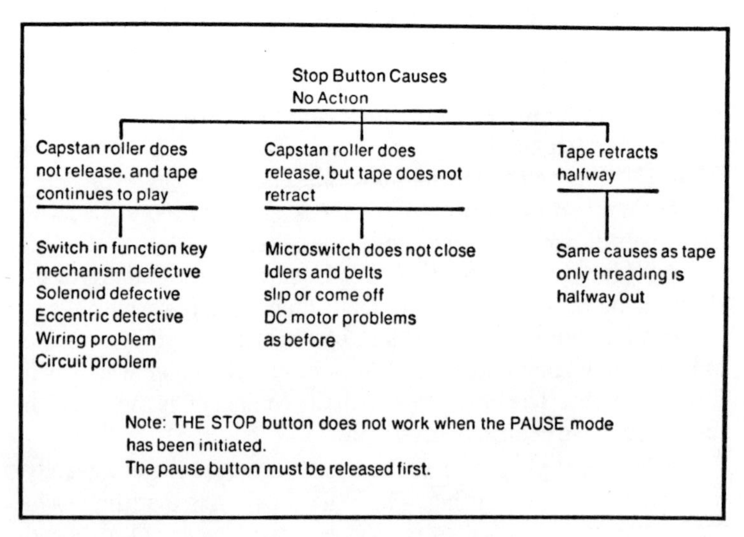

Fig. 17-2. Fault diagnosis- stop button depressed, no action.

The first four of these affect playback and record alike, but circuit problems tend to affect one mode or the other only. For this reason the two modes can be treated separately, and a mix of trees and lists are used to describe the faults and aid in determining the causes.

Playback Problems

Three main playback problems will be covered:

No video on the screen when a tape is played.
Noise only on the screen when the tape is played.
A bad quality picture is seen on the screen.

●**No picture on the screen in playback.** To logically determine which of the many possible causes would produce this fault, it is suggested the tree in Fig. 17-3 be followed. This assumes that there is no obvious mechanical fault with the machine. The steps are arranged with the most likely causes first, with instruction concerning the steps as follows:

Step 1. The often repeated obvious things should be checked first; this is usually some switch in the wrong position for the desired mode of operation. These are the most common—**machine**: channel select switch, color-monitor switch or pause. (Ensure the tape is run out and moving, this eliminates any mechanical problems.) **TV monitor**: TV-VTR switch, on-off switch, channel dial, brightness, or contrast. If none of these produce a picture, then go on the Step 2.

Step 2. Play the test tape. Do *not* play the manufacturer's alignment tape. If this produces a picture, then the system is in working order and the program tape is defective. This can be further checked by playing the program tape on another machine and using another monitor. If this step does not produce a picture, then continue checking with the test tape in the machine.

Step 3. As previously mentioned, the plugs and cables are constant sources of trouble. There are two main ways to check a cable: (1) replace the suspect item with a known good one, or (2) use a test meter and check the continuity of the cable. If the cables are good, go on to Step 4.

Step 4. It is possible that the cassette machine has no output. This is best checked with a known good monitor and a known good cable. If the RF output is being used, then change the channel and try the video output. If the video output is being used, try the RF output. If no output is produced in either case, then go through the list of items in Step 6.

Step 5. Occasionally the TV monitor will become defective. If the TV is used in the RF mode, change the channel and then try the video mode. If it is used in the video mode, try the RF mode.

Step 6. If there is no output from the machine, then any one of the following may be the cause (it is suggested that most of these be checked by a serviceman): dirty video heads, cassette micro-switch is not actuated, cassette is not fully in the machine, dirty control head, bad tape is not contacting heads, other microswitches do not operate, fuses, head relay is struck in wrong position, internal plugs and wires are defective, defective video heads, pre-amp dead, circuit problems on PC boards, rotating transformer is dead, muting circuit, or in RF mode check the fine tuning on the TV set.

- **Noise on screen but no picture.** This is very similar to the previous case, and the same line of tracing can be followed. The tree here presents an alternative which is often useful.

The first step is to change the mode of the TV monitor, if this is possible. If the RF mode is being used, change the channel and then try the video mode. If the video mode is in use, then try the RF mode. If these changes do not produce a picture, the simple tree shown in Fig. 17-4 can be followed.

In case the noise is fixed but no picture appears, revert to the previous case for checks.

- **Bad quality picture on the screen.** Bad quality covers a multitude of faults, each of which must be examined and corrected individually. It is almost impossible to categorize them. Below is a

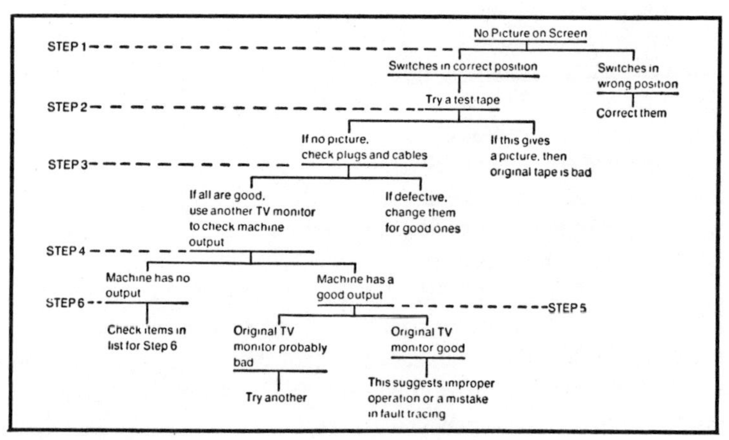

Fig. 17-3. Fault diagnosis—no picture on playback.

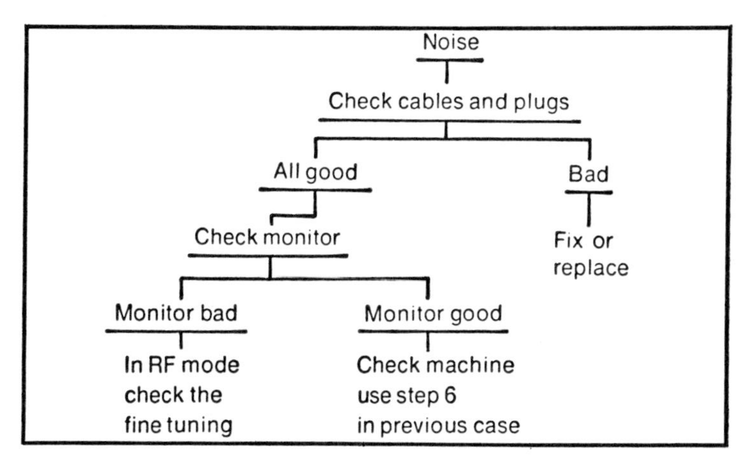

Fig. 17-4. Fault diagnosis—noise, but no picture.

list of typical problems with possible causes and the best corrective action.

- **The picture does not lock up on playback:**

Tracking off—Adjust tracking contol
Dirty guides—Clean
Dirty control head—Clean
Tape tension wrong—Adjust
Bad Tape—Try the tape on another machine and try a known good tape on this machine

- **Hooking at top of the picture:**

Tape tension wrong—Adjust
Tracking wrong—Adjust
Monitor horizontal wrong—Adjust horizontal hold

If these do not correct the picture, then servo problems are indicated.

- **Many dropouts:**

Tracking wrong—Adjust
Bad tape—Use an external dropout compensator
Internal D/O compensator defective—Adjust or fix

- **Overly white picture:**

Monitor controls wrong—Adjust
Overmodulated tape—Can do nothing about this; check machine on which tape was made.

- **Picture not sharp:**

Monitor out of focus—Adjust (monitor may not have a focus control)

Camera out of focus; if off air recording, then TV needs adjusting—Adjust if possible
Possible loss of HF response—Call serviceman for alignment

- **Noise all over picture:**

Heads becoming clogged—Clean
One head defective—Change heads

- **Low level picture with noise:**

Heads dirty—Clean
Heads getting old—Check age; replace
Head penetration insufficient—Change heads

- **Machine will not play tape from another machine:**

Tracking control—Adjust
Guides dirty—Clean
Needs servo align—Check with standard test tape
Control head dirty or defective—Clean and check
Tape path wrong—Remove top and look
Pinch roller problems—Remove top and look
Belts old or defective—Change
Tape tension wrong—Check
Motor brake problems—Check servo

- **Excessive hooking at bottom of picture:**

Head switching wrong—Check servo switching
Servo problems—Check and align

- **No color or bad color:**

Color-mono switch—Check
Maybe not a color tape—Check with another machine
Is the TV monitor a color set—Check and change
TV set fine tuning off—Adjust
Heads dirty—Check and clean
Tracking—Adjust
Check color controls on TV—Adjust if necessary
Color lock control—Check and adjust
Tension wrong—Check
Color killer circuit defective—Check with service manual
3.58 oscillator or crystal dead—Check with service manual

Recording Problems

Three main problems will be covered, and it should be noticed that these involve some sections of the playback circuitry:

No E-E picture in the record mode.

An E-E picture is on the screen, but nothing is recorded onto the tape.

A recording takes place, but there is no color.

● **No E-E picture.** The input signal to the machine passes through all the record electronics to the video heads and then passes through all the playback electronics to the output. This E-E signal is an indication that all the electronics in the machine are working. With no E-E picture, a test tape should be played. If this produces a picture, then the playback electronics and the heads have been established to be in working order, and the problem lies in the record side. If a playback picture is not produced, then the checks for no playback picture should be made first. When a playback picture does appear on the screen, then the record electronics can be checked.

Often the problem can be found in the most simple operational conditions, and these should be checked first. The following list shows the most likely items to give trouble:

Cassette red cap is not in place
Cassette is not dropped into its correct position
Cassette microswitch is not actuated
Record button is not properly down or interlock is defective
Microswitch on record button is defective
No video input
Input select switch is in wrong position
Defective cables at input
Head relay is stuck

● **An E-E picture, but no recording on the tape.** The fact that an E-E of the input signal is observed immediately eliminates all the operational errors and items like bad cables and plugs. The trouble here is either the tape does not make contact with the heads on record or some internal switching or electronics are wrong. Playback of the test tape will eliminate the heads as the offending item and will establish that the rotary transformer are good. Check the following:

Tape is not run out when record button is pressed
Defective microswitch in record mode
Bad connections to record electronics
Record amplifiers are dead

● **No color recorded.** If this fault appears, the first check is to play back the test tape and see if the machine is playing back color. If it is not, then the previous checks should be made. If the machine does play color, then the fault can be limited to one of three things:

Input signal not in color
Defective record amplifier
Defective record switching

AUDIO PROBLEMS

Videocassette machines have two identical audio channels. Each has two input sockets which are connected to the audio amplifier for that channel, and they both have several output points which are all fed from the output of the channel amplifier.

Both channels share the same bias and erase oscillator, and the difference between the channels is found in their operational use.

In all audio problems the obvious cables, power switches, and plugs should be checked first. If this fails to correct the fault, then the following sections should be regarded.

No Playback Audio

If no playback audio is heard from a tape, the first check to make is with the test tape. If this produces no audio, then check these:

Playback level on the monitor may be too low
Audio select lever may be in wrong position
The sound may be on the other track of the tape, so try the program tape using the other output
Check the TV-VTR switch on the monitor

If these do not produce an audio output, the substitution of cables and monitors should be tried to establish which of the items is giving the trouble. If it is found that the audio is not coming out of the machine, then these are the most likely causes:

Nothing on the tape—Play the test tape
Dirty audio head—Clean it
Audio amp is defective
Switch is defective
Audio head wiring is broken
Audio head is defective
Internal connections are broken
Tape may be out of tape path

The outputs of the machine appear at several places, but they are all fed from the same amplifier. Lack of audio at one output is likely to be common to all outputs.

No Audio Recording

The input audio always should appear at the output of the machine. If this is not so, then all the input cables, plugs, and switches should be checked. If audio at the output is produced, then the problem is due to internal causes.

If a recording does not appear to have been made, the playback should be checked first, as previously described.

If it appears that the machine has internal problems which occur solely in the record mode, these items should be checked:

The tape tracking
Heads may be dirty
The gain of the record amplifier
The record amplifier works
The bias oscillator works and the bias level
The head relay—it may be stuck

Other Audio Problems

- **Hum in the audio.** Usually this is caused by bad grounds or broken shield cables. They should be checked. Another cause can be that the control track shield has broken.
- **Buzz in the audio.** This is different from a hum. The main causes are that the antenna is misconnected, the TV tuning is off, the video is overmodulated on the tape, or the tuner AGC is not working.
- **Wow.** This is due to the usual causes found in audio machines. Check the capstan pressure roller, the tape tension, and the reel tables.
- **Audio distortion.** This could be due to bad audio on the tape or because of too high a level incoming and the AGC not working correctly.
- **Noisy playback.** If the record level was too low and the bias was wrong on the original tape, then this is a tape fault and cannot be corrected. Bad tape and dirty heads can cause this also in both playback and record.
- **Low level playback.** This is due to the playback amplifier gain being too low or to the level on the tape being too low. Head and amplifier problems also will cause this.
- **Overloaded audio.** The playback amplifier gain is too high, and the AGC may not be working. Also the audio may have been recorded at too high a level

SERVO PROBLEMS

Servo problems show up in playback more often than in any other mode. The symptom is simple: the tape will not play correctly, it will not track, and often the head switching is irregular. A faulty servo will prevent any form of playback and also will affect recording. The moment such problems are indicated, and immediate alignment of the servo system must be undertaken. Other internal problems which can affect the servo are as follows:

Dirty or defective control heads
Defective head-tach coils
Noncomposite video input
Broken wires to heads and PC boards
No control track on the tape being played back
Previous control track has not been erased, so the tape has two control tracks

Servo alignment should be undertaken only by trained personnel.

RF SYSTEM PROBLEMS

These are due either to faults with the TV tuner in the machine or the TV monitor, the RF modulator in the machine, or the antenna switching. Typical faults are no picture, a very noisy picture, a picture with beats and patterns across it, or no audio with the picture.

Almost all of these can be corrected by adjusting the channel or the fine tuning, or by selecting the correct position of the input switch. Cables also will cause trouble here and should be checked.

If a fault is traced to the RF modulator in the machine, this should be unplugged and another used. Do not attempt to service this item—return it to the manufacturer.

CONTROL SYSTEM PROBLEMS

Due to their diversity it is not possible to categorize the problems which can affect these circuits. They are a mixture of both mechanical and electronic faults which are best examined on their own individual merits. Typical faults are as follows:

Standby lamp stays on
Light on photosensing circuit goes out
Photosensing element goes bad
Microswitch operated by the cassette does not work
Auto stop and rewind do not work

The causes of these can be so diverse that it is impossible to list them or attribute them to any particular operational error or mechanical malfunction.

POWER SUPPLY PROBLEMS

These usually manifest themselves as no lights and no motion of the tape. Occasionally these will be in good order, but there will be no picture or audio.

In the case of no apparent power the first check is to insert a cassette and watch for or listen for the main motor. It should momentarily rotate and then slow down to a stop. If this does not occur, then all the fuses should be checked. A blown fuse should be replaced and the machine tried again. If the fuse immediately blow, then the machine should be unplugged from the wall and *no attempt should be made to operate it* until the fault has been rectified. Never put in a fuse which is above the rated size.

All troubleshooting of power supplies should be left to trained personnel. Electric shock is possible. Always have a shock treatment guide or chart available, and take all precautions.

The possible faults which can blow fuses are as follows:

AC cord is broken or shorted
Defective fuse holder
Defective on-off switch
Other power carrying switch is defective
Wiring is burned out or is touching chassis
Solenoid or relay coil is burned out
Solenoid or relay diode is burned out
A motor is defective
Noise filter on the motor is shorted
Electronic brake coil is shorted
Some other component is shorted or is out of position and is touching the chassis

Checking for these types of faults is not hard, but it usually requires experience in this type of work. It is best left to a qualified serviceman.

Further power problems can be caused by a drop in the incoming line voltage. This can cause malfunctions in both the electronic circuits and the mechanical parts which are driven by solenoids and motors. A good indication of this type of trouble is that the indicator lights may be dimmer than usual or may vary in brightness. In such cases, a variable transformer used in the power line can be a help. If the incoming line voltage is metered, the variac can be manually adjusted.

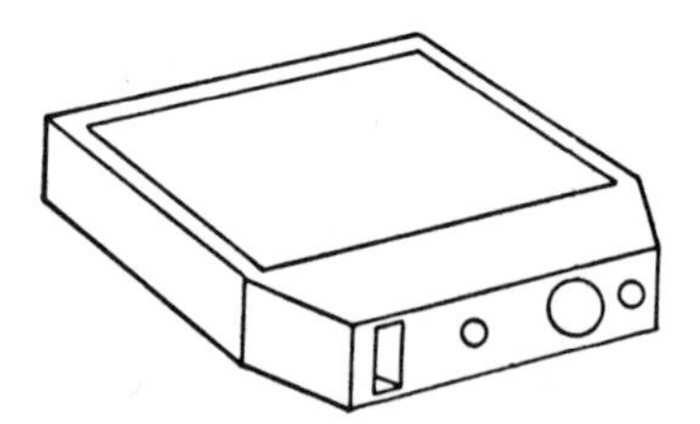

18 When and Why To Use Videocassettes

The practitioners of any science and technology have often tended to overlook the social implications and effects on the public, and television engineers are no exception. The producers of television programs, on the other hand, have always been aware of their public, but only as passive viewers of material sent over the air or played back in a controlled situation.

Passivity for a television audience has long been accepted as natural, and this had been suggested as one of the reasons why the open reel machines had so little use outside the TV studio. When this was coupled with the public attitude against having to use a highly technical and complex machine, it formed a reasonable hypothesis for the lack of use of videotapes as a widespread educational and informational medium. With the added difficulties of obtaining large numbers of copies and the difficulty of distribution, the growth of the open reel videotapes continued to be far less than hoped for and always remained below potential.

It was a consideration of these and other social factors by several companies in different parts of the world, rather than technical details, that led to the development of an alternative videotape system which would overcome or sidestep all the objections and place TV and video firmly in the realm of education, industrial training, art, and general public use. It began the development of television as a nonbroadcast audiovisual medium for information distribution.

To be usable audiovisual medium the system has to meet several requirements, the most important begin these:

- It must be easy to use, which in fact means some form of cassette to eliminate all tape handling and threading.
- No special equipment should be required other than the tape machine. And the viewing device should be a normal, unmodified, domestic TV set.
- At least one hour of time should be possible on the tape.
- It should be as inexpensive as possible.

- It should record and play back in both monochrome and color and do this automatically.
- No long or extensive training or education should be required for its use.
- The tapes should be easy to file in a library, easy to copy, and easy to distribute.

Several slightly different systems have been proposed and introduced which meet these requirements, and three social uses were considered to be important in their development and serve as pointers for the use of the cassettes: educational, industrial, and social uses.

EDUCATION

In education a major problem is the underprivileged and the undereducated. There are not enough teachers to give these persons the time they require to raise their level of knowledge, standards, or expectations. And so in general they continually fall behind and eventually drop out when they try to take classes. Various audiovisual aids have been tried to alleviate this problem, but with limited success. The medium was either wrong for the subject, required a prelearned ability such as reading, or was too complicated to use, such as an open reel VTR or a film projector.

The ease with which a videocassette can be used and the large number of plays possible can be a great help, because it enables repetition of lessons in an easily presentable form. This in fact has been a sufficient reason for many institutions to adopt the cassettes.

Another aspect of education is that complex subjects often need more than one exposure for a new student to understand what is going on. Having lessons on an easy-to-play medium can eliminate instruction time and can avoid losing time better utilized in lectures and laboratory classes. It also enables the students to repeatedly listen to and watch the top teachers in a given subject.

Modern learning is often referred to as an interactive process rather than a one-way flow of ideas from the teacher to a group of students. Self-instruction and seminars are more useful and more in vogue now than ever before and will probably remain the dominant teaching style in many areas for a long time. This attitude allows the student to proceed at his or her own pace and to work when feeling motivated rather than when regimented.

Thus the vidoecassettes are proving to be one of the greatest helps known. With the ease of stopping, starting, and rewinding the tape, the review and gradual assimilation of material is enormously facilitated. Another effect is that the brighter and self-motivated students are allowed to proceed at an accelerated pace and even to graduate in less time than that required by the traditional "in-step" classroom methods. Educators are saying that students who have used taped material now come to classes and seminars better prepared and with a higher level of understanding and ask much more pertinent questions than those who do not use tapes. The increase in knowledge with the continued use of videocassettes has been demonstrated beyond a doubt.

Cassettes are turning out to be a very useful new medium. They have some resemblance to open reel tapes but almost nothing in common with film, and are exhibiting some characteristics which are entirely their own. They are not yet perfectly or fully developed as an audiovisual medium, but the advances they have made are tremendous.

INDUSTRY

In industry the needs are similar to those in education. Training and retraining are a major part of most modern industries, and this includes both the undereducated as well as the highly educated. In both cases continued review of material and proceeding at one's own pace are important.

Modern industries are continually changing products as well as manufacturing and serving techniques. A major problem always has been the time and expense of bringing a limited number of persons over a long distance for training and then relying on them to train others at a later date. This process has been speeded up by sending videotapes to the outlying centers and training a few key individuals only in the central location. The results of this is a far easier and more thorough training for everybody in much less time.

Many new industrial products, especially in the science and engineering fields, are made for sale over a broad spectrum of companies, and along with these has come the necessary updating and increasing knowledge required by the end user. A good example of this are the advances in integrated circuit use and manufacture. Many companies in these areas produce videotapes for sale or rent to client companies with complex instructions and lessons on how to best use the new products.

In many of these areas the traditional audiovisual methods have been inadequate, and the open reel machines have had only limited success. Often a dynamic presentation was needed—which eliminated slides and audiotapes—and the reel-to-reel tapes were too expensive and cumbersome. A videocassette could be most useful and, in fact, is now gaining ground very fast.

SOCIAL USES

The development of a home video player has always been a desirable goal, but it was one of the areas in which the open reel machines made absolutely no impact. An examination of the reasons convinced the manufacturers that such a device would have to be a fully automatic cassette system, and this was in fact one of the dominating goals in the development phase. The desirability of having one's own favorite programs on tape is very attractive, but so also is the availability of the vast resources of the educational world. Here it is possible the cassette will succeed where the open reel machines fail, but it is more likely this area will fall to the videodiscs.

The end of the 1960s brought on the attitude of "need to know" especially among the young, and the desire for information distribution increased enormously. The proliferation of alternative or underground newspapers, magazines, and video became enormous. This added greatly to the acceptance of the small VTR as an information and art medium, and it was realized that a cassette-type format could be most valuable in this area. Cassettes are an almost ideal medium to work for the nontechnical person, and they are perfect for the safe transport of the tape, for producing quick and easy copies, and for playback many times to individuals or small groups. For the small, independent video producer they could be the ideal format, providing the easiest route to the most interested audience.

In a completely different vein, the managing of an audiovisual center—especially television—always has had to be a very technically oriented undertaking. This is due mainly to the technical complexity of the medium and the difficulties of keeping it operational. There is no doubt that this inhibits the creative spirit and productive attitude of many who do not understand TV technology, and it does become a limitation in many nonbroadcast areas.

A simple nontechnical cassette-type system would make a small studio more information and program oriented and would have less emphasis on the technical capabilities and limitations of the medium, which is a major consideration for general use. This means that the message and the content can become the dominant factors much more than previously, and the problems of achieving the recording can take second place. This does not mean the engineering can be ignored—it just changes the emphasis.

MEDICAL USES

The medical profession, probably more than most others, is subject to gaining new information continuously. The problem of keeping up is most acute for the doctor and the dentist. This is now being helped greatly by schools producing taped lectures and instructions in many areas and making them available to professionals, who use the tapes either singly or in small groups. The cassette provides an ideal medium for an easy-to-use information display system which can be "bicycled" around on a nearly continual basis.

DISTRIBUTION CONSIDERATIONS

The distribution problem encountered with videotapes is very real and has been a limiting factor in their widespread use. Much of the material viewed by the end users is not produced in his or her own studio or facility but is increasingly received from some outside production center which specializes in educational or other programs.

One of the greatest and most annoying problems faced by the producer is the proliferation of many different formats of videotape, none of which has been able to take a commanding lead and to be accepted as a standard. The result of this was that no really large mass distribution of tapes and certainly no large mass of copying centers have been developed. There are just too many formats to copy onto with insufficient sales in each of the separate formats.

Obviously a standard became necessary, and so the desire developed for better quality pictures and the use of more color. If this could be realized, then the widespread use of videotapes could be achieved, and this was a major factor in the consideration leading to the design of the cassette systems.

A new videocassette could be a standard format of playback tape with high quality color pictures which could be produced on any original production format and then transferred in large numbers—using semiautomatic machinery—to one standard format for distribution and play back by the end user. This also would allow the end user to receive tapes from many different centers with the knowledge that they could be played on in-house machines.

Recent experience with the cassette format suggests that this has been achieved with considerable success and presents a very good reason for using videocassettes.

ADVANTAGES AND DISADVANTAGES

Other than their extreme ease of operation, there are several advantages enjoyed by cassettes, but it should be understood that these only become apparent when the machines are used in a manner which capitalizes on these good points:

- They are easier than reel-to-reel VTRs to operate in all modes. They also are much easier to use than a film projector.
- The tape is protected against misuse and damage by its enclosed nature. This is most important when tapes have to be used many times in an uncontrolled situation.
- The tapes can be rerecorded when they are outdated—a definite advantage over film.
- They have a much longer life than film. Film will give about 100 plays before it becomes useless; a videotape will give about 1000.
- A TV screen can operate in ambient light; it does not need a darkened room like film. This means a cassette can be used in conjunction with other media or other activities.
- The mechanism is silent, so several can be used simultaneously in close proximity. In this they are ideal for multilanguage translations and for use in libraries.
- The two audio tracks allow tapes to be bilingual or to have two levels of commentary or instruction.
- The viewer can easily stop, restart, rewind, fast forward, or repeat whenever required and with no fuss. This is not so with film.
- The tapes are easy to library store, which aids in distribution and "networking."
- Most manufacturers use a container box which is suitable for protection, storage, and mailing.
- The machines are neat in appearance and do not look like monstrous pieces of technical equipment, so no one objects to their presence.

There are a few disadvantages which also should be considered:

- Most of the machines are rather heavy and bulky and are not easily portable.
- Although they produce a good picture, it has less resolution and "quality" than a film or broadcast tape. It is, however, good enough for most purposes.
- The initial cost of the machines is high.

- Basically they are a postproduction machine and are unsuitable for serious studio work.
- Their time base stability is not good, and they have nonphased or heterodyne color

There are certain other factors which should be considered, because the above lists are not complete.

Cost is an all-important factor. Although the initial cost of providing the machines is much higher than for film, the operating costs are very low. A cassette costs about $1 for each minute in length, and for this the tape can be used about 1000 times. This is much cheaper than anything else available.

The size of the machine and the location in which it will be used are very important. Most are not "rack mountable" in the way other electronic equipment is, and they do not lend themselves easily to remote control. If they are to be used by young people, then a safe cart on which they are mounted flat should be used. This will allow a TV set to be put on another shelf on the cart and permit the whole setup to be wheeled around. Most of the machines are small enough to be used in a library carrel or a small office.

Prime among the reasons for using cassettes is their complete acceptance in nonprofessional areas, mainly because they are easy to use and will work with an ordinary domestic TV set—thus making other expensive equipment unnecessary. Added to this is their reported high reliability and the fact they have spread so widely.

The choice of a cassette or a cartridge must be made by the user. The main advantage possessed by the cartridge is that they have a compatibility with an open reel format, but against this must be balanced the necessity of rewinding the tape onto the cartridge reel and then ensuring it will play without problems. Definite advantages of the cassettes are that there are more machines of this format, that it is made by more manufacturers than other formats, and that there are more places which specialize in mass copies for this format than any other.

THE STUDIO

The choice of equipment and format for the studio is important, but it is not necessarily central to the use of videocassettes. No hard and fast rules can be formulated about the type of equipment needed in a studio. The choice will always be a matter of conflicting requirements. Factors like the size and complexity of the productions are obviously important, but so also are the personnel who will run and use the studio, the size and type of audience served, the type of program material envisaged, the available budget, location, etc. All these interact to influence the decision, and often they are pulling in opposite directions and tend to make general advice impossible to give. In fact, it always becomes an individual decision for that particular production center.

In most professional and semiprofessional production studios the open reel machines will be used for the original recording. This is mainly because they have been built as recording machines for a studio and have superior time base stability and better record and playback electronics than the cassette machines. Also, editing is much easier, which is important in a studio.

A prime decision in the choice of studio equipment must be the number of copies which will be made and how much editing will be required before the final tape is ready. If the programs will be complex productions which are assembled from both the studio and previously recorded tapes, then the final master tape may be a second, third, and even fourth generation tape. If this is so, then the use of high quality helical machines with time-base correctors will be necessary, or even broadcasting equipment will have to be used.

If fairly simple productions are all that will be attempted, with no more than simple assemble edits used to build the program onto the original tape which is then copied directly onto cassettes, then almost any good quality, open reel, helical machine will be suitable.

If only the simplest cases of short, straight recordings are to be made directly from a studio, then the cassette machines could be used.

An important factor which must be considered is the type of information produced, the use to which it will be put, who will produce it, and who will view it.

For example, if the finished tape will be shown in a sales office for the play back of simple messages and instructions from a central office, then a simple playback deck and a monochrome TV set would suffice. This would enable the central office recording equipment to be a simple one or a two camera setup and may even be transportable.

For a school, where off air recording of educational TV programs will be made for play back later, a simple TV set and cassette machine with a timer will suffice. If the students are to put together simple material themselves, then a small studio will be required.

In an establishment which tends to undertake medical instruction and information gathering, a high quality color studio and tape machines to match will be required.

Assuming the decision about the level and type of production has been made, the question which now arises is the choice of tape format and the facilities required on the machines. In a closed system—that is, one in which no material is sent outside or received from elsewhere—the decision can be made on the money available, personal preference, etc. But for a system which must interact with or be part of a larger overall system, the choice can be both limited and very difficult. Often it is necessary to have more than one tape format, which immediately implies that copying will be a normal part of the operation. It is here that the videocassettes become very attractive, because they represent a nearly universal standard format.

EXAMPLES OF VIDEOCASSETTE USE

Many large and well-known establishments as well as many small places have gone into extensive videocassette use. The following short list of examples illustrates the thinking of the manufacturers when developing the cassettes and shows how well the cassettes have been accepted:

- Many prerecorded lessons are becoming available for schools and colleges. These are often produced in a broadcasting-type studio and then are copied onto cassettes for distribution.
- Many companies are now producing sales and technical information for their smaller divisions in the form of short video programs and then distributing them on videocassettes. Demonstrations of new techniques in the use and servicing of new equipment benefit greatly from videotaping. Copies of the tapes can be sent to all interested parties on a regular basis. The recipient of the tape then can play it as many times as desired and to as many people necessary for maximum effect.
- Many companies with new products are faced with the problem of educating their potential customers in the use of their products. This often means raising the level of the customer's knowledge in advanced technical and scientific areas. Prerecorded instructions have proven of great value here, and now several engineering companies have libraries of such material for general distribution.
- The problem of doctors and dentists in keeping up with new medical information is continual. Now many schools are producing tapes and distributing them on cassettes to groups of medical

personnel. The need for this has been greatly increased because many states have passed laws requiring a certain number of hours of continuing education for medical practitioners, and in many cases attendance in class is impossible.

- Public access television is an outgrowth of the cable television industry, and it is spreading in many places. It has been hampered by a lack of technical personnel and a scarcity of easy-to-use equipment. Because much of this type of TV relies on "bicycling" tapes around, the introduction of the cassettes has been most valuable.
- The Open University in Britain makes extensive use of video programs both over the air and on videocassettes. Taped lessons are available to students either in the home or at "media centers," where instruction by experts in a subject is readily available. The beauty of this is that one can enjoy and study a lecture many times over rather than just once, as is traditional in a college. These programs have been so successful many foreign universities are now using them.
- The Dental School of the University of Michigan uses extensive videocassette playback facilities for instructional purposes.

CONCLUSION

Videocassettes are not a new technology, they are a new use of a technology already developed and refined. The cassettes came about because of the realization by the manufacturers that society was at last ready to benefit from TV but had no device which was yet adequate enough for this purpose.

Most of the reasons for the use of the videocassette machines centers around their operational simplicity and the lack of technical complexity for the user. Their acceptance has been almost immediate and complete, and they have transformed the use of TV and video into a medium of active general purpose use. Any viewer can now select and insert a program and can choose the time and frequency of viewing and can even generate personal high quality material.

The relative ease of fast duplication and mass distribution of the cassettes has transformed them into one of the most widespread audiovisual media in use today, and the fact they are standardized in format has helped this enormously.

They are basically an information-distribution medium rather than another technical device to be used in a studio by engineers, and already they are beginning to revolutionize the communication systems in society.

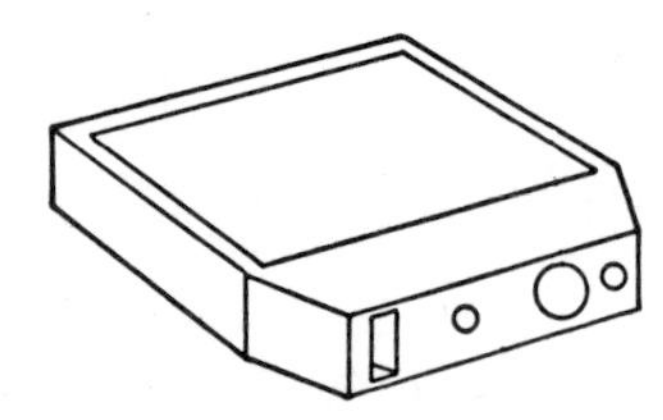

Index

Edited by Roland S. Phelps